2009 STATE OF THE WORLD

Into a Warming World

Other Norton/Worldwatch Books

State of the World 1984 through *2008*
(an annual report on progress toward a sustainable society)

Vital Signs 1992 through *2003* and *2005* through *2007*
(a report on the trends that are shaping our future)

Saving the Planet
Lester R. Brown
Christopher Flavin
Sandra Postel

How Much Is Enough?
Alan Thein Durning

Last Oasis
Sandra Postel

Full House
Lester R. Brown
Hal Kane

Power Surge
Christopher Flavin
Nicholas Lenssen

Who Will Feed China?
Lester R. Brown

Tough Choices
Lester R. Brown

Fighting for Survival
Michael Renner

The Natural Wealth of Nations
David Malin Roodman

Life Out of Bounds
Chris Bright

Beyond Malthus
Lester R. Brown
Gary Gardner
Brian Halweil

Pillar of Sand
Sandra Postel

Vanishing Borders
Hilary French

Eat Here
Brian Halweil

Inspiring Progress
Gary T. Gardner

2009

STATE OF THE WORLD

Into a Warming World

A Worldwatch Institute Report on
Progress Toward a Sustainable Society

Robert Engelman, Michael Renner, and Janet Sawin, *Project Directors*
Ambika Chawla, *Project Coordinator*

Jessica Ayers
David Dodman
Christopher Flavin
Gary Gardner
W. L. Hare
Saleemul Huq
Lisa Mastny
Alice McKeown
William L. Moomaw
Sara J. Scherr
Sajal Sthapit

Linda Starke, *Editor*

W · W · NORTON & COMPANY
NEW YORK LONDON

1776 Massachusetts Avenue, N.W.
Suite 800
Washington, DC 20036
www.worldwatch.org

Printed in the United States of America.

The text of this book is composed in Galliard, with the display set in Gill Sans. Book design, cover design, and composition by Lyle Rosbotham; manufacturing by Victor Graphics.

First Edition
ISBN 978-0-393-33418-0

W.W. Norton & Company, Inc., 500 Fifth Avenue, New York, N.Y. 10110
www.wwnorton.com

W.W. Norton & Company Ltd., Castle House, 75/76 Wells Street, London W1T 3QT

1 2 3 4 5 6 7 8 9 0

♻ This book is printed on recycled paper.

Acknowledgments

In the 20 years since the historic testimony by Goddard Institute scientist James Hansen, the science of climate change has come a long way, as manifested in the reports of the Intergovernmental Panel on Climate Change (IPCC). The politics is still lagging behind, but the urgency of constructive climate action is now clearer than ever. Hansen's work and courage has been a major inspiration in compiling this twenty-sixth edition of *State of the World*. This volume offers a range of informed perspectives on pathways for adapting to a warming world while avoiding catastrophic consequences.

This book is part of Worldwatch's Climate and Energy Program, which is dedicated to achieving a substantial reduction in the combustion of fossil fuels and a transformation of the global energy system in order to stabilize the climate and increase energy security. Worldwatch works to encourage businesses, governments, and individuals to adopt the policies and behaviors that are key to the transition to an efficient, low-carbon energy system based on the sustainable use of renewable resources.

Even more than previous editions, this State of the World is the product of a broadly collaborative effort, involving the expertise of dedicated individuals from around the world. All deserve our sincere thanks for their contributions to the book and to the Institute's work. In addition to this print edition, additional views on this vital topic can be found on our Web site: www.worldwatch.org.

We give special thanks to our Board of Directors for its tremendous support and leadership over the past year, particularly our new Chairman, Tom Crain; our new Vice Chairman, Robert Friese; and our new Treasurer, Geeta B. Aiyer.

This year we particularly wish to recognize and thank Øystein Dahle, who stepped down as Board Chairman in 2008 and is now a Board Member emeritus. Øystein is one of Norway's leading environmentalists, and we will miss his visionary leadership and wonderful Norwegian stories. We are pleased, however, that he will continue to chair our Scandinavian affiliate, Worldwatch Norden, and to inspire us with his deep commitment to ecological reform.

We are also grateful to a number of foundations and institutions whose support over the past year has made this book and Worldwatch's many other projects possible: the American Clean Skies Foundation; the Blue Moon Fund; the Casten Family Foundation; the Compton Foundation, Inc.; Ecos Ag–Basel; the Better World Fund on behalf of the Energy Future Coalition; the Food and Agriculture Organization of the United Nations; the Ford Foundation; the German Federal Ministry of Food, Agriculture and Consumer Protection; the Goldman Environmental Prize; the Richard and Rhoda

Goldman Fund; the Hitz Foundation; the W.K. Kellogg Foundation; the Steven C. Leuthold Family Foundation; the V. Kann Rasmussen Foundation; the Royal Norwegian Ministry of Foreign Affairs; the Shared Earth Foundation; the Shenandoah Foundation; the Sierra Club; Stonyfield Farm; the TAUPO Fund; the Flora L. Thornton Foundation; the United Nations Environment Programme; the United Nations Population Fund; the Wallace Genetic Foundation, Inc.; the Wallace Global Fund; the Johanette Wallerstein Institute; and the Winslow Foundation.

State of the World would not exist without the generous contributions of more than 3,000 Friends of Worldwatch, who fund nearly one third of the Institute's annual operating budget. Their faithful support is indispensable to our work and we thank them for their commitment to the Institute and its vision for a more sustainable world.

For the 2009 edition of *State of the World*, the Institute enlisted a record number of scholars and leading thinkers on climate change issues from a variety of organizations around the world. We are grateful for their commitment to this ambitious project in addition to the many demands of their own work and lives. Bill Hare, a visiting Australian scientist at the Potsdam Institute for Climate Impact Research in Germany and a member of the IPCC, contributed the chapter that summarizes the science of climate change and what must be accomplished to ensure a safe landing. The chapter on the potential for farming and land use to mitigate warming was written by Sara Scherr, President and CEO of Ecoagriculture Partners in Washington, DC, and her colleague, Nepali national Sajal Sthapit. William R. Moomaw, Director of the Center for International Environment and Resource Policy at Tufts University and a member of the IPCC, coauthored the chapter on building a new energy future with Worldwatch senior researcher Janet L. Sawin. The chapter on adaptation and building resilience was contributed by Jessica Ayers of the London School of Economics; David Dodman, a researcher with the Human Settlements and Climate Change Groups at the International Institute for Environment and Development (IIED); and Saleemul Huq, head of the Climate Change Group at IIED, who was a Coordinating Lead Author of the chapter on adaptation and mitigation in the IPCC's Fourth Assessment Report and founder of the Bangladesh Centre for Advanced Studies in Dhaka.

A diverse and large number of international experts contributed the shorter pieces entitled *Climate Connections*. We are grateful for their perspectives and insights into key challenges and solutions that lie ahead as we make the transition to a warming world. Contributors include Lorena Aguilar, Juan Almendares, Ken Caldeira, Edward Cameron, Dennis Clare, Kert Davies, David Dodman, Paul R. Epstein, Manfred Fischedick, Juan Hoffmaister, David Kanter, Tim Kasser, Robert K. Kaufmann, Thomas Lovejoy, Janos Maté, Malina Mehra, Balgis Osman-Elasha, K. Madhava Sarma, David Satterthwaite, Alifereti Tawake, Betsy Taylor, Daniel Vallentin, Peter Viebahn, Jennifer Wallace, Tao Wang, Jim Watson, and Durwood Zaelke. Two of these pieces were prepared by Worldwatch researchers Ambika Chawla and Yingling Liu, and a third was contributed by senior researcher Michael Renner along with Sean Sweeney and Jill Kubit of Cornell University. Worldwatch's Gary Gardner, who codirected the 2008 edition of *State of the World*, and research associate Alice McKeown prepared the Climate Change Reference Guide and Glossary.

This book owes a special thanks to Ambika Chawla, Worldwatch State of the World Fellow in 2008 and the able Project Coordinator for all the *Climate Connections* articles. Ambika hit the ground running the moment she arrived and took to heart her mission of assembling a group of outside authors as geographically diverse as the issue of global climate change. She then worked with individual authors to make their pieces as compelling as possible. Her passionate commitment that voices from the Global South be heard in this volume helped produce more-productive ideas for addressing climate change at every level. Ambika also helped in the research and preparation of Chapter 6.

We would also like to thank an energetic and dedicated team of research interns and fellows for their hard work over the past several months. With appreciation, we acknowledge the work of Stanford MAP Sustainable Energy Fellow Amanda Chiu on Chapters 1, 4, and 6; Jeff Harti for research assistance with Chapter 4; and Hannah Doherty and Jennifer Wallace for research help with Chapter 6. And a special thanks goes to the Institute's senior editor, Lisa Mastny, for her help with editing early versions of many of the *Climate Connections* and for her quick and thorough work in compiling the set of significant global events that appear in the book's Year in Review timeline.

Once research is complete and initial drafts are written, State of the World chapters undergo a rigorous review that includes a full day of in-house critique and comment. We are grateful to Worldwatch staff from all Institute departments who participated in the 2008 review process, including program researchers Erik Assadourian, Hilary French, and Brian Halweil and *World Watch* magazine editor Tom Prugh.

In addition, authors of this year's *State of the World* benefited from numerous specialists who provided guidance and critical information and from a distinguished international panel of reviewers who took time to read draft chapters and provide valuable feedback. For their helpful and penetrating insights and comments, we thank the following: Chapter 2 benefited from analysis and assistance from Malte Meinshausen; Chapter 3 benefited from the expertise of Meike Andersson, Beto Borges, Ruth DeFries, Svetlana Edmeades, Myles Fisher, Katherine Hamilton, Celia Harvey, Willem Janssen, Andrew Jarvis, Rattan Lal, Anna Lappe, Vanessa Meadu, Peter A. Minang, David Molden, Paulo Moutinho, Danielle Nierenberg, Thomas Oberthür, Nora Ourabah, Molly Phemister, Al Rotz, Kendra Sand, Seth Shames, and Kathy Soder; Chapter 4 was strengthened by invaluable input and extensive review by Edgar DeMeo and Kurt Yeager, and the expertise of additional reviewers Tom Crain, Wolfram Krewitt, Junfeng Li, Robert Pratt, Wilson Rickerson, and Wanxing Wang; Chapter 5 authors received feedback from Tim Forsyth; and Chapter 6 came together thanks to advice and review by Tom Athanasiou, Christoph Bals, and Alan Miller.

For their assistance with various *Climate Connections*, we thank the following: Neil Leary for contributions on building resilience in Sudan; Marek K. Kolodziej and Kathryn L. Schu on carbon taxes and cap and trade systems; Ashley Clark on biodiversity; Dolan Chatterjee and Nick Mabey on India's role in addressing climate change; Renske Mackor and Steve Rivkin on climate justice movements; Stephen O. Andersen, Ana Maria Kleymeyer, Sarah Knutson, Matthew Stilwell, Xiaopu Sun, and Alexandra Viets on technology transfer; and Larry Kohler and Peter Poschen on green jobs.

Linda Starke has been the editor of the *State of the World* series since its inception in 1984—a remarkable achievement that has ensured a high level of quality and consistency. Once again, Linda's skill and ability to work with dozens of far-flung authors in the face of tight deadlines was an essential ingredient in producing this book. We are grateful to Linda for her quarter-century service to the series.

Behind the scenes, Art Director Lyle Rosbotham, whose work this year was supplemented by his wife, Joan Wolbier, rapidly turned typescript into the beautifully designed book in your hands. We also thank Kate Mertes for preparing the index.

Once the book goes to press, the Worldwatch communications team of Darcey Rakestraw and Julia Tier swing into action to spread its core messages to a global audience. Meanwhile, Director of Publications and Marketing Patricia Shyne coordinates with our international publishing partners to ensure they have the necessary information for releasing *State of the World* in their respective countries and also manages promotions for this book and our other Worldwatch publications. Another Worldwatch veteran, Director of Finance and Administration Barbara Fallin, is now in her twentieth year of running our Washington office very smoothly. An equally important underpinning to our operations is found in the untiring efforts of our development staff. Mary Redfern manages our foundation relations with unparalleled dedication and thoroughness. Courtney Berner and Kimberly Rogovin apply their energy and enthusiasm to deepening our relationships with individual donors and friends of the Institute.

W. W. Norton & Company in New York has published *State of the World* in each of its 26 years. We are grateful to Amy Cherry, Erica Stern, Nancy Palmquist, and Devon Zahn for their work in producing the book and ensuring that it gets maximum exposure in bookstores and university classrooms across the United States.

State of the World would have a limited international audience were it not for our network of publishing partners, who provide advice, translation, outreach, and distribution assistance. We give special thanks to Eduardo Athayde of the Universidade Mata Atlântica in Brazil; Soki Oda of Worldwatch Japan; Benoit Lambert in Switzerland, who also connects us to readers in France and French-speaking Canada; Christoph Bals and Klaus Milke of Germanwatch and Bernd Rheinberg of the Heinrich Böll Foundation in Germany; Sylvia Shao of Science Environment Press in China; Anna Bruno Ventre and Gianfranco Bologna of WWF Italy; Maria Antonia Garcia of Fundación Hogar del Empleado for the Castilian version and Helena Cots of Centre UNESCO de Catalunya for the Catalan version in Spain; Yiannis Sakiotis of the Society of Political Analysis Nikos Poulantzas in Greece; Kartikeya Sarabhai and Kiran Chhokar of the Centre for Environment Education in India; Sang-ik Kim of the Korean Federation of Environmental Movement in South Korea; Øystein Dahle, Hans Lundberg, and Ivana Kildsgaard of Worldwatch Norden in Norway and Sweden; George Cheng of Taiwan Watch Institute in Taiwan; Yesim Erkan of TEMA in Turkey; Tuomas Seppa of Gaudeamus & Otatieto in Finland; Marcin Gerwin of Earth Conservation in Poland; Professor Marfenin and Anna Ignatieva of the Center of Theoretical Analysis of Environmental Problems at the International Independent University of Environmental and Political Sciences in Russia; Milan Misic of IP NARODNA KNJIGA in Serbia; and Jonathan Sinclair Wilson, Michael Fell, Rob West, and Alison Kuznets of Earthscan in the United Kingdom.

Our readers are ably served by the customer service team at Direct Answer, Inc. We are grateful to Katie Rogers, Ginger Franklin, Katie Gilroy, Lolita Harris, Cheryl Marshall, Valerie Proctor, Ronnie Hergett, Marta

Augustyn, Heather Cranford, Colleen White, Sharon Hackett, RJ Cranford, and Karen Piontkowski for providing first-rate customer service and fulfilling our customers' orders.

Like the world we are monitoring and observing, Worldwatch continues to be a dynamic place. We would also like to acknowledge the valuable service of several staff members who have moved on to new challenges this year: Zoe Chafe, Andrew Burnette, Raya Widenoya, and James Russell, who joined us as a Stanford MAP Sustainable Energy Fellow. Even as we miss their important contributions, we were joined by Ben Block, staff writer for our Eye on Earth online news service, and Amanda Chiu, the latest in a series of able MAP Fellows. And we were reminded of the needs and opportunities of this planet's next generation when we welcomed into the Worldwatch family Clio Halweil, daughter of Brian Halweil, and Charles and William O'Meara Sheehan, twins born to Molly O'Meara Sheehan.

Robert Engelman, Janet L. Sawin,
and Michael Renner
Project Directors

Worldwatch Institute
1776 Massachusetts Ave., NW
Washington, DC 20036
worldwatch@worldwatch.org
www.worldwatch.org

Contents

Climate Connections

Boxes

Tables

Figures

Units of measure throughout this book are metric unless common usage dictates otherwise.

Foreword

R. K. Pachauri
Director General, The Energy and Resources Institute
Chairman, Intergovernmental Panel on Climate Change

The Worldwatch Institute's *State of the World* reports have evolved into a remarkable source of intellectual wealth that provides understanding and insight not only on the physical state of this planet but on human systems as they are linked with ecosystems and natural resources around the world. It is especially heartening that the focus of *State of the World 2009* is on climate change.

The contents of this volume are of particular interest as they are based on the findings of the Fourth Assessment Report of the Intergovernmental Panel on Climate Change (IPCC) and provide a comprehensive overview of the policy imperatives facing humanity as we come to grips with this all-important challenge confronting the world today. The IPCC report provided the global community with up-to-date knowledge through an overall assessment of climate change that went substantially beyond its Third Assessment Report. On the basis of strong and robust scientific evidence, the IPCC stated clearly that "warming of the climate system is unequivocal, as is now evident from observations of increases in global average air and ocean temperatures, widespread melting of snow and ice, and rising global average sea level." The evidence from observations of the past 150 years or so leads to some profound conclusions. For instance, 11 of the last 12 years are among the 12 warmest years ever recorded in terms of global surface temperature.

This edition of *State of the World* brings out clearly the difference between inaction based on a business-as-usual approach and action to mitigate greenhouse gas (GHG) emissions in order to avoid the worst impacts of climate change. U.N. Secretary-General Ban Ki-moon has rightly called climate change "the defining challenge of our age." Several world leaders have made similar statements to highlight the importance of taking climate change seriously when developing initiatives and plans for the future. *State of the World 2009* has framed the challenge appropriately by emphasizing the importance of not only new technologies but also a very different approach in terms of human behavior and choices. An important element of future solutions is a different form of global governance—one that would create a high level of seriousness in the implementation of global agreements.

It is profoundly disappointing, for example, that although the United Nations Framework Convention on Climate Change (UNFCCC) came into existence in 1992 it took five more years to provide the convention with an agreement that could be implemented—the Kyoto Protocol. A further source of disappointment is the fact that the Kyoto Protocol, which required ratification by a minimum number of countries accounting for a specific share of greenhouse gas emissions, did not enter into force until 16 February 2005. All of this, unfortunately, provides a sad commentary on the importance that the global community has accorded the problem so far.

It was against this dismal record of inaction, and just after the release of the Synthesis Report of the recent IPCC report, that hopes were raised that the Thirteenth Conference of the Parties to the UNFCCC, held in Bali in December 2007, would finally agree on some firm action on an agreement beyond 2012, the final year covered by the Kyoto Protocol. The meeting was even rescheduled to four weeks after the Synthesis Report was due to be published, so that the delegates would have time to study the IPCC's findings. The Bali Action Plan that was adopted, following a great deal of debate and discussion, certainly provides hope for the future. It is gratifying that the discussions in Bali—and certainly the final declaration—were based predominantly on the assessment contained in the Synthesis Report, the final document in IPCC's Fourth Assessment Report.

State of the World 2009 has been structured logically into chapters that clearly explain the sequence that must guide our understanding of the problem and help set directions for taking action. Particularly relevant is the explanation of what would constitute a safe level of concentration of GHGs. Recall that the main objective of the UNFCCC is stabilization of GHGs in the atmosphere at a level that would prevent dangerous anthropogenic interference with Earth's climate system. Article 2 of the treaty notes that such a level should be achieved within a time frame sufficient to allow ecosystems to adapt naturally to climate change, ensure that food production is not threatened, and enable economic development to proceed in a sustainable manner. Unfortunately, understanding what level of emissions would actually be dangerous is still not clear in policymaking circles around the world.

Several commentators in recent months have expressed deep concern at the current imbalance in the global market for foodgrains, which has hurt some of the poorest people on Earth. There is now mounting evidence that foodgrain output would be threatened by climate change, particularly if the average temperature were to reach 2.5 degrees Celsius above preindustrial levels. Some regions of the world would, of course, be affected far more than others. In Africa, for instance, 75–250 million people would experience water stress as early as 2020 as a consequence of climate change. Some countries on that continent may also be suffering from a 50-percent decline in agricultural yields by then.

The definition of what constitutes dangerous anthropogenic interference is therefore directly related to specific locations, because not only are the impacts of climate change likely to vary substantially across the planet but the capacity to adapt is also very diverse in different societies. What could be labeled as a dangerous level of anthropogenic interference may have already been reached or even exceeded in some parts of the world. Some small island states, for instance, often with land areas not more than a meter or two above sea level, face serious risks from flooding and storm surges that represent a major threat to life and property even today.

Mitigation measures that can help stabilize the concentration of GHGs in the atmosphere have been assessed as generally very low in cost, and most of these carry large-scale cobenefits that in effect reduce the costs further quite significantly. *State of the World 2009* clearly explains the benefits of harnessing low-carbon energy on "a grand scale." The world has been slow in adopting some of these energy options simply because we have not as yet taken full advantage of economies of scale. Nor have we carried out adequate research and development that would allow new technologies to evolve effectively within a short period of time. One important way to develop and disseminate appropriate technologies would be to place a price on carbon, which would provide significant incentives to producers as well as consumers. But there is also an important role for regulatory measures, standards, and codes that can lay down appropriate benchmarks to be observed in different sectors of the economy. Government policy, therefore, will be an important driver of action in the right direction for mitigation of greenhouse gas emissions.

The strongest message from *State of the World 2009* is this: if the world does not take action early and in adequate measure, the impacts of climate change could prove extremely harmful and overwhelm our capacity to adapt. At the same time, the costs and feasibility of mitigation of GHG emissions are well within our reach and carry a wealth of substantial benefits for many sections of society. Hence, it is essential for the world to look beyond business as usual and stave off the crisis that faces us if we fail to act.

This publication comes at a time when governments are focused on reaching an agreement in Copenhagen at the end of 2009 to tackle the challenge of climate change. It will undoubtedly influence the negotiators from different countries to look beyond the narrow and short-term concerns that are far too often the reason for inaction. Indeed, we all need to encourage and join them in showing a determination and commitment to meet this global challenge before it is too late.

State of the World: A Year in Review

Compiled by Lisa Mastny

This timeline covers some significant announcements and reports from October 2007 through September 2008. It is a mix of progress, setbacks, and missed steps around the world that are affecting environmental quality and social welfare.

Timeline events were selected to increase awareness of the connections between people and the environment. An online version of the timeline with links to Internet resources is available at www.worldwatch.org/features/timeline.

NATURAL DISASTERS
Wildfires across drought-stricken southern California char some 2,000 square kilometers, destroying at least 1,500 homes and forcing more than half a million people to evacuate.

© 1986 Andrea Fisch/courtesy Photoshare

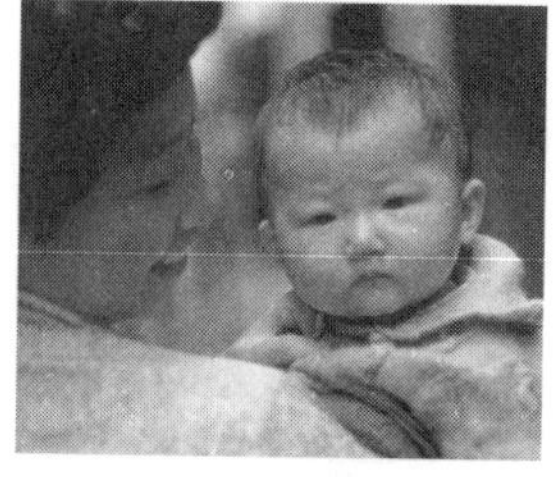

POLLUTION
Russian tanker spills 2,000 tons of heavy fuel oil near the Black Sea, affecting local fishing and bird populations and coating beaches with a thick black sludge.

CLIMATE
Former US Vice President Al Gore and the Intergovernmental Panel on Climate Change win the Nobel Peace Prize for galvanizing international action against climate change.

HEALTH
China reports that birth defects in the nation's infants have soared nearly 40 percent since 2001 due to pollution and worsening environmental degradation.

WILDLIFE
Conservation groups and the government of the Democratic Republic of the Congo create a vast new reserve to protect the endangered bonobo ape, the closest human relative.

OCTOBER — NOVEMBER

2007 STATE OF THE WORLD: A YEAR IN REVIEW

2 4 6 8 10 12 14 16 18 20 22 24 26 28 30 — 2 4 6 8 10 12 14 16 18 20 22 24 26 28

CLIMATE
Scientists say Arctic sea ice has declined to its lowest level since satellite assessments began in the 1970s, opening the Northwest Passage fully for the first time in memory.

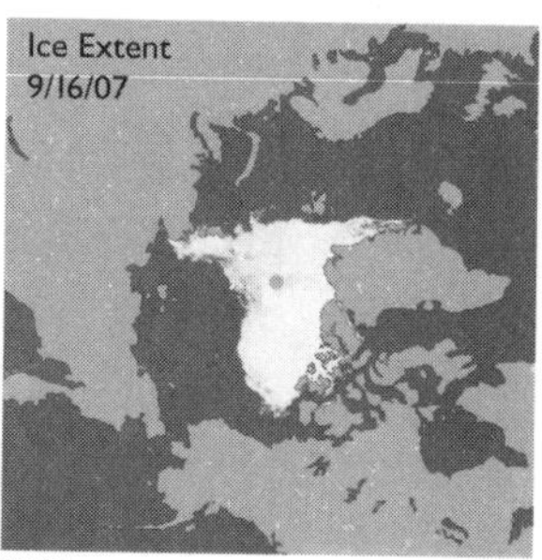

NASA

FISHERIES
Experts say Southeast Asia's oceans are rapidly running out of fish, threatening the livelihoods of some 100 million people and increasing the need for government protection of fish stocks.

WATER
ADB says developing countries in Asia could face an "unprecedented" water crisis in a decade due to climate change, population growth, and mismanagement of water resources.

HEALTH
In a one-day snapshot of obesity, doctors report that 24 percent of men and 27 percent of women worldwide are obese—nearing the obesity levels found in the United States.

NATURAL DISASTERS
Cyclone Sidr lashes Bangladesh, killing some 3,000 people and destroying an estimated 458,000 houses, 350,000 head of livestock, and 60,700 hectares of crops.

BIODIVERSITY
WWF reports that four Antarctic penguin populations are under pressure from climate change as habitat loss and overfishing disrupt breeding and feeding.

© Jenny Rollo

CLIMATE
Report warns that as many as 150 million people in the world's big coastal cities are likely at risk from flooding by the 2070s, more than three times as many as now.

NATURAL DISASTERS
Officials say China is suffering from its worst drought in a decade, leaving millions of people short of drinking water and shrinking reservoirs and rivers.

CLIMATE
Scientists demonstrate that recent warm summers have caused the most extreme Greenland ice melting in 50 years, providing further evidence of global warming.

DECEMBER 2008 JANUARY

2 4 6 8 10 12 14 16 18 20 22 24 26 28 30 2 4 6 8 10 12 14 16 18 20 22 24 26 28 30

ECONOMY
UN says climate change is creating millions of "green jobs" in sectors from solar power to biofuels that will slightly exceed layoffs elsewhere in the economy.

TRANSPORTATION
Indian auto manufacturer Tata unveils its $2,500 "people's car," the Nano, raising concerns about crowded roads and rising pollution.

CLIMATE
Group reports that trade in global carbon credits rose 80 percent in 2007, to $60 billion, up from $33 billion the previous year.

Tata Motors

CLIMATE
At UN climate talks in Bali, nearly 200 nations agree to launch negotiations on a new climate change treaty following a groundbreaking reversal of US position.

FORESTS
Brazilian scientist says Amazon deforestation is likely to increase in 2008 for the first time in four years, raising concerns about the effectiveness of national forest protection policies.

WILDLIFE
The eight South Asian nations agree to cooperate more in addressing wildlife trade problems in the region, one of the prime targets of organized wildlife crime networks.

© Lyle Rosbotham

ECONOMY
Report says global investments in renewable energy topped $100 billion for the first time in 2007, led by wind power and driven by supportive policies.

ENERGY
Price of oil passes the all-time inflation-adjusted peak of $103.76 set in April 1980 and is now three times what it was four years ago.

Photodisc

CLIMATE
Study reports that the US West is warming at nearly twice the rate of the rest of the world and is likely to face more drought conditions in many of its fast-growing cities.

FEBRUARY MARCH

2008 STATE OF THE WORLD: A YEAR IN REVIEW

2 4 6 8 10 12 14 16 18 20 22 24 26 28 2 4 6 8 10 12 14 16 18 20 22 24 26 28 30

AGRICULTURE
Global seed vault in Svalbard, Norway, opens with 100 million food crop seeds from more than 100 countries, the most comprehensive and diverse collection in the world.

Adzuki beans, Wikimedia

Switchgrass, NREL

ENERGY
Studies report that more greenhouse gases are released when clearing land to grow current biofuel crops than would be reduced when the biofuels displaced fossil fuels.

CLIMATE
UN reports that the world's glaciers are continuing to melt away, with record losses reported between 2004–05 and 2005–06 and the average rate of melting and thinning more than doubling.

PUBLIC EDUCATION
Some 50 million people worldwide participate in Earth Hour, switching off lights in some 370 cities in more than 35 countries to raise awareness of climate change.

CLIMATE
Report says capping carbon emissions would cost US households less than a penny on the dollar over 20 years, refuting claims that mandatory limits would damage the economy.

NATURAL DISASTERS
Cyclone Nargis kills some 78,000 people and leaves millions homeless in Myanmar, while critics blame mangrove destruction and a slow government response for the high fatality rate.

Sgt. Andres, USMC

FORESTS
Brazilian environment minister and rainforest activist Marina Silva resigns after facing ongoing struggles with the Lula administration over Amazonian forest policies.

CLIMATE
Study says financial incentives for cutting carbon emissions from deforestation could earn developing countries up to $13 billion in carbon credits per year.

Agência Brasil

APRIL

MAY

2 4 6 8 10 12 14 16 18 20 22 24 26 28 30 2 4 6 8 10 12 14 16 18 20 22 24 26 28 30

CONSUMPTION
San Francisco reports a 70 percent recycling rate—the highest in the United States—through measures such as recycling, composting, and reuse.

Vicky S

NATURAL DISASTERS
A 7.9 magnitude earthquake hits China's Sichuan province, killing some 70,000 people, injuring 374,000 more, and leaving 4.8 million homeless.

ENERGY
Texas oilman T. Boone Pickens places the largest-ever order for wind turbines, spending $2 billion for 667 turbines to develop the world's largest wind farm.

© Lyle Rosbotham

CONSUMPTION
China bans the production and use of plastic bags in supermarkets and retail shops as part of a campaign to fight "white pollution" in the country.

FOOD
World Food Programme announces it will provide $1.2 billion in additional food aid for the 62 countries hit hardest by the food and fuel crisis.

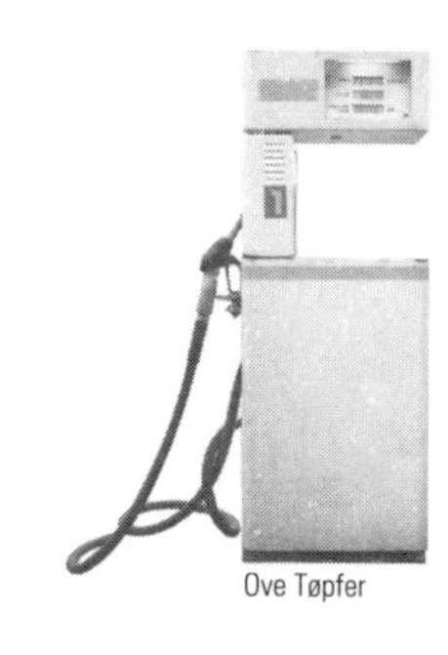
Ove Tøpfer

ENERGY
US average price for a gallon of regular gasoline tops $4 for the first time ever.

CLIMATE
Study reports that China's CO_2 releases accounted for two thirds of increased global emissions in 2007 and are 14 percent higher than those from the United States.

Photodisc

FOOD
FAO says rising land degradation reduces crop yields and may threaten the food security of 1.5 billion people, about a quarter of the world's population.

Photodisc

ENERGY
The price of oil hits a new all-time inflation-adjusted high of $147.27.

Photodisc

FORESTS
Reports say booming demand for food, fuel, and wood as world population surges will put unprecedented and unsustainable demands on remaining forests.

WILDLIFE
China wins the right to make a one-off purchase of registered elephant ivory stocks from four African countries under strict conditions.

GOVERNANCE
Internal review says World Bank investments fail to give enough attention to long-term sustainability and place uneven emphasis on economic benefits of environmental preservation.

JUNE — JULY

2008 STATE OF THE WORLD: A YEAR IN REVIEW

2 4 6 8 10 12 14 16 18 20 22 24 26 28 30 | 2 4 6 8 10 12 14 16 18 20 22 24 26 28 30

ENERGY
Pacific Gas and Electric agrees to purchase 800 megawatts of solar cells, the largest such sale ever, to be installed in two solar farms that can supply electricity to 239,000 homes.

Ewout Bos

MARINE SYSTEMS
Researchers say the number of "dead zones" in the world's oceans and coastal areas has nearly doubled every decade since the 1960s, to some 400, due mainly to fertilizer runoff.

ENERGY
Report says US installed wind capacity exceeds 20,000 megawatts, enough electricity to serve 5.3 million American homes and making the US the world leader in wind power capacity.

NREL

CLIMATE
Ten northeastern states hold the first US cap and trade auction of carbon dioxide emissions allowances, raising nearly $40 million for renewable energy technologies and energy efficiency programs.

AUGUST — SEPTEMBER

See page 205 for sources.

2 4 6 8 10 12 14 16 18 20 22 24 26 28 30 | 2 4 6 8 10 12 14 16 18 20 22 24 26 28

CONSUMPTION
China celebrates the opening of what it calls the first "green" Olympics, after spending some $20 billion on mass transit and the addition of new renewable energy systems in Beijing.

Joowwww

ENERGY
UN says abolishing some $300 billion in global subsidies for fossil fuels could cut world greenhouse gas emissions by up to 6 percent while also boosting economic growth.

CONSUMPTION
Study says exports now account for one third of China's CO_2 emissions as manufacturers there feed a growing global appetite for cheap goods.

DOE

ENERGY
U.S government lifts a longstanding ban on offshore drilling, opening most of the country's coastline to oil and gas leasing and exploration.

2 0 0 9

STATE OF THE WORLD

Into a Warming World

About This Book

It is New Year's Day, 2101. Somehow, humanity survived the worst of global warming—the higher temperatures and sea levels and the more intense droughts and storms—and succeeded in stabilizing Earth's climate. Atmospheric greenhouse gas concentrations peaked a few decades ago and are expected to continue their downward drift throughout the twenty-second century. Global temperatures are slowly returning to their pre-warming levels. The natural world is gradually healing. The social contract largely held. And humanity as a whole is better fed, healthier, and more prosperous today than it was a century ago. What did humanity do in the twenty-first century—and especially in 2009 and the years immediately following—to snatch a threatened world from the jaws of climate change catastrophe?

This is the scenario for success that the *State of the World 2009* Project Directors challenged each of the book's authors to address. The goal was to go beyond the short-term thinking about climate change that prevails today and to explore more deeply its implications for humanity and the planet. To do that, this edition departs in important ways from the 25-year tradition of Worldwatch Institute's annual book, gathering more than 40 authors—far more than in any previous edition. The talent represented in these pages is rich and diverse. More than a dozen authors are natives of or have firm roots in the developing countries so important to the book's theme: how to keep climate change at manageable levels and how to adapt to what is coming our way no matter how successful we are in reducing future emissions of greenhouse gases.

The first chapter in *State of the World 2009* presents the climate dilemma; the second, the emissions path needed to glide toward a safe landing. The third and fourth describe the needed transitions toward carbon-absorbing forestry and food production and toward a low-carbon and eventually a no-carbon energy future. The fifth lays out the importance of building resilience to climate change. The sixth proposes components of the agreement that nations must reach to begin stabilizing the climate, even while adapting to a warming world. And in another first for *State of the World*, the middle of the book features a large selection of short pieces called *Climate Connections.* These take on 22 critical topics on the theme of preventing and addressing climate change. The book ends with a Climate Change Reference Guide and Glossary that aims to be a useful primer for following the developments on climate change that will unfold this year.

State of the World 2009: Into a Warming World provides hope amidst the grim certainty that we are living in the early years of a vast unplanned change in the planet's climate. All the authors in this book agree that it is anything but too late to save the climate for an enduring human civilization. Yet the subtitle was chosen carefully and after much discussion: We are entering a warming world. Human alterations of the atmosphere and climate will without doubt outlive the readers of this book. But we are privileged to live in a brief window of time when human beings can act decisively to stop the warming before its impacts become impossible to reverse or to tolerate. How we handle the challenge ahead will make for history on an epic time scale.

CHAPTER 1

The Perfect Storm

Christopher Flavin and Robert Engelman

Something extraordinary happened at the top of our planet in the past three summers. For a few weeks each year—in the final days of the northern summer—a large stretch of open water appeared around the Arctic, making it briefly possible to pilot a ship from the Atlantic to the Pacific without going through the Panama Canal or around the Cape of Good Hope. Never before in recorded human history has it been possible to make that journey.[1]

As a barometer of global environmental change, the loss of the permanent ice cap at the North Pole is like a seismograph that suddenly jumps off the charts. For several decades now, Earth's heat balance has been severely out of equilibrium. Earth is absorbing more heat than it is emitting, and across the planet ecological systems are responding. The changes so far have been almost imperceptible, and even now they appear from the human viewpoint gradual.

But don't be fooled: the changes represented by melting glaciers, acidifying oceans, and migrating species are—on a planetary timescale—breaking all known speed limits. The planet that humans have known for 150,000 years (encompassing the Pleistocene and Holocene epochs, as geologists describe them) is changing irrevocably thanks to human actions. In 2000 the Nobel Prize-winning chemist Paul Crutzen and his colleague Eugene F. Stoermer concluded that these changes are so profound that the world has entered a new geological epoch—which they aptly named the Anthropocene.[2]

Changing Earth's climate is like sailing a massive cargo ship. Tremendous energy is required to get such a ship moving—and its forward progress is at first almost imperceptible—but once it is traveling at full speed, it is very hard to stop. It is now virtually certain that children born today will find their lives preoccupied with a host of hardships created by an inexorably warming world. Food supplies will be diminished, and many of the world's forests will be destroyed. Not just the coral reefs that nurture many fisheries but the chemistry of the oceans will face disruption. Indeed, the world's oceans are already acidifying rapidly. Coastlines will be rearranged, and so will the world's wetlands. Whether you are a farmer or an office worker, whether you live in the northern or southern hemisphere, whether you are rich or poor, you will be affected.[3]

Fiddling While the World Burns

Like a distant tsunami that is only a few meters high in the deep ocean but rises dramatically as it reaches shallow coastal waters, the great wave of climate change has snuck up on people—and is now beginning to break. Climate change was first identified as a potential danger by a Swedish chemist in the late nineteenth century, but it was not until the late 1980s that scientists had enough evidence to conclude that this transformation was under way and presented a clear threat to humanity.

An American scientist, James Hansen of the National Aeronautics and Space Administration, put climate change squarely on the agenda of policymakers on 23 June 1988. On that hot summer day, Hansen told a U.S. Senate Committee he was 99 percent certain that the year's record temperatures were not the result of natural variation. Based on his research, Hansen had concluded that the rising heat was due to the growing concentration of carbon dioxide (CO_2) and other atmospheric pollutants. "It's time to stop waffling so much and say that the evidence is pretty strong that the greenhouse effect is here."[4]

Hansen's words, joined with those of other scientists, echoed around the world. Within months government officials were beginning to consider steps to reduce greenhouse gas emissions, with much of the focus on the kind of international agreement that would be needed to tackle this most global of problems. In 1992 the United Nations Framework Convention on Climate Change was adopted by heads of state in Rio de Janeiro, and in 1997 the Kyoto Protocol, with its legally binding emissions limits for industrial countries, was negotiated.[5]

As the 1990s came to an end the world appeared to be moving to tackle the largest and most complex problem humanity has ever faced. But fossil fuel interests mobilized a counterattack—pressuring governments and creating confusion about the science of climate change. Taking advantage of the inevitable uncertainties and caveats contained in leading climate assessments, a handful of climate skeptics—many of them PhDs with oil industry funding—managed to position climate change as a scientific debate rather than a grim reality.

The climate change skeptics had their greatest influence in the United States, putting it at loggerheads with the European Union, which since the early 1990s has been the strongest advocate of action on climate change. In November 2000, in the waning days of the Clinton administration, climate negotiators met in The Hague with the intention of finalizing details of the Kyoto Protocol—which in principle had been agreed to three years earlier. Two weeks of intense discussions concluded with an agonizing all-night session that ended in failure. Distrust and miscommunication between American and European negotiators were at the heart of this historic diplomatic failure—a failure that became more significant a short time later when the U.S. Supreme Court decided that Al Gore would not be the next President of the United States.[6]

In the months that followed, many remained optimistic: before his election, President George W. Bush had indicated his support for addressing the climate problem and working cooperatively with other countries. Two months later—under heavy pressure from Vice President Cheney and the oil industry—he executed an abrupt U-Turn, rejecting the Kyoto Protocol outright and throwing negotiations into a tailspin. Europe, Canada, Japan, and Russia were shocked into completing and ultimately ratifying the Kyoto Protocol in the following years, but time and political momentum had been lost. More significantly, the unilateral actions of the U.S. government deepened North-South fissures on climate change—a divide that has now become the largest obstacle to progress.[7]

Storm Clouds Gather

The tragedy of these two wasted decades is that during this period the world has moved from a situation in which roughly a billion people in industrial countries were driving the problem—the United States, for example, has 4.6 percent of the world's population but accounts for 20 percent of fossil-fuel CO_2 emissions—to today's reality in which the far larger populations of developing countries are on the verge of driving an even bigger problem.[8]

Global emissions of carbon dioxide from fossil fuel combustion and cement production rose from 22.6 billion tons in 1990 to an estimated 31 billion tons in 2007—a staggering 37-percent increase. This is 85 million tons of carbon dioxide spilled into the atmosphere each day—or 13 kilograms on average per person. The annual increase in emissions shot from 1 percent a year in the 1990s to 3.5 percent a year from 2000 to 2007—with China accounting for most of that remarkable leap.[9]

Between 1990 and 2008 U.S. emissions of carbon dioxide from fossil fuel combustion grew by 27 percent—but emissions in China rose 150 percent, from 2.3 billion to 5.9 billion tons. More suddenly and dramatically than experts had expected, China and other developing countries are entering the energy-intensive stages of economic development, and their factories, buildings, power plants, and cars are consuming vast amounts of fossil fuels. As recently as 2004, the International Energy Agency projected that it would be 2030 before China passed the United States in emissions. It now appears that the lines crossed in 2006.[10]

Accelerating emissions are not the only factor driving increased concern. Tropical deforestation—estimated at 13 million hectares per year—is adding 6.5 billion tons of carbon dioxide to the atmosphere annually. The world's largest tropical forest, the Amazon, is disappearing at a faster pace as high agricultural prices encourage land clearing. More alarmingly, Earth's natural sinks—its oceans and biological systems—appear to be losing their ability to absorb a sizable fraction of those emissions. As a result, the increase in atmospheric CO_2 concentrations has accelerated to the fastest rate ever recorded.[11]

Scientists are reticent by nature, and the overwhelming complexity and inevitable uncertainty of the climate problem have led them to produce equivocal and hard-to-interpret studies that have given considerable comfort to those who argue it is too early to act on climate change. In the past year, however, a few brave scientists have cast reticence aside. Speaking in Washington on the twentieth anniversary of his historic testimony, James Hansen had a sharp warning for policymakers: "If we don't begin to reduce greenhouse gas emissions in the next several years, and get on a very different course, then we are in trouble....This is the last chance."[12]

Climate scientists have discovered a particularly inconvenient truth: by the time definitive predictions of climate change are adopted by scientific consensus, the climate system may have reached a tipping point at which climate change begins to feed on itself—and becomes essentially irreversible for centuries into the future. The loss of Arctic ice, for example, will allow more sunlight to heat the Arctic Ocean, accelerating the buildup of heat and putting the vast Greenland ice sheet at risk. And there are early indications that the rapid rise in Arctic temperatures is thawing the tundra and thereby releasing additional amounts of CO_2 and methane.

The political will for change is building, thanks to the strong base in science and widening public awareness of climate change and its risks.

These dramatic changes will affect the entire planet, but the world's poor will suffer first and suffer most. The latest climate models indicate particular vulnerability in the dry tropics, where the food supplies for hundreds of millions of people will be undermined by climate change. Hundreds of millions more who live in the vast Asian mega-deltas will be at risk from rising sea levels and increased storm intensity. Health threats from malaria, cholera, and other diseases that are likely to flourish in a warmer world will add to the burdens facing the world's poor. The fact that many of the 1.4 billion people who now live in severe poverty already face serious ecological debts—in water, soil, and forests—will exacerbate the new problems presented by climate change.[13]

When they were released in 2007, the latest findings of the Intergovernmental Panel on Climate Change were taken as an urgent warning of the dangers ahead. But the torrent of scientific data to emerge since then has led some scientists to sharpen their advice. James Hansen and W. L. Hare of Germany's Potsdam Institute are among those who have concluded that to prevent "dangerous climate change"—the goal that governments have already agreed to—global emissions must begin declining within the decade and then fall to no more than half the current level—and possibly even to zero—by the middle of this century. (See Chapter 2.)[14]

This is a tall order indeed. Some would call it impossible. But the resources, technologies, and human capacity for change are all in place. The missing ingredient is political will, and that is a renewable resource.

A New Political Climate

Over the past few years, political will to tackle the climate problem has grown in many countries around the world. The European Union has committed to reducing its emissions to 20 percent below the 1990 level in 2020—and to reaching 30 percent if other industrial countries join them in a strong international agreement. And the political will for change is building, thanks to the strong base in science and widening public awareness of climate change and its risks. In late 2007, Australians voted out a conservative government in part out of impatience with the Prime Minister's unwillingness to support the Kyoto Protocol; the new Prime Minister promptly secured its ratification. His first trip outside Australia was to a climate negotiation in Bali, and his government has been working to build a national climate plan ever since.[15]

In the United States, climate policy is raging like a prairie fire at the state level. By late 2008, some 27 states had adopted climate plans, and groups of eastern and western states are developing their own regional emissions cap and trade systems. In April 2008, the governors of 18 states gathered at Yale University to proclaim: "Today, we recommit ourselves to the effort to stop global warming, and we call on congressional leaders and the presidential candidates to work with us—in partnership—to establish a comprehensive national climate policy." And the U.S.

business community is responding as well: 27 major corporations, including Alcoa, Dow Chemical, General Motors, and Xerox, have announced their support for caps on national greenhouse gas emissions.[16]

Developing countries are joining in too. In June 2008, the prime minister of India released the much-anticipated National Action Plan on Climate Change. It focuses on eight areas intended to deliver maximum benefits in terms of domestic climate change mitigation and adaptation: solar energy, energy efficiency, sustainable habitat, water, sustaining the Himalayan ecosystem, green India, sustainable agriculture, and sustainable knowledge for climate change. China announced a new climate plan in 2007, and during the course of 2008 continued to strengthen its energy efficiency programs, including a new incentive system that ties promotion of local officials to their success in saving energy.[17]

These advances are welcome. But the world needs to change course much faster. To concentrate the attention of policymakers, a mass global movement is needed in support of a new climate treaty that picks up where the Kyoto Protocol leaves off in 2012. It is everyone's planet, after all, and everyone's climate. There are signs that such a public movement is now growing in industrial as well as developing countries, but it is not yet sufficiently strong or pervasive to counter the vested interests that stand on the other side.

One reason is that climate negotiations are numbingly hard to follow. Outside of a hard-working community of government negotiators, nongovernmental organizations, and academics, most people have little sense of what is happening. In a modest effort to help demystify the process, this book eschews terms of art and uses everyday language as much as possible. (See the Climate Change Guide following Chapter 6 for a glossary of terms used in the climate debate.)

Ten Key Challenges

Ten challenges must be met in order to create the world of zero net greenhouse gas emissions that will be needed to achieve climate stability.

Thinking Long-term. Human beings have evolved to be very good at focusing on an immediate threat—whether it is wild animals the first humans faced on the plains of Africa or the financial panic that gripped the world in late 2008. Climate change is a uniquely long-range problem: its effects appear gradual on a human time scale, and the worst effects will likely be visited on people not yet alive. To solve this problem, we must embrace the future as our responsibility and consider the impact of today's decisions on future generations. Just as Egyptians built pyramids and Europeans built cathedrals to last millennia, we need to start acting as if the future of the planet matters beyond our own short lives.

Innovation. The world needs to develop and disseminate technologies that maximize the production and use of carbon-free energy while minimizing cost and optimizing convenience. (Convenience matters: the ease of transporting, storing, and using carbon-based fuels is among their attractions, not captured in price alone.) An effective climate pact will offer incentives that accelerate technological development and ensure that renewable energy and other low-emission technologies are deployed in all countries regardless of ability to pay the costs. (See Chapter 4.) We need to dramatically increase the efficiency with which we use carbon-based energy and lower release into the atmosphere of land-based CO_2, methane, nitrogen oxides, and greenhouse gases stemming from cooling and various industrial processes. The opportunities for quick and inexpensive emissions reductions remain vast and mostly untapped.

Population. It is essential to reopen the global dialogue on human population and promote policies and programs that can help slow and eventually reverse its growth by making sure that all women are able to decide for themselves whether and when to have children. A comprehensive climate agreement would acknowledge both the impacts of climate change on vulnerable populations and the long-term contribution that slower growth and a smaller world population can play in reducing future emissions under an equitable climate framework. And it should renew the commitment that the world's nations made in 1994 to address population not by pressuring parents to have fewer or more children than they want but by meeting the family planning, health, and educational needs of women.[18]

Changing Lifestyles. The world's climate cannot be saved by technology alone. The way we live will have to change as well—and the longer we wait the larger the needed sacrifices will be. In the United States, the inexorable increase in the size of homes and vehicles that has marked the past few decades has been a major driver of greenhouse gas emissions and the main reason that U.S. emission are double those of other industrial countries. Lifestyle changes will be needed, some of which seem unattractive today. But in the end, the things we may need to learn to live without—oversized cars and houses, status-based consumption, easy and cheap world travel, meat with every meal, disposable everything—are not necessities or in most cases what makes people happy. The oldest among us and many of our ancestors willingly accepted such sacrifices as necessary in times of war. This is no war, but it may be such a time.

Healing Land. We need to reverse the flow of carbon dioxide and other greenhouse gases from destroyed or degraded forests and land. Soil and vegetation can serve as powerful net removers of the atmosphere's carbon and greenhouse gases. (See Chapter 3.) Under the right management, soil alone could absorb each year an estimated 13 percent of all human-caused carbon dioxide emissions. To the extent we can make the land into a more effective "sink" for these gases we can emit modest levels essential for human development and well-being. Like efficiency, however, an active sink eventually faces diminishing returns. And any sink needs to be secured with "drain stoppers" to prevent easy return of greenhouse gases to the atmosphere when conditions change.[19]

Strong Institutions. "Good governance" can be a cliché—until someone needs it to survive. The final months of 2008 laid painfully bare the dangerous imbalance between a freewheeling global economy and a regulatory system that is a patchwork of disparate national systems. And if there was ever a global phenomenon, the climate is it. In fact it is not hard to imagine the climate problem driving a political evolution toward global governance over the long term, but given the public resistance to that idea the next most effective climate-regulating mechanism will be the strength and effectiveness of the United Nations, multilateral banks, and major national governments. New institutions and new funds will be needed, but it could take a major public awakening or a dramatically deteriorating climate to overcome the obstacles to inventing and establishing them.

The Equity Imperative. A climate agreement that can endure and succeed will find mechanisms for sharing the burden of costs and potential discomforts. Per capita fossil fuel CO_2 emissions in the United States are almost five times those in Mexico and more than 20 times the levels in most of sub-Sahara. An effective climate agreement will acknowledge the past co-optation of Earth's greenhouse-gas absorbing capacity by the wealthiest and most industrialized countries and the corresponding need to reserve most of what little absorbing capacity is left for countries in development. Most people live in such countries, and they bear little responsibility for causing this problem—though it is worth recalling that a small but growing share of their populations already have large carbon footprints.[20]

Economic Stability. In the fall of 2008 the global economy foundered, raising the obvious question: can a world heading into hard economic times add to its burdens the costs of switching from fossil to renewable fuels or managing precious land for carbon sequestration? Any climate agreement built on an assumption of global prosperity is doomed to failure. And as growing and increasingly affluent populations demand more of the resources of a finite planet, we may have to balance the future of climate against present realities of hunger, poverty, and disease. A robust international climate regime will need to design mechanisms that will operate consistently in anemic as well as booming economic times. And a strong pact will be built on principles and innovations that acknowledge and accommodate the problem of cost—while building in monitoring techniques to ensure that efficiency is not achieved at the expensive of effective and enduring emission cuts and adaptation efforts.[21]

Political Stability. A world distracted by major wars or outbreaks of terrorism will not be able to stay focused on the more distant future. And just such a focus is needed to prevent future changes in climate and adapt to the ones already occurring. A climate pact could encourage preemptive action to diminish insecurity caused or exacerbated by climate change. But unless nations can find ways to defuse violent conflict and minimize the chance that terrorism will distract and disrupt societies, climate change prevention and adaptation (along with development itself) will take a back seat. On the bright side, negotiating an effective climate agreement offers countries an opportunity, if they will only seize it, to practice peace, to look beyond the narrowness of the interests within their borders at their dependence on the rest of the world, to see humanity as a single vulnerable species rather than a collection of nations locked in pointless and perpetual competition.

Solving the climate problem will create the largest wave of new industries and jobs the world has seen in decades.

Mobilizing for Change. As fear of climate change has grown in recent years, so has political action. But opponents of action have repeatedly pointed to the vast costs of reducing emissions. At a time of serious economic problems, the power of that argument is growing, and some of those who are persuaded are going straight from denial to despair. The most effective response to both of those reactions is, in the words of Common Cause founder John Gardner, to see global warming as "breathtaking opportunities disguised as insoluble problems." Solving the climate problem will create the largest wave of new industries and jobs the world has seen in decades. Michigan, Ohio, and Pennsylvania in the United States are among those that have devoted enormous efforts to attracting new energy industries—with a glancing reference to climate change and a major focus on creating new jobs to revive "rustbelt" economies.[22]

In November 2009, the world faces a test. Will the roughly 200 national governments that meet in Copenhagen to forge a new climate agreement come up with a new protocol that provides both vision and a roadmap, accelerating action around the globe? The challenges are many: Will the global financial crisis and conflict in the Middle East distract world leaders? Will the new U.S president have time to bring his country back into a leadership position? Will the global North-South divide that has marked climate talks in recent years be overcome?

State of the World 2009 presents some potential answers to these challenges. One vital theme stands out from the rest: climate change is not a discrete issue to be addressed apart from all the others. The global economy fundamentally drives climate change, and economic strategies will need to be revised if the climate is ever to be stabilized—and if we are to satisfy the human needs that the global economy is ultimately intended to meet.

We cannot afford to have the Copenhagen climate conference fail. The outcome of this meeting will be written in the world's history books—and in the lasting composition of our common atmosphere.

CHAPTER 2

A Safe Landing for the Climate

W. L. Hare

Our climate system is in trouble. It has warmed by over 0.7 degrees Celsius in the last 100 years. Most of the warming since at least the mid-twentieth century is very likely due to human activities. Warming's impacts on human and natural systems are now being observed nearly everywhere—perhaps most obviously in the recent loss of Arctic sea ice, which in 2007 and 2008 reached record low levels at the end of the northern summer. In spite of nearly 20 years of international attention, emissions of greenhouse gases (GHGs)—principally carbon dioxide (CO_2) from the burning of fossil fuels—continue to grow rapidly. As a consequence, the concentration of carbon dioxide in the atmosphere has increased faster during the last 10 years than at any time since continuous measurements began in 1960.[1]

Unabated, current increasing trends in emissions can be expected to raise Earth's temperature by a further 4–6 degrees Celsius (7.2–10.8 degrees fahrenheit), if not more, by the end of this century. If even half that much warming occurs, it will bring huge damages and potentially catastrophic problems. The Fourth Assessment Report of the Intergovernmental Panel on Climate Change (IPCC), which was released at the end of 2007, predicted serious risks and damages to species, ecosystems, human infrastructure, societies, and livelihoods in the future unless warming is reduced. The report's projected risks and damages are larger and more serious than previously estimated and threaten development in several regions of the world. The IPCC also found that reducing greenhouse gas emissions would lower the global temperature increase and consequently lessen the risks and damages. Yet it is also important to note at the outset that even reducing emissions 80 percent by 2050 will not eliminate all serious risks and damages.[2]

One of the great icons of the modern world,

W. L. Hare is a scientist in Earth System Analysis at the Potsdam Institute for Climate Impact Research in Germany and advises Greenpeace International on climate policy and science.

the jet aircraft, provides a telling metaphor for what the world faces in terms of climate change. Jet aircraft burn prodigious quantities of fossil fuels in order to move passengers and freight across vast distances in relative safety and luxury. Yet like the climate system, the rules of operating these machines are not widely understood by anyone except the few people whose job it is to know about such things. The climate system is like a jet aircraft that has become airborne safely but is now facing grave difficulty and must land as a matter of urgency before disaster becomes inevitable. If we do not reduce emissions fast enough and bring the warming of the climate system to a halt, we risk a major catastrophe.

This chapter is about how much and how fast the world needs to reduce greenhouse gas emissions in order to prevent or limit serious damage—in other words, to bring the climate system to a safe landing. But first it is important to review the current state of scientific knowledge on the risks, damages, and impacts estimated for different levels of warming in order to see what level might prevent dangerous changes and thus be "safe."

Preventing dangerous climate change is the universally agreed ultimate goal of climate policy established in the 1992 U.N. Framework Convention on Climate Change (UNFCCC). (See Box 2–1.) Once a dangerous level of change has been defined, scientists can calculate with reasonable confidence an emission pathway that can limit warming and other changes to this level, taking into account continuing uncertainties in their understanding of the climate system.[3]

Projected Climate Change and Sea Level Rise

In the latest IPCC report the projected levels of global warming in the absence of efforts to reduce emissions are not dramatically different from those made in earlier reports: warming by 2100 is projected to be in the range of 1.1–6.4 degrees Celsius above the average in the 1980–99 period. Given that emissions, warming, and sea level rise during the current decade have all been at the upper end of projected ranges, it would be prudent to assume that the likely warming in the absence of major emission reductions over the next century will be toward the mid or upper end of the range projected by the IPCC.[4]

The main reference point for greenhouse gas concentrations and temperature increases is typically preindustrial times. This is usually taken as 1750, so preindustrial CO_2 concentration levels are given as 278 parts per million (ppm) CO_2. Increases in greenhouses gases (taken together as CO_2-equivalent (CO_{2eq}) concentrations) are generally related to this number. A doubling of GHG concentrations means an increase that is equivalent to the effect of about 556 ppm CO_2 (often just rounded to 550 ppm CO_2).

As far as possible, global temperature increases here are referred to as increases above the preindustrial level. Given that a global instrumental temperature series only exists for the period after 1850, the preindustrial period is defined as the 30-year average from this year. (The average global mean temperatures between 1750 and the 1850s were quite similar, so this is considered satisfactory.) From the 1850s to the five-year period ending in 2007, global mean temperature increased by more than 0.7 degrees Celsius. In the IPCC report, projections are often stated with respect to the period 1980–99 (with 1990 used as the midpoint), which was a bit over 0.5 degrees Celsius warmer than the preindustrial period. So the IPCC's projected increase for the twenty-first century of 1.1–6.4 degrees Celsius above 1980–99 levels would be about 1.6–6.9

Box 2–1. Preventing Dangerous Climate Change

The guiding principles of international efforts to deal with climate change were established in 1992 in the United Nations Framework Convention on Climate Change, which was adopted in Rio de Janeiro at the Earth Summit: "The ultimate objective of this Convention and any related legal instruments...is to achieve...stabilization of greenhouse gas concentrations in the atmosphere at a level that would prevent dangerous anthropogenic interference with the climate system. Such a level should be achieved within a time-frame sufficient to allow ecosystems to adapt naturally to climate change, to ensure that food production is not threatened and to enable economic development to proceed in a sustainable manner."

This is a powerful statement, as it contains a legally binding requirement to prevent dangerous changes. In practice, however, exactly what this means remains undefined in international law. The article is ambiguous, as it leaves open core questions such as dangerous to whom and to what. What if food production increases in some regions due to global warming and increased CO_2 concentration, as is projected for the northern high latitudes, but decreases perhaps dangerously in other regions, as is projected for low-latitude tropical regions such as Africa? Is that dangerous within the meaning of the convention? Answering such questions is fundamental to the development of a fair and equitable global approach to climate change.

While most attention in debates about climate change has focused on changes in climate, it needs also to be noted that under Article 2 "dangerous anthropogenic interference" relates to the climate system as whole: changes in ocean acidity due to human-induced CO_2 increases that result in adverse changes in the oceans and marine ecosystems could also be deemed dangerous. University of Toronto climatologist Danny Harvey has pointed that there are important differences between terms such as dangerous interference and dangerous climate change. (For simplicity's sake, however, these are used synonymously in this chapter.)

Decisions as to what is "dangerous" fundamentally affect the rate, timing, and scale of emissions reductions required regionally and globally in the coming years and decades. If "dangerous interference" is considered to begin only once the global average temperature exceeds 4 degrees Celsius above the preindustrial level, then it will be hard to justify urgent and stringent mitigation action in the next 10–30 years, as greenhouse gas emissions would not need to peak until well after the 2050s before dropping. If, on the other hand, warming of more than 2 degrees above preindustrial is deemed dangerous, then there is acute and urgent emphasis on near-term emission actions leading to large global emissions reductions of 80 percent or more by 2050.

degrees Celsius above preindustrial level. Since 1980–99, the climate system has already warmed about 0.25 degrees Celsius.[5]

For projected sea level rise the IPCC was unable to estimate fully all the contributions of global warming, as numerical computer models of the ice sheets of Greenland and Antarctica cannot yet adequately project the effects. So the range of sea level rise estimated by the IPCC—between 0.18 and 0.59 meters by 2100 above 1980–99 levels—was heavily qualified, given that the possible future rapid loss of ice from Greenland and Antarctica could not be quantified. The already observed rapid loss of ice in response to recent warming of the atmosphere and ocean around Greenland and West Antarctica indicates that these ice sheets could be more vulnerable to warming than implied by ice sheet models and hence could add significantly to future sea level rise. As a consequence, the IPCC could not give a "best estimate" or upper bound for sea level rise.[6]

After the writing of the IPCC science

report was completed, Stefan Rahmstorf of the Potsdam Institute for Climate Impact Research projected future sea level rise based on the observed relationship between sea level and temperature over the last century. Using a similar range of emission and climate projections, he estimated a sea level rise in the range of 0.5–1.4 meters above 1990 levels by 2100. More recent work indicates that the increase during this century could be even higher. In short, the evidence points to a likelihood of meter-scale sea level rise by 2100, well above the top end of the range quantified by the IPCC. Thus, much larger risks to coastal zones and small islands seem likely during this century than had previously been estimated.[7]

There is much greater confidence now than in earlier IPCC assessments in the regional changes that can be expected in a warmer world. Warming will be greatest in the high north and in the interiors of the continents. Reduction in snow cover, a thawing of permafrost, and decreases in the extent of sea ice in both hemispheres can be expected.[8]

Weather extremes and water availability are two of the most important projections in terms of impacts on human and natural systems. More-frequent heat extremes and heat waves, more-intense tropical cyclones, and heavier precipitation and flooding can be expected in many regions. Recent projections confirm that extreme high surface temperatures will rise faster than global warming and indicate a 10 percent chance of "dangerously high" surface temperatures over 48 degrees Celsius every decade in much of the world by 2100 if the global temperature exceeds 4 degrees Celsius above the preindustrial level.[9]

Precipitation can be expected to decrease in most subtropical land regions but to increase in the high latitudes. The IPCC assessment found with "high confidence that many semi-arid areas (e.g. Mediterranean basin, western United States, southern Africa and northeast Brazil) will suffer a decrease in water resources due to climate change." By the 2050s it is projected that there will be less annual river runoff and water availability in dry regions in the mid-latitudes and tropics but an increase in high-latitude regions and in some tropical wet areas.

Especially Affected Systems, Sectors, and Regions

For the first time the systems, sectors, and regions most likely to suffer adverse effects were identified in the latest IPCC report, providing important details of risks, impacts, and vulnerabilities at different levels of future warming. The especially affected ecosystems identified were tundra, boreal forest and mountain regions, Mediterranean types, tropical rainforests where precipitation declines, coral reefs, mangroves and salt marshes, and systems dependent on sea ice. A sector identified as of special concern is the health of vulnerable populations who have a low capacity to adapt. As Hurricane Katrina and the European heat wave of 2003 showed, even in high-income countries the poor, the elderly, and young children can be particularly at risk from climatic extremes.[10]

For sea ice, the IPCC projected a decrease in both the Arctic and Antarctic under every unmitigated emissions scenario, with summer sea ice in the Arctic disappearing almost entirely toward the end of this century. This would have far-reaching adverse consequences for ice-dependent species and ecosystems as well as speeding up the warming far into the interior of the bordering continental regions of Russia, Canada, and Alaska.[11]

Large losses of sea ice threaten the continued existence of polar bears. Based on the projections available for the latest assessment, the IPCC predicted that this risk would occur for a global warming of 2.5–3.0 degrees Celsius above the preindustrial level. But it seems clear that this threshold could be much lower, as the observed rapid loss of summer ice (about 9.1 percent a year for the 1979–2006 period) exceeds the projections in nearly all the latest IPCC models.[12]

In already dry regions in the mid-latitudes, in drier parts of the tropics (predominantly developing countries), and in regions that depend on melting snow and ice for river and stream flows, water resources will be adversely affected. Glaciers in regions such as central Asia and the Himalaya and Tibetan plateau are melting faster than expected. Large adverse effects on water supply availability are predicted, threatening billions of people with water insecurity. Developing countries are not the only ones at risk. Serious water supply impacts have been seen in Australia from the 2001–07 drought—the most extreme and hottest drought recorded for this continent. Water inflows into Australia's largest and most important river basin, the Murray-Darling, are expected to decline 15 percent for each 1 degree Celsius of warming, and dramatic and adverse impacts are forecast for the water supply for large cities in southeast Australia.[13]

Agriculture and food supply in low-latitude regions, which are predominantly poor developing countries, are projected to be adversely affected even at low levels of warming. Recent climate trends, some of which can be attributed to human activities, appear to have had a measurable negative impact on global production of several major crops. In India, for example, it is clear that agricultural production has suffered due to a combination of climate change and air pollution.[14]

Substantial to sometimes severe adverse effects on food production, water supply, and ecosystems are projected for sub-Saharan Africa and small island developing states if the average temperature reaches 1.5 degrees Celsius above preindustrial level. Large river deltas, such as those of the Nile in Africa and of the Mekong and Ganges-Brahmaputra in Asia, are particularly at risk as they are home to large vulnerable populations and have a high exposure to sea level rise, storm surges, and river flooding.[15]

Tipping Points

Levels of warming that can trigger changes in large-scale components of the climate system, that can be irreversible for all practical purposes, and that have large-scale adverse consequences are often called tipping points. If a tipping point is passed, then a subsequent cooling of the climate system would likely not reverse the change. In some cases, such as disintegration of the West Antarctic ice sheet, the process would continue until a new equilibrium is reached.[16]

Elements of the climate system that are susceptible to "tipping" include Arctic summer sea ice (possible complete loss), the Greenland ice sheet (a meltdown would raise sea level 6–7 meters over many centuries to millennia), the West Antarctic ice sheet (disintegration would raise sea level 4–5 meters over several centuries), the circulation of the major Atlantic Ocean currents (risks of complete shutdown, with cooling of Europe and other adverse impacts), and the Amazon rainforest (risk of collapse due to warming and rainfall reductions).

A recent assessment indicates that a significant number of tipping points could be approached if the climate warms more than 3 degrees Celsius over the preindustrial level. Loss of the West Antarctic ice sheet is one such element. Other tipping points could be approached at warming levels over 1.5–2 degrees Celsius, such as the loss of the Greenland ice sheet. Arctic summer sea ice could be lost at even lower levels of warming (0.8–2.6 degrees Celsius), and its rapid loss would amplify warming in the adjacent continents, accelerating permafrost decay.[17]

What Levels of Warming Might Be Safe?

Deciding what level of climate change is dangerous and what might be safe is not a purely scientific question. It involves normative and political judgments about acceptable risks. Science has, however, a fundamental role to play in providing information and analysis relevant to this question and has contributed to policy and political debates on acceptable levels of climate change since the 1980s.[18]

By the late 1980s the scientific community had begun to recognize that a warming of much more than 1–2 degrees Celsius over the preindustrial level could lead to rapid and adverse changes to many human and natural systems. In 1986 the U.N. Environment Programme set up an Advisory Group on Greenhouse Gases, which in 1990 reported that a 2-degree warming could be "an upper limit beyond which the risks of grave damage to ecosystems, and of non-linear responses, are expected to increase rapidly." Also in the late 1980s the Enquete Komission, a joint committee of German parliamentarians and scientists, sought to define acceptable limits. Warming more than 0.1 degree Celsius per decade was seen as especially risky to forest ecosystems, with an overall acceptable maximum warming estimated to be 1–2 degrees Celsius. In 1995 the German government's Global Change Advisory Council found that 2 degrees Celsius should be the upper limit of "tolerable" warming.[19]

Efforts to define acceptable limits to warming at a political level started in the European Union and among its member states. Based on the IPCC's Second Assessment Report at the end of 1995, the European Union's Council of Environment Ministers in 1996 called for warming to be limited to 2 degrees Celsius above the preindustrial level. Nearly a decade later this position was confirmed by European Union Heads of Government after consideration and debate over the findings of the IPCC's 2001 Third Assessment Report, as well as more recent scientific developments. Since 2005 other countries have joined in calling for global mean warming to be limited to 2 degrees: Chile, Iceland, Norway, Switzerland, the Least Developed Countries, and Small Island Developing States. The latter two groups of countries have argued that 2 degrees may in fact be too much warming if their safety and survival are to be guaranteed.[20]

From the nongovernmental sector, the Climate Action Network, which has worked on climate change since 1989, has called for warming to be limited to a peak increase as far below 2 degrees Celsius as possible. It also calls for warming to be reduced as fast as possible from this peak. In 1997, based on a review of risks identified in mid-1990s, Greenpeace International called for the long-term committed increase of temperature to be limited to less than 1 degree Celsius above preindustrial and for warming rates to be less than 0.1 degree Celsius per decade.[21]

Several groups of scientists who have attempted to define a safe limit have also endorsed the need to stop before warming by 2 degrees Celsius. In a 2007 paper, NASA's James Hansen and colleagues argued for a limit of 1.7 degrees Celsius above preindustrial on the basis that potential changes above this level—including irreversible loss of the Greenland and Antarctic ice sheets and species extinction—would be "highly disruptive." Following further analysis of ongoing climate changes and of Earth's sensitivity to climate changes in the past, Hansen and his colleagues called for an "initial" CO_2 stabilization level of 350 ppm, significantly below present levels of close to 390 ppm. This would produce a warming in the long term of around 1 degree Celsius if the climate sensitivity were close to the IPCC best estimate of 3 degrees Celsius. The present CO_2 level, they argued, "is already too high to maintain the climate to which humanity, wildlife, and the rest of the biosphere are adapted." One implication of Hansen's reasoning is that warming may need to be lowered even from this level in centuries to come in order to reduce the risks of large-scale loss of ice from the ice sheets.[22]

Taking into account uncertainties in the sensitivity of the climate system to greenhouse gas increases, climatologist Danny Harvey of the University of Toronto has argued that even the present GHG concentration levels may constitute dangerous interference with the climate system. This would mean that a "safe" warming limit would be below 1.3–1.4 degrees Celsius above the preindustrial level, given that the present GHG concentration levels would likely warm the planet by about this amount once the world ocean and climate systems fully respond to these concentrations.[23]

The findings of the latest IPCC assessment and more-recent studies strongly reinforce the conclusions reached by all these different groups that "safe" levels of warming lie at 2 degrees Celsius or below. Table 2–1 summarizes salient examples of highly significant projected risks both below and above that level of warming.[24]

It is clear from this overview that substantial risks, dangers, and damages are likely across multiple sectors should global temperatures warm 1.5–2 degrees Celsius above the preindustrial level. Risks of extinction and major ecosystem disruption are evident at the low end of this range and increase rapidly with the rising temperature. While scientists are uncertain of the probability that a warming in the range of 1.5–2 degrees Celsius would destabilize the Greenland or West Antarctic ice sheets, this would have very large consequences if it did happen and hence qualifies as a high risk that "is something that should rather be avoided."[25]

A warming of 2 degrees Celsius is clearly not "safe" and would not prevent, with high certainty, dangerous interference with the climate system.

It is hard to avoid the conclusion that even a warming of 2 degrees Celsius poses unacceptable risks to key natural and human systems. It is clearly not "safe" and would not prevent, with high certainty, dangerous interference with the climate system. From thermal expansion of sea water alone, a meter or more of sea level rise over centuries cannot be excluded if there is a 2 degrees Celsius warming.

Furthermore, there is no "magic number" lower than 2 degrees Celsius that would limit warming to safe levels with high confidence. Warming in the range of 1.5–2 degrees Celsius clearly contains a significant risk of dangerous changes. Thus the amount of time the climate system remains in this temperature region should be minimized if it cannot be prevented. Below 1.5 degrees Celsius, there still appears to be a risk of dangerous changes. And at even a 1 degree Celsius warming there remains a risk of significant loss of ice from the ice sheets as well as large damages to vulnerable ecosystems.

Thus it does not appear possible to define at present an ultimate warming limit that is unambiguously safe or that undoubtedly would prevent dangerous interference with

Table 2–1. Risks and Impacts at Different Warming Levels above Preindustrial Level

System	1.5–2.0 Degrees Celsius	2.0–2.5 Degrees Celsius	>2.5 Degrees Celsius
Ecosystems and biodiversity	• 10–15 percent of species assessed committed to extinction, and significant risks for many biodiversity hotspots • Sharply accelerating risk of extinction for land birds, with loss of 100–500 species per degree of warming • Evidence from observed amphibian and reptile declines "portend a planetary-scale mass extinction"	• Major losses of endemic plants and animals in Southern Africa, northeastern Australia	• 20–30 percent of plant and animal species assessed at increased risk of extinction • Loss of 20–80 percent of Amazon rainforest and its biodiversity
	• Widespread damages to coral reef systems due to bleaching	• Increasing damage to coral reefs	• Widespread mortality of corals
	• Observed larger-than-expected losses of Arctic sea ice indicate increasing risk of extinction for the polar bear • High extinction risk projected for the King Penguin, with a reduction in adult survival of about 30 percent per degree of warming		• High risk of extinction for the polar bear due to projected loss of Arctic sea ice
Food Production	• Decreases in cereal production for some crops in low-latitude poor regions • Risk of highly adverse and severe impacts on food production in some African countries • Substantial risks to rice production in Java and Bali	• Significant decreases in crop production of around 5 percent for wheat and maize in India and rice in China • Agriculture losses of up 20 percent of GDP in low-lying island states • Recent review indicates that increases in productivity projected in IPCC report for warming of up to 2 degrees Celsius may not occur	• Risk of decline in crop yield globally
Coastal regions	• Increased damages from storms and floods, with up to 3 million additional people at risk of coastal flooding	• Increasing damages	• Increasing damages
Health	• Increasing burden from malnutrition and from diarrheal, infectious, and cardiovascular	• Increasing damages	• Increasing damages

Table 2–1. continued

System	1.5–2.0 Degrees Celsius	2.0–2.5 Degrees Celsius	> 2.5 Degrees Celsius
Health, continued	diseases, with increased mortality from heat waves, floods, and droughts		
Water	• Many hundreds of millions at risk of increased water stress in Africa, Asia, and Latin America	• Increasing number at risk of water stress	• 2 billion at risk of increased water stress for warming over 2–2.5 degrees Celsius
	• Glacial area in the Himalaya and Tibetan plateau regions could be reduced by 80 percent, adversely affecting billions of people • Transition to a more arid climate in southwestern North America	• Colorado River flow reduced to unprecedented levels that cannot be compensated by increased reservoir capacity or operating policies for water supplies	
Sea level rise	• Greenland ice sheet risk of irreversible meltdown for warming of 1.9–4.6 degrees Celsius • New data from the last interglacial period, 125,000 years ago, indicates that average rates of sea level rise in this period were rapid, around 1.6 meters per century	• Increasing risk of Greenland meltdown raising sea level; rapid sea level rise from this "cannot be excluded"	• Loss of ice sheet would raise sea level by some 2–7 meters over centuries to millennia
	• Accelerating ice loss from the West Antarctic ice sheet indicates risk of significant sea level rise at low levels of warming	• Increasing risk	• Increasing likelihood of partial or complete loss of the West Antarctic ice sheet, raising sea level 1.5–5 meters over several centuries to millennia
	• Commitment to minimum sea level rise of 0.3–1.2 meters over many centuries due to thermal expansion (0.2–0.6 meters per degree Celsius of global average warming)	• New projections indicate likely well above 0.5 meters of sea level rise by 2100 • Commitment to minimum sea level rise over many centuries of 0.4–1.5 meters due to thermal expansion irrespective of loss of the ice sheets and glaciers, which would only add to this risk	• New sea level rise projections of 0.5–1.4 meters above 1990 levels

Source: See endnote 25.

the climate system. It would seem safest and most prudent to reduce emissions fast enough in the coming decades so that global warming can be stopped soon and as far below 2 degrees Celsius as possible. The warming would then also need to be reduced as rapidly as possible, aiming to get it below 1 degree Celsius above preindustrial level—in other words, to at most about one fifth of a degree Celsius from where it is today.

Emission Pathways That Could Limit Warming to "Safe" Levels

Working out an emission path that would limit warming to any particular level involves accounting for a wide range of uncertainties in the causal chain from emissions to concentration to radiative forcing (the warming effect of changed concentrations in GHGs and aerosols, gaseous suspensions of fine solid or liquid particles that are associated with most CO_2 emissions, on the energy balance of the lower atmosphere) to climate change. Major uncertainties include the sensitivity of the climate system to changes in GHG concentration, the rate at which the ocean takes up heat from the atmosphere, the effects of aerosols on radiative forcing, and the response of the carbon cycle to changes in climate.[26]

In addition to scientific and technical uncertainties, it is important to decide how much confidence there needs to be that a warming limit will be achieved—in other words, how certain to be that specific risks and damages will be avoided or prevented. The emission pathways that are consistent with limiting warming to, say, 2 degrees Celsius or below with a 50 percent confidence are very different, and higher, than those that would do so with 90 percent confidence. (See Box 2–2 for how GHG concentration levels change for different probabilities of limiting warming to 2 degrees Celsius.) Before turning to the specific question of "safe" levels of emissions, this section reviews some of the important scientific aspects of generating an emission pathway.[27]

The important greenhouse gases have long lifetimes in the atmosphere, with large fractions of emissions remaining there for decades to centuries—and in some case, such as CO_2, for a thousand years or longer. Cutting emissions of these long-lived gases therefore leads to only slow reductions in their warming effect. Aerosols from human activities (principally sulfate compounds, organic and black carbon, nitrates, and dust) have a net cooling effect on the lower atmosphere and offset some of the warming effect of the long-lived GHGs.[28]

Aerosols have short lifetimes in the air, on the order of days or weeks. Reducing aerosol emissions thus has a rapid effect on temperatures since aerosol concentrations can drop quickly. The effect is so large that if all combustion and other activities that emit CO_2 and lead to the production of aerosols were cut to zero overnight, there would be a sharp warming spike before temperatures began to decline. The rapid drop in the concentration of aerosols would lead to a sudden loss of their cooling effect, which would occur faster than the slow reduction in the warming effect due to the much more slowly declining greenhouse gas concentrations. In realistic scenarios, when GHG emissions are reduced, air pollutants are also reduced. This leads to a more rapid reduction in aerosol concentrations, including those related to black carbon (suspended particles that absorb heat and contribute to warming), than in the greenhouse gas concentrations. As a consequence, the drop in aerosol cooling leads to a delay in the reduction of warming that would otherwise occur.[29]

Box 2–2. Greenhouse Gas Concentrations and Global Warming

Converting greenhouse gas concentrations to temperature cannot be done with certainty, as scientific knowledge of the sensitivity of the climate system is uncertain. Climate sensitivity is defined as the global mean temperature increase that would result in the long term after a doubling of CO_2 concentration above the preindustrial level of about 278 ppm. This temperature would be reached after a few hundred years, when the climate system comes into balance with the increased greenhouse gas concentration. The IPCC Fourth Assessment Report found that this was higher than previously estimated. It increased the "best" estimate from 2.5 to 3 degrees Celsius and the lower bound estimate from 1.5 to 2 degrees Celsius, and it kept the upper bound of 4.5 degrees Celsius unchanged from earlier assessments. There is some possibility that the climate sensitivity could be higher than 4.5 degrees.

Climate sensitivity is a vital number: if it were low (1 degree Celsius), then CO_2 levels could perhaps be doubled to around a concentration of 550 ppm CO_2 without causing large risks to many systems. If it were high (4.5 degrees), a doubling of CO_2 concentration could lead to a potentially catastrophic level of warming.

For stabilization of greenhouse gas concentrations at 550 ppm CO_2, the best estimate of the warming at equilibrium would be 3 degrees Celsius. The uncertainty in scientists' knowledge of climate sensitivity means that there is a chance the warming would be lower or higher than this. Taking into account this uncertainty, there is about a 75 percent risk that stabilizing greenhouse gas concentrations at 550 ppm would lead to warming exceeding 2 degrees Celsius.

Stabilizing at a lower level, 475 ppm CO_{2eq}, would reduce the risk to about 50 percent: in other words, there would be about an even chance that warming would stabilize at 2 degrees. Risks of dangerous changes to the climate system at this level could not be avoided with any confidence. Finally, for a concentration pathway that peaks at 475 ppm CO_{2eq} and then drops to stabilize at 400 ppm CO_{2eq}, there would be about a 20 percent chance of exceeding 2 degrees Celsius. If concentrations were reduced further, the risk of exceeding 2 degrees would be lower still.

In 2005, atmospheric CO_2 concentrations were 379 ppm, and they are now over 382 ppm. The IPCC best estimate of the total CO_2-equivalent concentration in 2005 for all long-lived GHGs was about 455 ppm—and at the end of 2007 it was 460 ppm. For 2005, the most recent year for which comprehensive figures are available, the "net" forcing, after taking into account aerosols and other human-induced climate forcing agents, was around 375 ppm CO_{2eq}, or about the same as the CO_2 concentration. Aerosols are short-lived; hence reductions of these lead to rapid reductions in the net cooling effect, whereas reductions in long-lived GHGs produce only a slow reduction in the warming effect.

Source: See endnote 27.

A further important property of the climate system that has to be accounted for in devising a safe emission pathway is inertia. Although the atmosphere responds quickly to changes in greenhouse gas forcing, a substantial component of the overall response is linked to the very long time scales of hundreds to thousands of years that the ocean takes to respond fully to the same climate forcing changes. Once GHG concentrations are stabilized, global mean temperature would very likely also begin to stabilize after several decades, though a further slight increase is likely to occur over several centuries. For sea level rise the inertia is even larger, as thermal expansion of the ocean continues for many centuries after GHG concentrations have stabilized due the ongoing heat uptake by oceans.[30]

The response of the carbon cycle to additions of fossil CO_2 is also very long. Of 1,000 tons of fossil CO_2 emitted now, after one century less than 500 tons would remain in the atmosphere. The rate of uptake of fossil CO_2 emissions by the world's oceans slows rapidly after a century or so; 1,000 years from now, 170–330 tons would remain in the atmosphere—and even after 10,000 years some 100–150 tons would remain. As David Archer, a geologist from the University of Chicago puts it, the lifetime of a fossil CO_2 emission in the atmosphere might best be described as "300 years, plus 25% that lasts forever."[31]

Limiting the peak warming to less than 1 degree Celsius will require a multicentury commitment to action.

All these climate system processes and factors need to be brought within an integrated system model that accounts for the interactions between emissions of greenhouse gases and aerosol pollutants and the responses and interactions among the different components of the climate system. For the analysis here, a new version of the simple climate-carbon cycle model MAGICC has been used to comprehensively capture current scientific knowledge and uncertainties in the response of the climate system. MAGICC 6.0 has been calibrated against, and can emulate, the higher complexity Atmospheric Ocean General Circulation models and carbon cycle models reviewed in the latest IPCC assessment. And it includes enhanced representations of aerosol forcing, carbon cycle feedbacks, ocean heat uptake, and climate sensitivity behavior over time. Reduced complexity models such as MAGICC are used as it is not practical to run many different emissions scenarios through a full climate system model. And further, as no specific model is a perfect representation of the climate system, doing so would not describe the scientific uncertainties in the response to a given emissions path.[32]

The greenhouse gases covered in the UNFCCC and the Kyoto Protocol, as well as the ozone-depleting substances (also greenhouse gases) covered in the Montreal Protocol, need to be accounted for in devising emission pathways. The phaseout of these latter gases also has a positive benefit for the climate. Emissions of air pollutants affect aerosol concentrations, including black carbon, and also affect concentrations of tropospheric ozone, a short-lived GHG. All these key climate forcings are included in MAGICC 6.0.[33]

Emission pathways are usually expressed in terms of CO_2-equivalent emissions, where the effects of non-CO_2 gases—methane, nitrous oxide, hydrofluorocarbons, perfluorocarbons, and sulfur hexafluoride—are compared using global warming potentials calculated over a 100-year time frame, as in the UNFCCC and Kyoto Protocol. This convention is followed here.

Limiting the peak warming to less than 2 degrees Celsius will not be easy, and getting it back below 1 degree Celsius will be even harder, requiring a multicentury commitment to action. The inertia in the response of the climate system to rapid reductions in GHGs and aerosols means that even stringent short-term reductions in greenhouse gas emissions will not ensure that peak warming stays below about 1.6–1.8 degrees Celsius. Reducing emissions fast enough to actually lower the level of warming ultimately will be as difficult as it is essential.[34]

The goal of the pathway in this chapter is to show what might be a safe emission scenario without requiring a full demonstration of technical feasibility. It could be argued that the "safest" pathway is one that immediately cuts emissions to very low levels. But since that lacks technical and economic feasibility, such a pathway has little meaning. The jet plane metaphor is again helpful. Faced with a dire in-flight emergency, it would be safest to be on the ground immediately. In the real world, however, it takes time to prepare the aircraft, get into a safe configuration for descent and landing, and find a safe runway to land on. Otherwise the outcome would be an unmitigated disaster—the plane would crash.

The approach taken here is to construct a pathway whose achievement in practice is plausible technically. It goes beyond the technically and economically feasible pathways published elsewhere so far. No pathway published to date brings warming below 1 degree Celsius. A few pathways could get warming below 1.5 degrees Celsius by the twenty-third century if the negative CO_2 emissions at the end of the twenty-first century in these scenarios were sustained for at least 100 years.[35]

Recent research has demonstrated that it is technically and economically feasible to reduce CO_2 emissions fast enough so that GHG concentrations can be limited to around 400 ppm CO_{2eq}, or to lower in the longer term. Under these scenarios it is likely that peak warming would occur close to, if not below, 2 degrees Celsius. And in some cases temperatures might slowly decline beyond the twenty-first century. All these scenarios require rapid fossil fuel CO_2 emission reductions, approaching zero emissions between 2050 and 2100, along with rapid reductions in deforestation.[36]

One very important finding is that in order to reach low stabilization levels of GHG concentrations, nearly all these scenarios require negative CO_2 emissions by the last quarter of the twenty-first century at the latest. Without this it is impossible to draw down atmospheric CO_2 concentrations, owing to the long lifetime of this gas. Without this key component, CO_2 concentrations would drop only slowly, and warming would likely remain well above 1.5 degrees Celsius for many centuries.[37]

The possible need to stabilize CO_2 at low concentration levels to avoid dangerous climate changes has been recognized for a long time, as has the need for negative CO_2 emissions if low CO_2 stabilization levels are to be reached. But evaluation of the implications of the technologies required to achieve this is only just beginning. In the low stabilization studies, models rely on the capture of CO_2 from biomass-fired power plants to essentially draw CO_2 out of the air so it can be stored underground in stable geological reservoirs (referred to often as biomass energy with carbon capture and storage, or BECS, technology). Biofuel plantations grow plants that take up CO_2 from the air as they grow, and if much of this is captured when the plants are burned, the process effectively pumps CO_2 out of the air. The environmental and sustainability consequences of such a strategy have yet to be fully evaluated. Air capture technology—taking CO_2 out of the air and storing it underground—has also been proposed as a feasible technology.[38]

While reducing emissions from deforestation is important, the scale of potential uptake of carbon in forests and agricultural soils is unlikely to be sufficient to draw atmospheric CO_2 concentrations down significantly. Recent results using the LPJ (Lund-Potsdam-Jena) land biosphere model—with scenarios of population increase, deforestation, land use change, and agriculture from the Dutch IMAGE 2.2 integrated assessment model—indicate that under high environmental sustainability assumptions (taking into account the effects of increased CO_2 and climatic changes) the net uptake of carbon over the twenty-first century would not increase the additional carbon stored in terrestrial ecosystems due to human activities enough to outweigh the need for negative emissions from the energy sector.[39]

Recent scenarios that keep warming below 2 degrees Celsius and get to concentrations of around 400 ppm CO_{2eq} or lower reduce CO_2 emissions 60–70 percent below 1990 levels and cut total GHGs around 40–60 percent by 2050. And they have negative CO_2 emissions in the range of 1 billon to 8 billion tons of carbon per year in 2100. All would require BECS.[40]

Figure 2–1. CO_2 Emissions from Fossil Fuels through 2100, IPCC SRES (High) Scenario and the Below 1 Degree Celsius Scenario

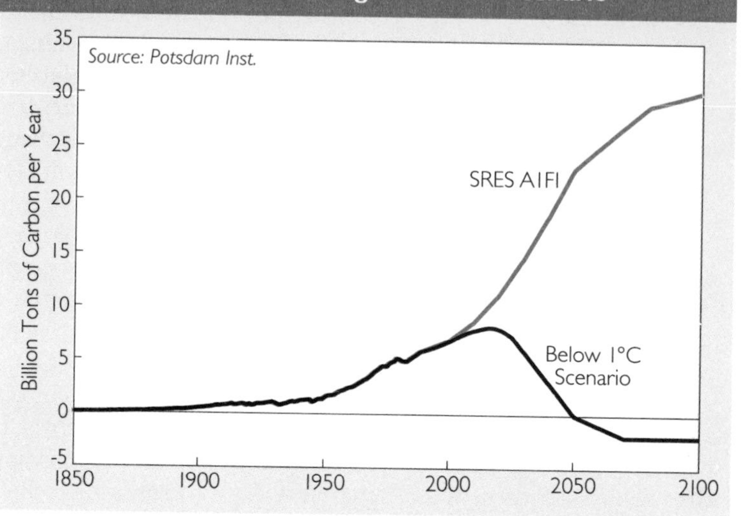

The emission pathway required to limit warming to below 2 degrees Celsius with higher confidence and at the same time reduce warming rapidly to below 1 degree Celsius (see Figure 2–1) would require a more rapid reduction in emissions by 2050 than in the most recent scenarios, which have already been at the limits of what models indicate is feasible based on present technological assessments. Plausible additional measures to achieve this include a more rapid reduction in fossil fuel emissions.

Getting fossil CO_2 emissions down to close to zero in 2050—which would be 25 years earlier than in most low-stabilization scenarios—would require an earlier and more massive global deployment of renewable energy systems, accelerated energy efficiency measures, and a limit to the lifetime of coal power plants. (See Chapter 4.) Deploying as-yet-unproven carbon capture and storage (CCS) technology after the mid-2020s may also help. However, the expected large life-cycle energy and emissions costs of CCS technology indicate that it cannot be relied on to reduce fossil CO_2 emission to zero. The faster that renewable energy systems can be scaled up and deployed, the less will be needed of CCS coal and gas power plants.[41]

In addition to action on fossil fuels, deforestation would need to be halted well before 2030, and there would need to be large-scale efforts to store carbon in soils through progress toward sustainable agriculture and regrowing forests. (See Chapter 3.) The reductions assumed here for emissions of methane and nitrous oxide, two powerful greenhouse gases, from agriculture and industry are not taken significantly further than can be found in the literature for low scenarios. And the emission pathway is relatively insensitive to the phaseout schedules for emissions of ozone-depleting substances, hydrofluorocarbons, perfluorcarbons, and sulfur hexafluoride.[42]

The resulting pathway has Kyoto GHG reductions of around 85 percent from 1990 levels by 2050 after peaking before 2020. (See Figure 2–2.) GHG atmospheric concentrations drop below today's levels by the mid twenty-second century and toward the preindustrial level by the twenty-fourth century. After the 2050s this pathway also would require the capture from the atmosphere and permanent storage of initially around 2.5 billion tons of carbon a year (about 9 billion tons of CO_2 per year) for more than 200 years in order to draw total GHG concentrations down to below 300 ppm CO_{2eq}. Global temperatures should peak below 2 degrees Celsius around mid-century and begin a slow decline, dropping to present levels by the last half of the twenty-third century and to 1990s levels by the end of the twenty-fourth century. (See Figure 2–3.) In Figures 2–2 and 2–3, there are bands of projected levels due to

Figure 2–2. Greenhouse Gas Atmospheric Concentration through 2100, IPCC SRES (High) Scenario and the Below 1 Degree Celsius Scenario

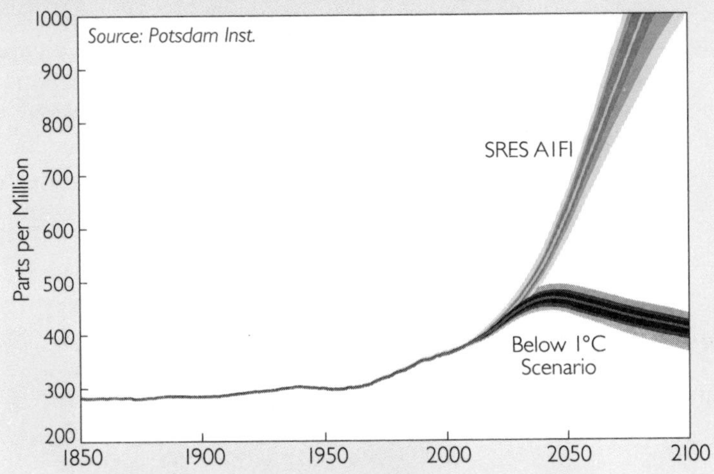

Note: The darker parts show the higher confidence range of changes for the emission pathway shown.

Figure 2–3. Global Mean Surface Temperature through 2100, with Uncertainty, IPCC SRES (High) Scenario and the Below 1 Degree Celsius Scenario

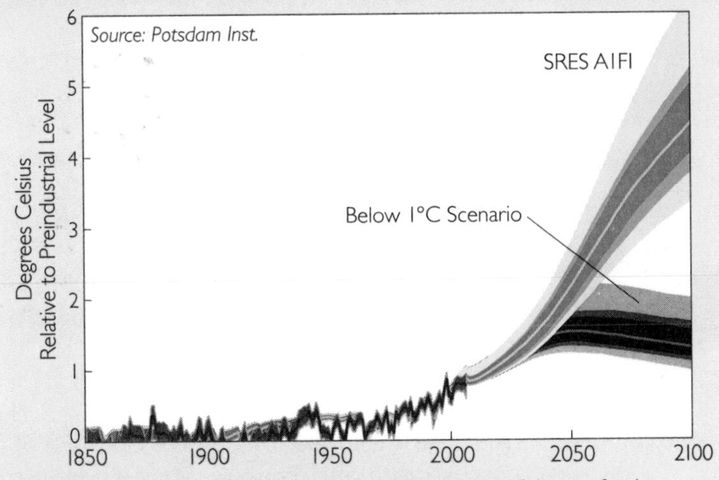

Note: The darker parts show the higher confidence range of changes for the emission pathway shown.

uncertainties in the science.

The amount of carbon that would need to be captured and stored to achieve all this would be on the same order as that emitted since the nineteenth century. As the amount of additional carbon that can be taken up and stored by the terrestrial biosphere due to human activities is limited—assumed here to be about 0.5 billion tons a year during much of the latter part of the twenty-first century and dropping to zero by 2200—the extraction of CO_2 from the atmosphere would have to be largely done using technologies similar to those just mentioned.

Just as the effects of climate change pose enormous long-term problems, a safe resolution of the problem will require a commitment to action that spans centuries. Returning to warming levels significantly below 2 degrees Celsius implies the need for large long-term extraction of CO_2 from the air and the storage of the captured carbon in secure underground reservoirs, which will need to be watched and managed over many centuries, perhaps millennia. Extracting CO_2 from the air appears to be a necessity that must be confronted within the next 50 years.

From any perspective the consequences of following an emissions pathway that keeps the temperature increase below 1 degree Celsius are quite radical and may be seen as technologically, economically, and politically close to impossible. But this needs to placed against the also quite radical risks that global warming poses if emissions are not reduced to low levels.[43]

As difficult as this emissions pathway seems, it is important to note that the low-emissions scenarios reviewed by the IPCC (all consistent with limiting warming to about 2 degrees Celsius) start out much like this one. In the lowest scenarios, global emissions need to peak before 2020. After that it may not be possible for technologies to be introduced fast enough to lower emissions at the rate required to keep warming below 2 degrees Celsius. Delay in acting entails faster rates of emissions reduction and significantly increased costs to reach the goal. And it might totally foreclose the ability to reduce GHG concentrations to low levels once societies are locked into emission-intensive energy sources and other infrastructure as well as development pathways that are carbon-intensive. Delay obviously also increases the risk of more-severe climate change impacts.[44]

Once a global emission pathway is defined, the next key question is, How much GHG can countries emit and still be consistent with global emissions limits? There are many possible ways to allocate emissions to countries to meet global limits, and review of these is beyond the scope of this chapter. Nevertheless it is useful here to point out the broad implications of the 2 degrees Celsius emissions pathway for different groups of countries in the next decade and beyond.[45]

An indication of the required reductions can be seen from the IPCC Fourth Assessment Report, where the reductions for different regions from a range of models are reviewed for different GHG stabilization scenarios. The lowest scenario was for stabilization at 450 ppm CO_{2eq}, far higher than the CO_{2eq} stabilization levels that would provide a higher probability of keeping warming below 2 degrees Celsius. Industrial-country GHG reductions in 2020 were generally required to be 25–40 percent below 1990 levels. By 2050, reductions for these countries would need to be 80–95 percent below 1990 levels. (The reductions refer to emission allowances and hence do not necessarily indicate the physical emissions levels of the countries in 2020 or 2050.) The exact reduction for each industrial country depends on the emission allocation system, individual circumstances, and other assumptions in the different models assessed.[46]

For developing countries, by 2020 there would need to be a substantial reduction in the growth of emissions in Latin America, the Middle East, and East Asia (China and others), but not in South Asia (including India) or Africa. By then the wealthier developing countries would need to reduce significantly the growth in emissions from their business-as-usual emissions. By the 2050s, all these regions would have to substantially reduce the growth in emissions. For the few scenarios available that stabilize GHG concentrations at 400 ppm CO_{2eq}, which would provide around a 75 percent chance of limiting warming to below 2 degrees Celsius, the emissions reductions in 2020 and 2050 are quite similar to but a little lower than in the higher scenarios.

Critical Priorities for the Next 10 Years

Halting the increase in global warming at far below 2 degrees Celsius is possible, and lowering global warming as rapidly as possible to below an increase of 1 degree Celsius appears critical if there is to be a high probability of preventing dangerous climate change. The emissions reduction actions required to achieve this are massive and appear to be at the outer edge of what is technically and economically feasible. Scenarios that can start to get within reach of these temperature goals require GHG emissions to peak before 2020 and then to drop toward 85 percent below 1990 levels by 2050, with further reductions beyond this time.

To return to the metaphor of the heavy jet aircraft facing an emergency, wherever the aircraft is going to land it needs to start preparing a long way out. Altitude needs to be lost without excessive speed buildup, fuel needs to be dumped and systems checked and prepared for a landing, and all this must be done quickly and expeditiously. As for climate policy, the vital preparation for a safe landing—whether the final safe landing place is a 2 degrees Celsius runway or a below 1 degree Celsius runway—is to halt the rise in global emissions by 2020 and to start to put in place the policies that can lower emissions. For policymakers, these are the decisions that must urgently be made at the end of 2009 when governments gather in Copenhagen for the next Conference of the Parties to the climate change convention.

CHAPTER 3

Farming and Land Use to Cool the Planet

Sara J. Scherr and Sajal Sthapit

For more than a decade, thousands of low-income farmers in northern Mindanao, the Philippines, who grow crops on steep, deforested slopes, have joined landcare groups to boost food production and incomes while reducing soil erosion, improving soil fertility, and protecting local watersheds. They left strips of natural vegetation to terrace their slopes, enriched their soils, and planted fruit and timber trees for income. And their communities began conserving the remaining forests in the area, home to a rich but threatened biodiversity. Yet these farmers achieved even more—their actions not only enriched their landscapes and enhanced food security, they also helped to "cool" the planet by cutting greenhouse gas emissions and storing carbon in soils and vegetation. If their actions could be repeated by millions of rural communities around the world, climate change would slow down.[1]

Indeed, climate change and global food security are inextricably linked. This was made abundantly clear in 2008, as rioters from Haiti to Cameroon protested the global "food crisis." The crisis partly reflected structural increases in food demand from growing and more-affluent populations in developing countries and short-term market failures, but it was also in part a reaction to increased energy costs, new biofuel markets created by legislation promoting alternative energy, and climate-induced regional crop losses. Moreover, food and fiber production are leading sources of greenhouse gas (GHG) emissions—they have a much larger "climate footprint" than the transportation sector, for example. Degradation and loss of forests and other vegetative cover puts the carbon cycle further off balance. Ironically, the land uses and management systems that are accelerating GHG emissions are also undermining the ecosystem services upon which long-term food and fiber production depend—healthy

Sara J. Scherr is President and CEO of Ecoagriculture Partners (EP) in Washington, DC. **Sajal Sthapit** is a Program Associate at EP.

watersheds, pollination, and soil fertility.[2]

This chapter explains why actions on climate change must include agriculture and land systems and highlights some promising ways to "cool the planet" via land use changes. Indeed, there are huge opportunities to shift food and forestry production systems as well as conservation area management to mitigate climate change in ways that also increase sustainability, improve rural incomes, and ease adaptation to a warming world.

The Need for Climate Action on Agriculture and Land Use

Land is one fourth of Earth's surface and it holds three times as much carbon as the atmosphere does. About 1,600 billion tons of this carbon is in the soil as organic matter and some 540–610 billion tons is in living vegetation. Although the volume of carbon on Earth's surface and in the atmosphere pales in comparison to the many trillions of tons stored deep under the surface as sediments, sedimentary rocks, and fossil fuels, surface carbon is crucial to climate change and life due to its inherent mobility.[3]

Surface carbon moves from the atmosphere to the land and back, and in this process it drives the engine of life on the planet. Plants use carbon dioxide (CO_2) from the atmosphere to grow and produce food and resources that sustain the rest of the biota. When these organisms breathe, grow, die, and eventually decompose, carbon is released to the atmosphere and the soil. Carbon from this past life provides the fuel for new life. Indeed, life depends on this harmonized movement of carbon from one sink to another. Large-scale disruption or changes on land drastically alter the harmonious movement of carbon.

Land use changes and fossil fuel burning are the two major sources of the increased CO_2 in the atmosphere that is changing the global climate. (See Box 3–1.) Burning fossil fuel releases carbon that has been buried for millions of years, while deforestation, intensive tillage, and overgrazing release carbon from living or recently living plants and soil organic matter. Some land use changes affect climate by altering regional precipitation patterns, as is occurring now in the Amazon and Volta basins. Overall, land use and land use changes account for around 31 percent of total human-induced greenhouse gas emissions into the atmosphere. Yet other types of land use can play the opposite role. Growing plants can remove huge amounts of carbon from the atmosphere and store it in vegetation and soils in ways that not only stabilize the climate but also benefit food and fiber production and the environment. So it is imperative that any climate change mitigation strategy address this sector.[4]

Extensive action to influence land use is also going to be essential to sustain food and forest production in the face of climate change. Agricultural systems have developed during a time of relatively predictable local weather patterns. The choice of crops and varieties, the timing of input application, vulnerability to pests and diseases, the timing of management practices—all these are closely linked to temperature and rainfall. With climate changing, production conditions will change—and quite radically in some places—which will lead to major shifts in farming systems.

Climate scenarios for 2020 predict that in Mexico, for example, 300,000 hectares will become unsuitable for maize production, leading to estimated yearly losses of $140 million and immense socioeconomic disruption. And in North America, the areas with the optimum temperature for producing syrup from maple trees are shifting northward, leaving farmers in the state of Vermont at risk of

Box 3–1. Greenhouse Gas Emissions from Land Use

Carbon dioxide (77 percent), nitrous oxide (8 percent), and methane (14 percent) are the three main greenhouse gases that trap infrared radiation and contribute to climate change. Land use changes contribute to the release of all three of these greenhouse gases. (See Table.) Of the total annual human-induced GHG emissions in 2004 of 49 billion tons of carbon-dioxide equivalent, roughly 31 percent—15 billion tons—was from land use. By comparison, fossil fuel burning accounts for 27.7 billion tons of CO_2-equivalent emissions annually.

Deforestation and devegetation release carbon in two ways. First the decay of the plant matter itself releases carbon dioxide. Second, soil exposed to the elements is more prone to erosion. Subsequent land uses like agriculture and grazing exacerbate soil erosion and exposure. The atmosphere oxidizes the soil carbon, releasing more carbon dioxide into the atmosphere. Application of nitrogenous fertilizers leads to soils releasing nitrous oxide. Methane is released from the rumens of livestock like cattle, goats, and sheep when they eat and from manure and water-logged rice plantations.

Naturally occurring forest and grass fires also contribute significantly to GHG emissions. In the El Niño year of 1997–98, fires accounted for 2.1 billion tons of carbon emissions . Due to the unpredictability of these events, annual emissions from this source vary from year to year.

Land Use	Annual Emissions (million tons CO_2 equivalent)	Greenhouse Gas Emitted
Agriculture	6,500	
Soil fertilization (inorganic fertilizers and applied manure)	2,100	Nitrous oxide*
Gases from food digestion in cattle (enteric fermentation in rumens)	1,800	Methane*
Biomass burning	700	Methane, nitrous oxide*
Paddy (flooded) rice production (anaerobic decomposition)	600	Methane*
Livestock manure	400	Methane, nitrous oxide*
Other (e.g., delivery of irrigation water)	900	Carbon dioxide, nitrous oxide*
Deforestation (including peat)	8,500	
For agriculture or livestock	5,900	Carbon dioxide
Total	15,000	

* *The greenhouse gas impact of 1 unit of nitrous oxide is equivalent to 298 units of carbon dioxide; 1 unit of methane is equivalent to 25 units of carbon dioxide.*

Source: See endnote 4.

losing not only their signature product but generations of culture and knowledge.[5]

The Gangotri glacier in the Himalayas, which provides up to 70 percent of the water in the Ganges River, is retreating 35 meters yearly. Once it disappears, the Ganges will become a seasonal river, depriving 40 percent of India's irrigated cropland and some 400 million people of water. The frequency, intensity, and duration of rainfall are also likely to change, increasing production risks, especially in semiarid and arid rainfed production areas. Monsoons will be heavier, more variable, and with greater risk of flooding. An increased incidence of drought threatens nearly 2 billion people who rely on livestock grazing for part of their livelihoods, particularly the 200 million who are completely dependent on pastoral systems. The incidence and intensity of natural fires is predicted to increase. [6]

The poorest farmers who have little insurance against these calamities often live and farm in areas prone to natural disasters. More-frequent extreme events will create both a humanitarian and a food crisis.

On the other hand, climatic conditions may improve in some places. In the highlands of East Africa, for example, rains may become more reliable and growing seasons for some crops may expand. The growing season in northern latitudes in Canada and Russia will extend as temperatures rise. Even in these situations, however, there will be high costs for adapting to new conditions, including finding crop varieties and management that are adapted to new climate regimes at this latitude. The impacts on pest and disease regimes are largely unknown and could offset any benefits. For instance, the Eastern spruce budworm is a serious pest defoliating North American forests. Changing climate is shifting the geographic range of the warblers that feed on the budworms, increasing the odds for budworm outbreak.[7]

Many of the key strategies described in this chapter for agricultural, forest, and other land use systems to mitigate climate change—that is, to reduce GHG emissions or increase the storage of carbon in production and natural systems—also will help rural communities adapt to that change. Mobilizing action for adaptation in these directions rather than relying only on other types of interventions, such as seed varieties or shifts in market supply chains, could have significant success in slowing climate change.

Making Agriculture and Land Use Climate-friendly and Climate-resilient

An agricultural landscape should simultaneously provide food and fiber, meet the needs of nature and biodiversity, and support viable livelihoods for people who live there. In terms of climate change, landscape and farming systems should actively absorb and store carbon in vegetation and soils, reduce emissions of methane from rice production, livestock, and burning, and reduce nitrous oxide emissions from inorganic fertilizers. At the same time, it is important to increase the resilience of production systems and ecosystem services to climate change.[8]

Many techniques are already available to achieve climate-friendly landscapes. None is a "silver bullet," but in combinations that make sense locally they can help the world move decisively forward. This chapter describes five strategies that are especially promising: enriching soil carbon, creating high-carbon cropping systems, promoting climate-friendly livestock production systems, protecting existing carbon stores in natural forests and grasslands, and restoring vegetation in degraded areas. (See Figure 3–1.) Many other improvements will also be needed for production systems to adapt to climate change while meeting growing food needs and commercial demands, such as adapted seed varieties. But these five strategies are highlighted because of their powerful advantage in mitigating climate change as well as contributing broadly to more-sustainable production systems and other ecosystem services.[9]

Moreover, these strategies can help mobi-

Figure 3–1. Multiple Strategies to Productively Absorb and Store Carbon in Agricultural Landscapes

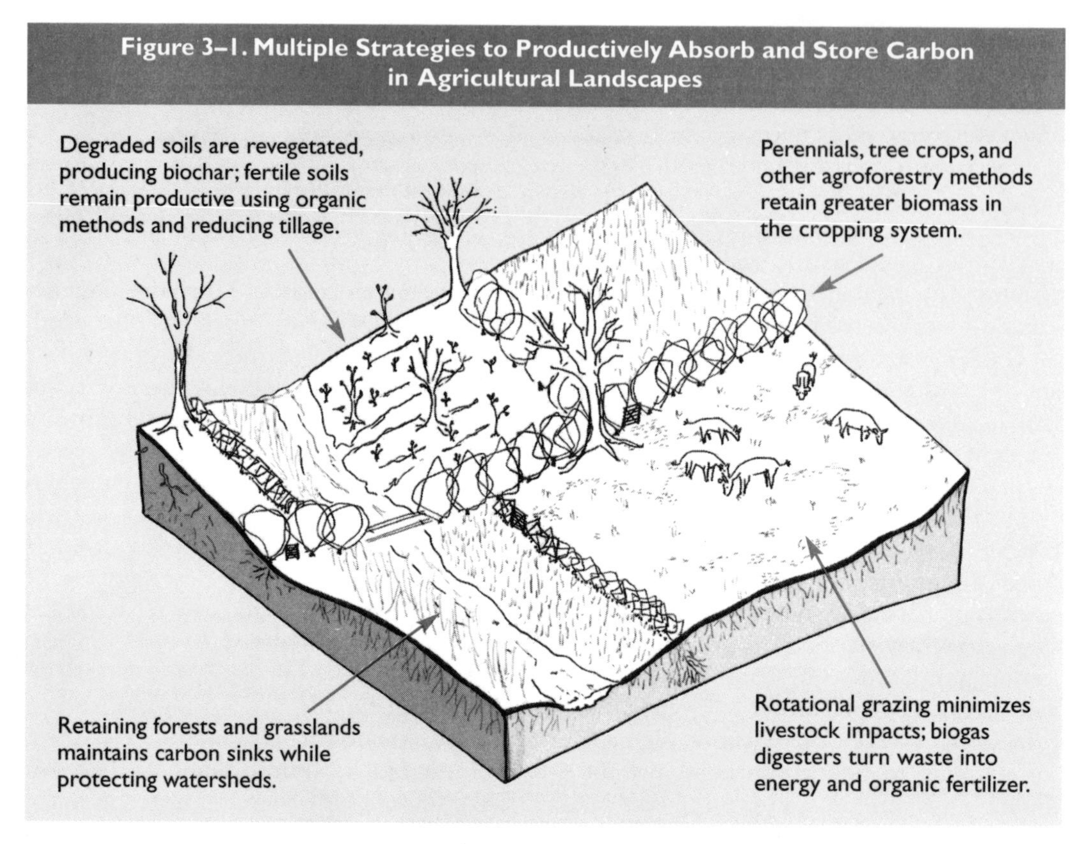

lize a broad political coalition to support climate action by meeting the urgent needs of farmers, grazers and rural communities, the food industry, urban water users, resource-dependent industries, and conservation organizations. They can help meet not only climate goals but also internationally agreed Millennium Development Goals and other global environmental conventions.

Many of these approaches will be economically self-sustaining once initial investments are made. It is important to implement this agenda on a large scale in order to have significant impacts on the climate. Key roles that governments need to play are to mobilize the financing and social organization needed for these initial investments, develop additional incentives for activities that are more time-consuming or costly yet offer no particular benefits to farmers or land managers, and invest in the development of technologies and management systems that are especially promising but not yet ready for widespread use.

Enriching Soil Carbon

Soil has four components: minerals, water, air, and organic materials—both nonliving and living. The former comes from dead plant, animal, and microbial matter while the living organic material is from flora and fauna of the soil biota, including living roots and microbes. Together, living and nonliving organic materials account for only 1–6 percent of the soil's volume, but they contribute much more to its productivity. The organic materials retain air and water in the soil and provide nutrients that the plants and the soil fauna depend on for life. They are also reservoirs of carbon in the soil.[10]

In fact, soil is the third largest carbon pool on the planet. In the long term, agricultural practices that amend soil carbon from year to year through organic matter management rather than depleting it will provide productive soils that are rich in carbon and require fewer chemical inputs. New mapping tools, such as the 2008 Global Carbon Gap Map produced by the Food and Agriculture Organization, can identify areas where soil carbon storage is greatest and areas with the physical potential for billions of tons of additional carbon to be stored in degraded soils.[11]

Enhance soil nutrients through organic methods. Current use of inorganic fertilizers is estimated at 102 million tons worldwide, with use concentrated in industrial countries and in irrigated regions of developing nations. Soils with nitrogen fertilizers release nitrous oxide, a greenhouse gas that has about 300 times the warming capacity of carbon dioxide. Fertilized soils release more than 2 billion tons (in terms of carbon dioxide equivalent) of greenhouse gases every year. One promising strategy to reduce emissions is to adopt soil fertility management practices that increase soil organic matter and siphon carbon from the atmosphere.[12]

Numerous technologies can be used to substitute or minimize the need for inorganic fertilizers. Examples include composting, green manures, nitrogen-fixing cover crops and intercrops, and livestock manures. Even improved fertilizer application methods can reduce emissions. In one example of organic farming, a 23-year experiment by the Rodale Institute compared organic and conventional cropping systems in the United States and found that organic farming increased soil carbon by 15–28 percent and nitrogen content by 8–15 percent. The researchers concluded that if the 65 million hectares of corn and soybean grown in the United States were switched to organic farming, a quarter of a billion tons of carbon dioxide could be sequestered.[13]

The economics and productivity implications of these methods vary widely. In some very intensive, high-yield cropping systems, replacing some or all inorganic fertilizer may require methods that use more labor or require costlier inputs, but there is commonly scope for much more efficient use of fertilizer through better targeting and timing. In moderately intensive systems, the use of organic nutrient sources with small amounts of supplemental inorganic fertilizer can be quite competitive and attractive to farmers seeking to reduce cash costs.[14]

Improvements in organic technologies over the past few decades have led to comparable levels of productivity across a wide range of crops and farming systems. The question of whether organic farming can feed the world, as some claim, remains controversial. And more research is needed to understand the potentials and limitations of biologically based soil nutrient management systems across the range of soil types and climatic conditions. But there is little question that farmers in many production systems can already profitably maintain yields while using much less nitrogen fertilizer—and with major climate benefits.

Minimize soil tillage. Soil used to grow crops is commonly tilled to improve the conditions of the seed bed and to uproot weeds. But tilling turns the soil upside down, exposing anaerobic microbes to oxygen and suffocating aerobic microbes by working them under. This disturbance exposes nonliving organic matter to oxygen, releasing carbon dioxide. Keeping crop residues or mulch on the surface helps soil retain moisture, prevents erosion, and returns carbon to the soil through decomposition. Hence practices that reduce tillage also generally reduce carbon emissions.[15]

A variety of conservation tillage practices accomplish this goal. In nonmechanized systems, farmers might use digging sticks to plant seeds and can manage weeds through mulch and hand-weeding. Special mechanized systems have been developed that drill the seed through the vegetative layer and use herbicides to manage weeds. Many farmers combine no-till with crop rotations and green manure crops. In Paraná, Brazil, farmers have developed organic management systems combined with no-till. No-till plots yielded a third more wheat and soybean than conventionally ploughed plots and reduced soil erosion by up to 90 percent. No-till has the additional benefit of reducing labor and fossil fuel use and enhancing soil biodiversity—all while cycling nutrients and storing carbon.[16]

In Paraná, Brazil, no-till plots yielded a third more wheat and soybean than conventionally ploughed plots and reduced soil erosion by up to 90 percent.

Worldwide, approximately 95 million hectares of cropland are under no-till management—a figure that is growing rapidly, particularly as rising fossil fuel prices increase the cost of tillage. The actual net impacts on greenhouse gases of reduced emissions and increased carbon storage from reduced tillage depend significantly on associated practices, such as the level of vegetative soil cover and the impact of tillage on crop root development, which depends on the specific crop and soil type. It is projected that the carbon storage benefits of no-till may plateau over the next 50 years, but this can be a cost-effective option to buy time while alternative energy systems develop.[17]

Incorporate biochar. Decomposition of plant matter is one way of enriching soil carbon if it takes place securely within the soil; decomposition on the surface, on the other hand, releases carbon into the atmosphere as carbon dioxide. In the humid tropics, for example, organic matter breaks down rapidly, reducing the carbon storage benefits of organic systems. Another option, recently discovered, is to incorporate biochar—burned biomass in a low-oxygen environment. This keeps carbon in soil longer and releases the nutrients slowly over a long period of time. While the burning does release some carbon dioxide, the remaining carbon-rich dark aromatic matter is highly stable in soil. Hence planting fast-growing trees in previously barren or degraded areas, converting them to biochar, and adding them to soil is a quick way of taking carbon from the atmosphere and turning it into an organic slow-release fertilizer that benefits both the plant and the soil fauna.

Interestingly, between 500 and 2,500 years ago Amerindian populations added incompletely burnt biomass to the soil. Today, Amazonian Dark Earths still retain high amounts of organic carbon and fertility in stark contrast to the low fertility of adjacent soils. There is a global production potential of 594 million tons of carbon dioxide equivalent in biochar per year, simply by using waste materials such as forest and milling residues, rice husks, groundnut shells, and urban waste. Far more could be generated by planting and converting trees. Initial analyses suggest that it could be quite economical to plant vegetation for biochar on idle and degraded lands, though not on more highly productive lands.[18]

Most crops respond with improved yields for biochar additions of up to 183 tons of carbon dioxide equivalent and can tolerate more without declining productivity. Advocates calculate that if biochar additions were applied at this rate on just 10 percent of the world's cropland (about 150 million hectares), this method could store 29 billion tons of CO_2-equivalent, offsetting nearly all the emissions from fossil fuel burning.[19]

Creating High-carbon Cropping Systems

Plants harness the energy of the sun and accumulate carbon from the atmosphere to produce biomass on which the rest of the biota depend. The great innovation of agriculture 10,000 years ago was to manage the photosynthesis of plants and ecosystems so as to dependably increase yields. With 5 billion hectares of Earth's surface used for agriculture (69 percent under pasture and 28 percent in crops) in 2002, and with half a billion more hectares expected by 2020, agricultural production systems and landscapes have to not only deliver food and fiber but also support biodiversity and important ecosystem services, including climate change mitigation. A major strategy for achieving this is to increase the role of perennial crops, shrubs, trees, and palms, so that carbon is absorbed and stored in the biomass of roots, trunks, and branches while crops are being produced. Tree crops and agroforestry maintain significantly higher biomass than clear-weeded, annually tilled crops.[20]

Although more than 3,000 edible plant species have been identified, 80 percent of world cropland is dominated by just 10 annual cereal grains, legumes, and oilseeds. Wheat, rice, and maize cover half of the world's cropland. Since annual crops need to be replanted every year and since the major grains are sensitive to shade, farmers in much of the world have gradually removed other vegetation from their fields. But achieving a high-carbon cropping system, as well as the year-round vegetative cover required to sustain soils, watersheds, and habitats, will require diversification and the incorporation of a far greater share of perennial plants.[21]

Perennial grains. Currently two thirds of all arable land is used to grow annual grains. This production depends on tilling, preparing seed beds, and applying chemical inputs. Every year the process starts over again from scratch. This makes production more dependent on chemical inputs, which also require a lot of fossil fuels to produce. Furthermore, excessive application of nitrogen fertilizer is a major source of nitrous oxide emissions, as noted earlier.[22]

In contrast, perennial grasses retain a strong root network between growing seasons. Hence, a good amount of the living biomass remains in the soil instead of being released as greenhouse gases. And they help hold soil organic matter and water together, reducing soil erosion and GHG emissions. Finally, the perennial nature of these grasses does away with the need for annual tilling that releases GHGs and causes soil erosion, and it also makes the grasses more conservative in the use of nutrients. In one U.S. case, for example, harvested native hay meadows retained 179 tons of carbon and 12.5 tons of nitrogen in a hectare of soil, while annual wheat fields only retained 127 tons of carbon and 9.6 tons of nitrogen. This was despite the fact that the annual wheat fields had received 70 kilograms of nitrogen fertilizer per hectare annually for years.[23]

Researchers have already developed perennial relatives of cereals (rice, sorghum, and wheat), forages (intermediate wheatgrass, rye), and oilseeds (sunflower). In Washington state, some wheat varieties that have already been bred yield over 70 percent as much as commercial wheat. Domestication work is under way for a number of lesser known perennial native grasses, and many more perennials offer unique and exciting opportunities.[24]

Shifting production systems from annual to perennial grains should be an important research priority for agriculture and crop breeding, but significant research challenges remain. Breeding perennial crops takes longer than annuals due to longer generation times. Since annuals live for one season only, they give priority to seeds over vegetative growth, making yield improvement in annuals easier than in perennials that have to allocate more resources to vegetative parts like roots in order to ensure survival through the winter. But in the quest for high-carbon agricultural systems, plants that produce more biomass are a plus. Through breeding, it may also be possible to redirect increased biomass content to seed production.

A Billion Tree Campaign launched in 2006 shattered initial expectations and mobilized the planting of 2 billion trees in more than 150 countries.

Agroforestry intercrops. Another method of increasing carbon in agriculture is agroforestry, in which productive trees are planted in and around crop fields and pastures. The tree species may provide products (fruits, nuts, medicines, fuel, timber, and so on), farm production benefits (such as nitrogen fixation for crop fertility, wind protection for crops or animals, and fodder for animals), and ecosystem services (habitat for wild pollinators of crops, for example, or micro-climate improvement). The trees or other perennials in agroforestry systems sequester and store carbon, improving the carbon content of the agricultural landscape.

Agroforestry was common traditionally in agricultural systems in forest ecosystems and is being newly introduced into present-day subsistence and commercial systems. The highest carbon storage results are found in "multistory" agroforestry systems that have many diverse species using ecological "niches" from the high canopy to bottom-story shade-tolerant crops. Examples are shade-grown coffee and cocoa plantations, where cash crops are grown under a canopy of trees that sequester carbon and provide habitats for wildlife. Simple intercrops are used where tree-crop competition is minimal or where the value of tree crops is greater than the value of the intercropped annuals or grazing areas, or as a means to reduce market risks. Where crops are adversely affected by competition for light or water, trees may be grown in small plots in mosaics with crops. Research is also under way to develop low-light-tolerant crop varieties. And in the Sahel, some native trees and crops have complementary growth patterns, avoiding light competition all together.[25]

While agroforestry systems have a lower carbon storage potential per hectare than standing forests do, they can potentially be adopted on hundreds of millions of hectares. And because of the diverse benefits they offer, it is often more economical for farmers to establish and retain them. A Billion Tree Campaign to promote agroforestry was launched at the U.N. climate convention meeting in Nairobi in 2006. Within a year and a half the program had shattered initial expectations and mobilized the planting of 2 billion trees in more than 150 countries. Half the plantings occurred in Africa, with 700 million in Ethiopia alone. By taking the lead from farmers and communities on the choice of species, planting location, and management, and by providing adequate technical support to ensure high-quality planting materials and methods, these initiatives can ensure that the trees will thrive and grow long enough and large enough to actually store a significant amount of carbon.[26]

Tree crop alternatives for food, feed, and fuel. In a prescient book in 1929, Joseph Russell Smith observed the ecological vulnerabilities of annual crops and called for "A Permanent Agriculture." This work highlighted the diversity of tree crops in the United States that could substitute for annual crops in producing starch, protein, edible and industrial oils, animal feed, and other goods as well as edible fruits and nuts—if only concerted efforts were made to develop genetic selection, management, and processing technologies. Worldwide, hundreds of indigenous species of perennial trees, shrubs, and palms are already producing useful products for regional markets but have never been subject to systematic efforts of tree domestication and improvement or to market development. Since one third of the world's annual cereal production is used to feed livestock, finding perennial substitutes for livestock feed is especially promising.[27]

Exciting initiatives are under way with dozens of perennial species, mainly tapping intra-species diversity to identify higher-yielding, higher-quality products and developing rapid propagation and processing methods to use in value-added products. For example, more than 30 species of trees, shrubs, and liane in West Africa have been identified as promising for domestication and commercial development. Commercial-scale initiatives are under way to improve productivity of the Allanblackia and muiri (*Prunus africanus*) trees, which can be incorporated into multistrata agroforestry systems to "mimic" the natural rainforest habitat. Growing trees at high densities is not, however, recommended in dry areas not naturally forested, as this may cause water shortages, as has happened with eucalyptus in some dry areas of Ethiopia.[28]

Shifting biofuel production from annual crops (which often have a net negative impact on GHG emissions due to cultivation, fertilization, and fossil fuel use) to perennial alternatives like switchgrass offers a major new opportunity to use degraded or low-productivity areas for economically valuable crops with positive ecosystem impacts. But this will require a landscape approach to biofuels planning in order to use resources sustainably, enhance overall carbon intensity in the landscape, and complement other key land uses and ecosystem services.[29]

Promoting Climate-friendly Livestock Production

Domestic livestock—cattle, pigs, sheep, goats, poultry, donkeys, and so on—account for most of the total living animal biomass worldwide. A revolution in livestock product consumption is under way as developing countries adopt western diets. Meat consumption in China, for example, more than doubled in the past 20 years and is projected to double again by 2030. This trend has triggered the rise of huge feedlots and confined dairies around most cities and the clearing of huge areas of land for low-intensity grazing. Livestock also produce prodigious quantities of greenhouse gases: methane (from fermentation of food in the largest part of an animal's stomach and from manure storage), nitrous oxide (from denitrification of soil and the crust on manure storage), and carbon (from crop, animal, and microbial respiration as well as fuel combustion and land clearing).[30]

Livestock now account for 50 percent of the emissions from agriculture and land use change. Remarkably, annual emissions from livestock total some 7.1 billion tons (including 2.5 billion tons from clearing land for the animals), accounting for about 14.5 percent of emissions from human activities. Indeed, a cow/calf pair on a beef farm are responsible for more GHG emissions in a year than someone driving 8,000 miles in a mid-size car.[31]

Serious action on climate will almost certainly have to involve reducing consumption of meat and dairy by today's major consumers and slowing the growth of demand in developing countries. No such shift seems likely, however, without putting a price on the cost of emissions. Meanwhile, some solutions are at hand to reduce emissions of greenhouse gases by existing herds.

Intensive rotational grazing. Innovative grazing systems offer alternatives to both extensive grazing systems and confined feedlots and dairies, greatly reducing net GHG emissions while increasing productivity. Conventional thinking says that the current number of livestock far exceeds the carrying capacity of a typical grazing system. But in many circumstances, this reflects poor grazing management practices rather than numbers.

Research shows that grasslands can sustainably support larger livestock herds through intensive management of herd rotations to allow proper regeneration of plants after grazing. By letting the plants recover, the soil organic matter and carbon are protected from erosion, while livestock productivity is maintained or increased. For example, a 4,800-hectare U.S. ranch using intensive rotational grazing tripled the perennial species in the rangelands while almost tripling beef production from 66 kilograms to 171 kilograms per hectare. Various types of rotational grazing are being successfully practiced in the United States, Australia, New Zealand, parts of Europe, and southern and eastern Africa. Large areas of degraded rangeland and pastures around the world could be brought under rotational grazing to enable sustainable livestock production.[32]

Rotational grazing also offers a viable alternative to confined animal operations. A major study by the U.S. Department of Agriculture compared four temperate dairy production systems: full-year confinement dairy, confinement with supplemental grazing, outdoor all-year and all-perennial grassland dairy, and an outdoor cow-calf operation on perennial grassland. The overall carbon footprint was much higher for confined dairy than for grazing systems, mainly because carbon sequestration in the latter is much higher even though carbon emissions are also higher. The researchers concluded that some of the best ways to improve the GHG footprint of intensive dairy and meat operations are to improve carbon storage in grass systems, use higher-quality forage, eliminate manure storage, cover manure storage, increase meat or milk production per animal, and use well-managed rotational grazing.[33]

Feed supplements to reduce methane emissions. Methane produced in the rumen (the first stomach of cattle, sheep, and goats and other species that chew the cud) account for about 1.8 billion tons of CO_2-equivalent emissions. Nutrient supplements and innovative feed mixes have been developed that can reduce methane production by 20 percent, though these are not yet commercially viable for most farmers. Some feed additives can make diets easier for animals to digest and reduce methane emissions. These require fairly sophisticated management, so they are mainly useful in larger-scale livestock operations (which are, in any case, the main sources of methane emissions).[34]

Advanced techniques being developed for methane reduction also include removing specific microbial organisms from the animal's rumen or adding other bacteria that actually reduce gas production there. Research is also under way to develop vaccines against the organisms in the stomach that produce methane.[35]

Biogas digesters for energy. Manure is a major source of methane, responsible for some 400 million tons of CO_2-equivalent. And poor manure management is a leading source of water pollution. But it is also an opportunity for an alternative fuel that reduces a farm's reliance on fossil fuels. By using appropriate technologies like an anaerobic biogas digester, farmers can profit from their farm waste while helping the climate. A biogas digester is basically a temperature-controlled air-tight vessel. Manure (or food waste) is fed into this vessel, where microbial action breaks it down into methane or biogas and a low-odor, nutrient-rich sludge. The biogas can be burned for heat or electricity, while the sludge can be used as fertilizer.[36]

Some communities in developing countries are already using manure to produce cooking fuel. By installing anaerobic digesters, a large pile of manure can be used to produce biogas as well as fertilizer for farms. Even collecting the methane and burning it to convert it to carbon dioxide will be an improvement, as methane has 25 times the global warming potential of carbon dioxide, molecule for molecule, over a 100-year period. And the heat this generates can be used to produce electricity. By thinking creatively, previously undervalued and dangerous wastes can be converted into new sources of energy, cost savings, and even income. Biogas digesters involve an initial cash investment that often needs to be advanced for low-income producers, but lifetime benefits far outweigh costs. This technology could be extended to millions of farmers with benefits for the climate as well as for human well-being through expanded access to energy.[37]

Biogas can even contribute to commercial energy. In 2005, for instance, the Penn England dairy farm in Pennsylvania invested $141,370 in a digester to process manure and $135,000 in a combined heat and power unit, with a total project cost of $1.14 million to process the manure from 800 cows. Now the farm makes a profit by using the biogas to generate 120 kilowatt-hours of electricity to sell back to the local utility, at 3.9¢ per kilowatt-hour. The system also produces sufficient heat to power the digester itself, make hot water, and heat the barns and farm buildings.[38]

Protecting Existing Carbon Stores in Natural Forests and Grasslands

The world's 4 billion hectares of forests and 5 billion hectares of natural grasslands are a massive reservoir of carbon—both in vegetation and root systems. As forests and grasslands continue to grow, they remove carbon from the atmosphere and contribute to climate change mitigation. Intact natural forests in Southeast Australia hold 640 tons of carbon per hectare, compared with 217 tons on average for temperate forests. Thus avoiding emissions by protecting existing terrestrial carbon in forests and grasslands is an essential element of climate action.[39]

Reduce deforestation and land clearing. Massive deforestation and land clearing are releasing stored carbon back into the atmosphere. Between 2000 and 2005, the world lost forest area at a rate of 7.3 million hectares per year. For every hectare of forest cleared, between 217 and 640 tons of carbon are added to the atmosphere, depending upon the type of vegetation. Deforestation and land clearing have many different causes—from large-scale, organized clearing for agricultural use and infrastructure to the small-scale movement of marginalized people into forests for lack of alternative farming or employment opportunities or to the clearing of trees for commercial sale of timber, pulp, or woodfuel. In many cases the key drivers are outside the productive land use sectors—the result of public policies in other sectors, such as construction of roads and other infrastructure, human settlements, or border control.[40]

Current international negotiations are exploring the possibility of compensating developing countries for leaving their forests intact or improving forest management.

Unlike many of the other climate-mitigating land use actions described in this chapter, protecting large areas of standing natural vegetation typically provides fewer short-term financial or livelihood benefits for landowners and managers, and it may indeed reduce their incomes or livelihood security. The solution sometimes lies in regulation, where there is strong enforcement capacity, as with Australia's laws restricting the clearance of natural vegetation. But in many areas the challenge is to develop incentives for conservation for the key stakeholders.

Several approaches are being used. One is to raise the economic value of standing forests or grasslands by improving markets for sustainably harvested, high-value products from those areas or by paying land managers directly for their conservation value. Current international negotiations are exploring the possibility of compensating developing countries for leaving their forests intact or improving forest management. During the Conference of the Parties to the climate convention in Bali in December 2007, governments agreed to a two-year negotiation process that would lead to adoption of a mechanism for Reducing Emissions from Deforestation and Degradation (REDD) after 2012. Implementation of any eventual REDD mechanism will pose major methodological, institutional, and governance challenges, but numerous initiatives are already under way to begin addressing these.[41]

A second incentive for conservation is product certification, whereby agricultural and forest products are labeled as having been produced without clearing natural habitats or in mosaic landscapes that conserve a minimum area of natural patches. For example, the Biodiversity and Agricultural Commodities Program of the International Finance Corporation seeks to increase the production of sustainably produced and verified commodities (palm oil, soy, sugarcane, and cocoa), working closely with commodity roundtables and their members, regulatory institutions, and policymakers. While the priority focus is on conservation of biodiversity, this initiative will have significant climate impacts as well, due to its focus on protecting existing carbon vegetative sinks from conversion, developing standards for sustainable biofuels, and establishing certification systems.[42]

A third approach is to secure local tenure rights for communal forests and grasslands so that local people have an incentive to manage these resources sustainably and can protect them from outside threats like illegal commercial logging or land grabs for agriculture. A study in 2006 of 49 community forest management cases worldwide found that all the initiatives that included tenure security (admittedly a small number) were successful but that only 38 percent of those without it succeeded. Diverse approaches and legal arrangements are being used to strengthen tenure security and local governance capacity.[43]

Reduce uncontrolled forest and grassland burning. Biomass burning is a significant source of carbon emissions, especially in developing countries. Controlled biomass burning in the agricultural sector, on a limited scale, can have positive functions as a means of clearing and rotating individual plots for crop production; in some ecosystems, it is a healthy means of weed control and soil fertility improvement. In a number of natural ecosystems, such as savanna and scrub forests, wild fires can help maintain biotic functions, as in Australia. In many tropical forest ecosystems, however, fires are mostly set by humans and environmentally harmful—killing wildlife, reducing habitat, and setting the stage for more fires by reducing moisture content and increasing combustible materials. Even where they can be beneficial from an agricultural perspective, fires can inadvertently spread to natural ecosystems, opening them up for further agricultural colonization.[44]

Systems are already being put in place to track fires in "real time" so that governments and third-party monitors can identify the people responsible. In the case of large-scale ranchers and commodity producers, better regulatory enforcement is needed, along with alternatives to fire for management purposes. For small-scale, community producers, the most successful approaches have been to link fire control with investments in sustainable intensification of production, in order to develop incentives within the community to protect investments from fire damage. These "social controls" have been effectively used to generate local rules and norms around the use of fire, as in Honduras and The Gambia.[45]

Manage conservation areas as carbon sinks. Protected conservation areas provide a wide range of benefits, including climate regulation. Just letting these areas stand not only helps the biodiversity within, it also stores the carbon, avoiding major releases in greenhouse gas emissions. Moreover, due to some early effects of climate change, important habitats for wildlife are shifting out of protected areas. Plants are growing in higher altitudes as they seek cooler temperatures, while birds have started altering their breeding times. Larger and geographically well distributed areas thus need to be put under some form of protection.

This need not always be through public protected areas. At least 370 million hectares of forest and forest-agriculture landscapes outside official protected areas are already under local conservation management, while half of the world's 102,000 protected areas are in ancestral lands of indigenous and other communities that do not want to see them developed. Conservation agencies and communities are finding diverse incentives for protecting these areas, from the sustainable harvesting of foods, medicines, and raw materials to the protection of locally important ecosystem services and religious and cultural values as well as opportunities for nature tourism income. Supporting these efforts to develop and sustain protected area networks, including public, community, and private conservation areas, can be a highly effective way to reduce and store greenhouse gases.[46]

Restoring Vegetation in Degraded Areas

Extensive areas of the world have been denuded of vegetation from large-scale land clearing for annual crops or grazing and from overuse and poor management in community and public lands with weak governance. This is a tragic loss, from multiple perspectives. People living in these areas have lost a potentially valuable asset for the production of animal fodder, fuel, medicines, and raw materials. Gathering such materials is an especially important source of income and subsistence for low-income rural people. For example, researchers found in Zimbabwe that 24 percent of the average total income of poor farmers came from gathering woodland products. At the same time, the loss of vegetation seriously threatens ecosystem services, particularly watershed functions and wildlife habitat.[47]

Efforts to restore degraded areas can thus be "win-win-win" investments. Although there may be fewer tons of CO_2 sequestered per hectare from restoration activities, millions of hectares can be restored with low opportunity costs and strong local incentives for participation and maintenance.

Revegetate degraded watersheds and rangelands. Hydrologists have learned that "green water"—the water stored in vegetation and filtrating into soils—is as important as "blue water" in streams and lakes. When rain falls on bare soils, most is lost as runoff. In many of the world's major watersheds, most of the land is in productive use. Poor vegetative cover limits the capacity to retain rainfall in the system or to filter water flowing into streams and lakes and therefore accelerating soil loss. From a climate perspective, lands stripped of vegetation have lost the potential to store carbon. Landscapes that retain year-round vegetative cover in strategically selected areas and natural habitat cover in critical riparian areas can maintain most, if not all, of various watershed functions, even if much of the watershed is under productive uses.[48]

With rapid growth in demand for water and with water scarcity looming in many countries (in part due to climate change), watershed revegetation is now getting serious policy attention. Both India and China have large national programs targeting millions of hectares of forests and grasslands for revegetating, and they see these as investments to reduce rural poverty and protect critical watersheds. In most cases, very low-cost methods are used for revegetation, mainly temporary protection to enable natural vegetation to reestablish itself without threat of overgrazing or fire. For example, in Morocco 34 pastoral cooperatives with more than 8,000 members rehabilitated and manage 450,000 hectares of grazing reserves. On highly degraded soils, some cultivation or reseeding may be needed. Two keys to success in these approaches are to engage local communities in planning, developing, and maintaining watershed areas and to include rehabilitation of areas of high local importance, such as productive grazing lands, local woodfuel sources, and areas like gullies that can be used for productive cropping.[49]

In Rajasthan, India, for example, community-led watershed restoration programs have reinstated more than 5,000 traditional *johads* (rainwater storage tanks) in over 1,000 villages, increasing water supplies for irrigation, wildlife, livestock, and domestic use and recharging groundwater. In Niger, a "regreening" movement, using farmer-managed natural regeneration and simple soil and water conservation practices, reversed desertification, increased tree and shrub cover 10- to 20-fold, and reclaimed at least 250,000 hectares of degraded land for crops. Over 25 years, at least a quarter of the country's farmers were involved in restoring about 5 million hectares of land, benefiting at least 4.5 million people through increased crop production, income, and food security. Extending the scale of such efforts could have major climate benefits, with huge advantages as well for water security, biodiversity, and rural livelihoods.[50]

Reestablish forest and grassland cover in biological corridors. Loss and fragmentation of natural habitat are leading threats to biodiversity worldwide. Conservation biologists have concluded that in many areas conservation of biodiversity will require the establishment of "biological corridors" through production landscapes, to connect fragments of natural habitat and protected areas and to give species access to adequate territory and sources of food and water. One key strategy is to reestablish forest or natural grassland cover (depending on the ecosystem) to play this ecological role. These reforestation efforts also have major climate benefits.

In Brazil's highly threatened Atlantic Forest, for example, conservation organizations working in the Desengano State Park struck a deal with dairy farmers to provide technical assistance to improve dairy-farm productivity in exchange for the farmers reforesting part of their land and maintaining it as a conservation easement. Milk yields tripled and farmers' incomes doubled, while a strategic buffer zone was established for the park.[51]

In northwestern Ecuador, two thirds of coastal rainforests have been lost due to logging and agricultural expansion, risking the survival of 2,000 plant and 450 bird species. The Chocó-Manabí corridor reforestation project is attempting to improve wild species' access to refuge habitats by restoring connectivity between native forest patches through reforestation efforts. This project is restoring 265 hectares of degraded pastures with 15 native trees species and as a result sequestering 80,000 tons of carbon dioxide. The opportunity for such investments is mobilizing new partnerships between wildlife conservation organizations, the climate action community, farmers, and ranchers.[52]

Market Incentives for Climate-friendly Agriculture and Land Use

All the strategies described in the preceding sections are already available or are well within technological reach at far lower cost than many climate solutions being discussed (such as geological storage of carbon). The challenge is shifting policy and investment priorities and supporting institutions to create incentives for farmers, pastoralists, forest owners, agribusiness, and all other stakeholders within the agriculture and forestry supply chains to scale up best practices and continue to innovate new ones. This will require concerted action by consumers, farmers' organizations, the food industry, civil society, and governments.

The central players in any response to climate change are the farmers and communities—those who actually manage land—and the food and fiber industry that shapes the incentives for the choice of crops, quality standards, and profitability. Some innovators are already showing the way. For example, the Sustainable Food Lab, a collaborative of 70 businesses and social organizations from throughout the world, has assembled a team of member companies, university researchers, and technical experts to develop and test ways to measure and provide incentives for low-carbon agricultural practices through the food supply chain, mainly by increasing soil organic matter, improving fertilizer application, and enhancing the capacity of crops and soil to store carbon.[53]

A key driver is consumer and buyer awareness. Consumers will take the needed steps once they realize that their choice of meat and dairy products, and their support for natural forests and grassland protection, can have as great an impact on the climate as how far they drive their cars. One immediate action is for consumers, processors, and distributors to adopt greenhouse gas footprint analysis for food and fiber products, addressing their full "life cycle," including production, transport, refrigeration, and packaging, to identify strategic intervention points.

In 2007, for instance, the Dole Corporation committed to establishing by 2021 a carbon-neutral product supply chain for its bananas and pineapples in Costa Rica. Their first step in this process was to purchase forest carbon offsets from the Costa Rican government equal to the emissions of its inland transport of these fruits. GHG impact is a key metric that can be used for evaluating new food and forest production technologies and for allocating resources and investments. Policymakers can then include incentives for reducing carbon emissions in cost structures throughout the food and land use systems, using various market and policy mechanisms.[54]

Product markets are also beginning to recognize climate values. The last 20 years have seen the rise of a variety of "green" certified products beyond organic, such as "bird-friendly" and "shade-grown," that have clear biodiversity benefits. Various certification options already exist for cocoa and coffee (through the Rainforest Alliance, Starbucks, and Organic, for example). The Forest Stewardship Council's certification principles "prohibit conversion of forests or any other natural habitat" and maintain that "plantations must contribute to reduce the pressures on and promote the restoration and conservation of natural forests," supporting the use of forests as carbon sinks. New certification standards are starting up that explicitly include impacts on climate, which will for the first time send clear signals to both producers and consumers.[55]

The rise of carbon emission offset trading could potentially provide a major new source of funding for the transition to climate-friendly agriculture and land use. (See Box 3–2.) A great deal can be done in the short term through the voluntary carbon market, but in the long run it will be essential for the international framework for action on climate change to fully incorporate agriculture and land use.[56]

Public Policies to Support the Transition

Governments can take specific steps immediately to support the needed transition by integrating agriculture, land use, and climate action programs at national and local landscape levels. Costa Rica is a leader in these efforts. The government has committed to achieving "climate neutrality" by 2021, with an ambitious agenda including mitigation through land use change. Costa Rica is a participant in the Coalition for Rainforest Nations, a group encouraging avoided-deforestation programs, and has already increased its forest cover from 21 percent in 1986 to 51 percent in 2006. The country is taking advantage of markets that make payments for ecosystem services and ecotourism to support these efforts.[57]

Currently, governments spend billions of dollars each year on agricultural subsidy payments to farmers for production and inputs, primarily in the United States ($13 billion in 2006, which was 16 percent of the value of agricultural production) and Europe ($77 billion, or 40 percent of agricultural production value) but also in Japan, India, China, and elsewhere. Most of these payments exacerbate chemical use, the expansion of cropland to sensitive areas, and overexploitation of water and other resources while distorting trade and reinforcing unsustainable agricultural practices. Some countries are beginning to redirect subsidy payments to agri-environmental payments for all kinds of ecosystem services, and these can explicitly include carbon storage or emissions reduction.[58]

Growth in commercial demand for agricultural and forest products from increased populations and incomes in developing countries and demand for biofuels in industrial

Box 3–2. Paying Farmers for Climate Benefits

Paying farmers and land managers to reduce carbon emissions or store greenhouse gases is a critical way to both mitigate climate change and generate ecosystem and livelihood benefits. The carbon market for land use has three main components: carbon emissions offsets for the regulatory market, as established by the Kyoto Protocol; offset activities in emerging U.S. regulatory markets operating outside the Kyoto Protocol; and the sale of voluntary carbon offsets coming from land use, land use change, and forestry, primarily to individual consumers, philanthropic buyers, and the private sector.

Developing countries can implement afforestation and reforestation projects that count toward emission reduction targets of industrial countries through the Clean Development Mechanism of the Kyoto Protocol. The treaty authorizes afforestation and reforestation but excludes agricultural or forest management, avoided deforestation or degradation, and soil carbon storage. However, each CDM project must address thorny issues of nonpermanence of carbon uptake by vegetation and soil, risks of potential displacement of emissions as deforestation just moves elsewhere, and sustainable development prospects in the host country that can limit implementation.

There is more innovation in the voluntary market, where buyers value multiple benefits. The value of forestry plus land use projects more than doubled from $35 million in 2006 to $72 million in 2007. Work is proceeding to lend more credibility, transparency, and uniformity in methods used for creating land-based carbon credits.

There are several ongoing initiatives to promote diverse types of land-use-based payments:

- The World Bank's $91.9-million BioCarbon Fund is financing afforestation, reforestation, REDD, agroforestry, and agricultural and ecosystem-based projects that not only promote biodiversity conservation and poverty alleviation but also sequester carbon.
- The Regional Greenhouse Gas Initiative in the northeastern United States will include afforestation and methane capture from U.S. farms.
- The trading system in New South Wales, Australia—the world's first—provides for carbon sequestration through forestry, including on-farm forest regeneration.
- The New Zealand government is investing more than $175 million over five years in a Sustainable Land Management and Climate Change Plan to help the agriculture and forestry sectors adapt to, mitigate, and take advantage of the business opportunities of climate change. This scheme will include specific cap-and-trade allocations to the dairy sector and will incorporate cash grants to encourage new plantings by landowners, increased research funding, technology transfer, and incentives to use more wood products and bio-energy.
- Rabobank, the world's largest agricultural financier, will pay farmers $83,000 to reforest, which will be sold as carbon offsets; the bank may use some of these credits to offset its own activities. This is the first transaction of its kind in Brazil's Xingu province, which has the country's highest deforestation rates. Soy and cattle farmers are targeted, and replanting is planned for riparian stretches through the region.
- REDD payments for avoided deforestation in Mato Grosso state alone in Brazil are estimated at $388 million annually.

Much larger initiatives are needed now to link carbon finance with investments to achieve rural food security by "re-greening" degraded watersheds, promoting agroforestry, restoring soil organic matter, rehabilitating degraded pastures, controlling fires, or protecting threatened forests and natural areas important for local livelihoods. If low-income landowners and managers are to benefit from payments for ecosystem services, they need secure rights, clear indicators of performance, and systems for aggregating buyers and sellers to keep transaction costs low.

Source: See endnote 56.

nations is stimulating investments by both private and public sectors. In 2003, African governments committed to increase public investment in agriculture to at least 10 percent a year, although only Rwanda and Zambia have done this so far. The World Bank and the Bill & Melinda Gates Foundation have committed to large increases in funding in the developing world. There is a major window of opportunity right now to put climate change adaptation and mitigation at the core of these strategies.[59]

This is beginning to happen in small steps. Brazil is crafting a diverse set of investment programs to support rural land users who invest in land use change for climate change mitigation and adaptation. The U.N. Environment Programme is initiating dialogues on "greening" the international response to the food crisis, linking goals of international environmental conventions with the Millennium Development Goals.[60]

Many available technologies and management practices could lighten the climate footprint of agriculture and other land uses and protect existing carbon sinks in natural vegetation.

But much more comprehensive action is needed. If not, this otherwise positive trend could seriously undermine climate action programs. A new vision is needed to respond to this food crisis that not only provides a short-term Band-Aid to refill next year's grain bins but also puts the planet on a trajectory toward sustainable, climate-friendly food systems.

National policy, however, is not enough. It is essential to invest in building capacity at local levels to manage ecoagricultural landscapes—to enable multistakeholder platforms to plan, implement, and track progress in achieving climate-friendly land use systems that benefit local people, agricultural production, and ecosystems.

Taking Action for Climate-friendly Land Use

Human well-being is wrapped up with how food is produced. Ingenious systems were developed over the past century to supply food, with remarkable reliability, to a good portion of the world's 6.7 billion people. But these systems need a fundamental restructuring over the next few decades to establish sustainable food systems that both slow and are resilient to climate change. Land-manager and private-sector action will determine the response, but public policy and civil society will play a crucial role in providing the incentives and framework for communities and markets to respond effectively.[61]

Food production and other land uses are currently among the highest greenhouse gas emitters on the planet—but that can be reversed. Although recent food price riots may discourage any actions that could raise costs, if action is not taken costs will rise anyway as local food systems are disrupted and as higher energy costs ripple through a system that has not been prepared with alternatives.

The strategy for reducing greenhouse gas emissions from agriculture and other land use sectors also must recognize the need for major increases in food and fiber production in developing countries to feed adequately the 850 million people currently hungry or undernourished, as well as continually growing populations. Investments must be channeled so that increased production comes from climate-friendly production systems rather than from systems that clear large areas of natural forest and grasslands, mine organic matter from the soil, strip vegetative cover from riparian areas, or leave soils bare for many months of the year.[62]

As described in this chapter, many available technologies and management practices could lighten the climate footprint of agriculture and other land uses and protect existing carbon sinks in natural vegetation and soils. Many more could become operational fairly quickly with proper policy support or adaptive research and with a more systematic effort to analyze the costs and benefits of different strategies in different land use systems. Other innovative ideas will emerge if leading scientists and entrepreneurs can be inspired to tackle this challenge. And many of the actions most needed in land use systems to adapt to climate change and mitigate GHG emissions will bring positive benefits for water quality, air pollution, smoke-related health risks, soil health, energy efficiency, and wildlife habitat. These tangible benefits can generate broader political support for climate action.

It is heartening that there are already so many initiatives to address climate change in the food and land use sectors, and these efforts have established a rich foundation of practical, implementable models. But the scale of action so far is dishearteningly small. With the exception of the recent REDD initiatives to save standing forests through intergovernmental action, which are still in an early stage, there are no major international initiatives to address the interlinked challenge of climate, agriculture, and land use.

A worldwide, networked movement for climate-friendly food, forest, and other land-based production is needed. This calls for forging unusual political coalitions that link consumers, producers, industry, investors, environmentalists, and communicators. Food, in particular, is something that the public understands. By focusing on food systems, climate action will become more real to people. It is realistic to expect that the prices of food and other land-based products will rise in a warming world, at least for a time. This must not be the result of scarcity caused by climate-induced system collapse but rather because new investment has been mobilized to create sustainable food and forest systems that also cool the planet.

Climate Connections

The Risks of Other Greenhouse Gases

Janos Maté, Kert Davies, and David Kanter

As the stability of the world's climate is increasingly at risk and governments grapple with the monumental task of cutting emissions, there is a group of little known but powerful greenhouse gases that, left unchecked, could ultimately undermine the best efforts to tackle the climate crisis. Fluorocarbons, or F-gases, are the quintessential greenhouse gases, since chemical engineers designed them to trap heat and to be stable and durable.

The Intergovernmental Panel on Climate Change calculated that the cumulative build-up of these gases in the atmosphere was responsible for at least 17 percent of global warming due to human activities in 2005. And the use of these chemicals worldwide is on the rise, with increased consumption in developing countries like China and India.[1]

Several chemical cousins make up the F-gas family: chlorofluorocarbons (CFCs), hydrochlorofluorocarbons (HCFCs), hydrofluorocarbons (HFCs), perfluorocarbons (PFCs), and sulfur hexafluoride (SF_6). The major applications of these chemicals today are in refrigeration and air-conditioning (including in cars), which account for 80 percent of F-gas use. The chemicals are also used as solvents, as blowing agents in foams, as aerosols or propellants, and in fire extinguishers. The most commonly used F-gases at the moment are the HFCs, a class of powerful greenhouse gases whose consumption is rising exponentially. HFCs were developed by the chemical industry in response to the discovery of damage to Earth's ozone layer due to CFC use. But this development ignored the known global warming effects of the newer chemicals.[2]

F-gases have incredibly strong global warming potential (GWP) relative to carbon dioxide (CO_2) pound per pound, or gram for gram, because they were built to trap heat very effectively. GWP is calculated relative to carbon dioxide, which is assigned a GWP of 1. Global warming potential depends on the ability of a molecule to trap heat and its "atmospheric lifetime"—how long the chemical stays in the atmosphere before it is broken down or is absorbed or settles out into the ocean, soil, or biosphere, for instance.[3]

GWPs are generally averaged over 100 years, as a baseline for comparison to carbon dioxide. The most popular HFC in use today has an atmospheric lifetime of about 14 years and a 100-year GWP of 1,400, but a 20-year GWP of 3,830. This means that one pound of this HFC is the same as 3,830 pounds of CO_2 in terms of global warming impact for 20 years after it is released into the atmosphere. Of course, there is one benefit to this high short-term GWP: phasing

Janos Maté is Ozone Policy Consultant for the Political Business Unit of Greenpeace International, based in Vancouver, Canada. ***Kert Davies*** *is Research Director at Greenpeace US, based in Washington, DC.* ***David Kanter*** *is a Research Fellow for Greenpeace International, based in the United Kingdom.*

out HFCs has an immediate global warming benefit. Cutting HFC use slows global warming right now, when that is most needed.[4]

The chemical industry argues that HFCs can be safely contained and prevented from leaking into the atmosphere, but so far containment has been an unqualified failure—more than 50 percent of all HFCs produced to date have already found their way into the atmosphere.[5]

Greenpeace estimates that HFCs will become an ever-increasing portion of the global warming pollution load exactly when scientists say greenhouse gases need to be reduced rapidly. The popular HFC-134a alone, which currently accounts for about 1 percent of greenhouse gases, is expected to account for 10 percent of emissions by 2050—and releases will still be on the rise. Growing HFC emissions could significantly hamper efforts to keep global temperatures from exceeding crucial tipping points in climate change that have been identified by scientists.[6]

Fortunately, there are environmentally safe, efficient, technologically proven, and commercially available alternatives to F-gases in almost all domestic and commercial applications. These use natural substances, such as simple hydrocarbons, carbon dioxide, ammonia, or even straight water. Typically, systems using natural refrigerants are at least as energy-efficient as those using HFCs or even more efficient. So they are less expensive to operate and create fewer greenhouse gases through reduced electricity use, and thus less load on dirty power plants.

In addition, there are novel technological solutions for air conditioning and refrigeration such as solar adsorption, which uses solar heat as the engine for compression of a liquid refrigerant, most commonly water and ammonia or water and lithium bromide salt. There are also refrigerators that operate with sound waves, and in certain climates simple evaporation devices provide air-conditioning.

The chemical industry does not profit from any of these alternatives. Getting rid of HFCs would put an end to the industry's global, long-term hold on the multibillion-dollar monopoly it has enjoyed with CFCs and other F-gases. The industry is therefore fighting any replacements, especially natural refrigerants, using its global lobbying efforts to soften—or stop—strong legislation and regulations.

There are many examples in all sectors of using natural refrigerants instead of HFCs. One outstanding example is the hydrocarbon domestic Greenfreeze refrigerator (safe for both the ozone layer and the climate) developed in 1993 by Greenpeace and a tiny East German company. Greenfreeze refrigerators are typically more efficient than their HFC counterparts. There are an estimated 300 million Greenfreeze-type refrigerators in the world today, built by leading manufacturers and accounting for approximately 40 percent of the 80 million refrigerators produced annually. This technology dominates the domestic refrigeration markets of Europe and is prominent in Japan and China, but it is conspicuously unavailable in North America due to obsolete regulatory obstacles.[7]

The search for HFC alternatives has been taken up by many large multinational corporations that use refrigeration and cooling technology. A technology-sharing coalition called Refrigerants Naturally was founded in 2004, set up by Greenpeace and the United Nations Environment Programme with the goal of replacing HFCs with natural refrigerants in vending machines, freezers, and fridges. The current partners include Unilever, Coca-Cola, PepsiCo, McDonald's, IKEA, and Carlsberg.[8]

The success of Refrigerants Naturally has

been broad. By 2008 Unilever had placed up to 275,000 climate-friendly retail ice-cream freezers in the field. And in late 2008 Ben & Jerry's ice cream, a Unilever brand, started using non-HFC freezers in the United States.

greentechnica.com

One of Ben & Jerry's non-HFC freezers freshly installed

The new units save at least 10 percent of the energy used by identical HFC freezers. Coca-Cola has developed a new, high-efficiency compressed carbon dioxide technology for its vending machines and planned to have up to 30,000 CO_2 vending machines in the field by 2008, which will increase to 100,000 by 2010. All Coca-Cola vending machines at the 2008 Beijing Olympics were HFC-free.[9]

Supermarkets and other retail stores are also making the switch from HFCs and F-gas technology. In March 2006, several major U.K. supermarket chains—including ASDA, Marks & Spencer, Sainsbury's, Somerfield, Tesco, and Waitrose—announced their decision to phase out their use of HFCs in cooling equipment and to convert to natural refrigerants such as carbon dioxide.[10]

In another crucial development, progress is being made in phasing out potent F-gases from automobile air-conditioning. The European Union (EU) moved to phase out HFC-134a in mobile air-conditioning by 2011. In response, the German car industry decided in August 2007 to deploy compressed carbon dioxide as the replacement refrigerant.[11]

HFCs and other non-ozone depleting F-gases are currently included in the Kyoto Protocol's "basket" of greenhouse gases, but the parties to the treaty have yet to address HFCs specifically and proactively. (CFCs and HCFCs were not included in the Kyoto Protocol because they were already being banned and phased out under the earlier Montreal Protocol on the ozone layer.) In the meantime, a few jurisdictions around the world are already taking action in a variety of ways.

Individual countries and jurisdictions—including Denmark, Switzerland, Austria, the EU Commission, and most recently the state of California—have all moved to curtail releases of F-gases by variously banning new uses, phasing out old ones, and providing incentives to reduce leakage of high-GWP F-gases. Governments have made these moves by passing regulations, providing incentives for technology upgrades, levying taxes on imports of HFCs based on their warming impact, and providing refunds for destruction of captured used F-gas. These bold moves have led to increased awareness and adoption of alternatives.[12]

The global solutions to this potentially devastating problem are quite clear and available. The Kyoto Protocol and the Montreal Protocol both could, in complementary ways, act swiftly to reduce and eliminate uses of F-gases. For example, the Montreal Protocol's practice of pushing developing countries to replace HCFCs with HFCs should be stopped immediately. And the treaty could be used to stimulate the recovery of millions of tons of "banked" HFCs, CFCs, and HCFCs sitting in old cooling equipment that need to be safely recovered at the end of the equipment's life and destroyed.

Meanwhile, the parties to the Kyoto Protocol are in the midst of serious and complex negotiations leading up to their next conference, in Copenhagen in late 2009, when new emissions reduction targets should be on the table. Negotiators are having a difficult time setting targets and commitments for reducing carbon dioxide, the largest greenhouse gas, and shaping a renewed commitment to limit deforestation-related greenhouse gas emissions. As of late 2008 the parties to the treaty had not found a path to develop strong specific incentives and actions on HFCs.

Some observers are now proposing that HFCs be regulated under the Montreal Protocol instead of the Kyoto Protocol. They call for HFCs to be phased out of production over time just as other F-gases were and simultaneously be removed from the Kyoto Protocol basket entirely. The argument is made that Montreal Protocol participants have more expertise with F-gases and assessment of alternatives and that developing countries already participate strongly in the treaty.[13]

This approach—removing agreed greenhouse gases or sectors from the climate change treaty at this point—risks weakening the Kyoto agreement. In addition, it would kill a strong financial incentive to reduce F-gas emissions by removing HFCs from the rapidly evolving GHG emissions trading schemes, erasing potentially lucrative carbon credits that adopters of non-HFC technology could earn and trade under the Kyoto Protocol. It is hoped that an effective approach will evolve during the treaty negotiations of 2009 and beyond, making efficient use of the best capacities of both treaties. Until then, the Kyoto Protocol arena remains the best place to kick-start and deal with this growing threat to the climate.

Reducing Black Carbon

Dennis Clare

Black carbon, a component of soot, is a potent climate-forcing aerosol and may be the second-leading cause of global warming after carbon dioxide (CO_2). Unlike CO_2, however, black carbon remains in the atmosphere for only a few days or weeks. Therefore reducing these emissions will have an almost immediate climate mitigation impact. While substantial reductions of greenhouse gas emissions should remain the anchor of overall climate stabilization efforts, dealing with black carbon may be the fastest means of near-term climate mitigation and could be critical in forestalling climate tipping points.[1]

Black carbon is a product of the incomplete combustion of fossil fuels, biofuels, and biomass. The main sources are open burning of biomass, diesel engines, and the residential burning of solid fuels such as coal, wood, dung, and agricultural residues. (See Figure.) Black carbon contributes to climate change in two ways: It warms the atmosphere directly by absorbing solar radiation and converting it to heat and indirectly by darkening the surfaces of ice and snow when deposited on them. This reduces albedo, the ability to reflect light, and thereby increases heat absorption and accelerates melting. As large masses of both land and sea ice disappear, they reflect less and less solar radiation, so heat is increasingly absorbed at the surface. Thus not only do sea and land ice face tipping points of irreversible melting, but this melting can create positive feedbacks leading to even further warming.[2]

Dennis Clare is a Law Fellow at the Institute of Governance & Sustainable Development in Washington, DC.

Combustion Sources of Black Carbon

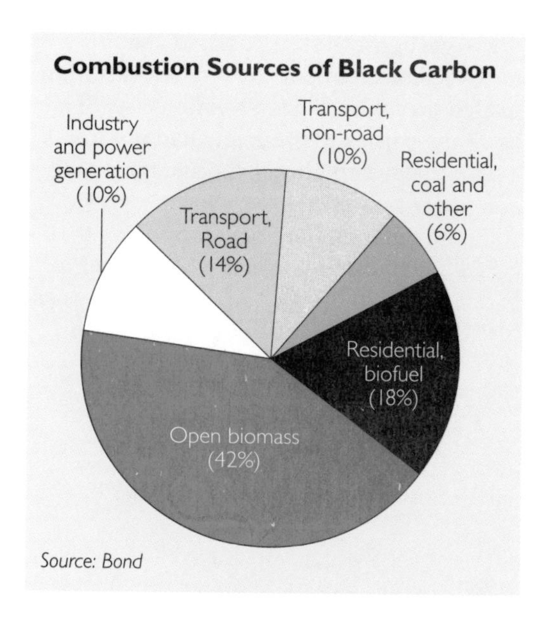

Source: Bond

The ability of black carbon to accelerate snow and ice melt makes it a particular concern in Arctic and glacial areas, where tipping points for melting are considered among the most imminent. According to Charles Zender of the University of California, Irvine, black carbon on snow warms Earth about three times as much as an equal forcing of CO_2. Melting glaciers are a concern not only because of the feedback warming they can cause but also because they feed rivers that supply water to hundreds of millions of people. As glaciers retreat due to melting, these water sources become threatened. V. Ramanathan and Y. Feng of the Scripps Institution of Oceanography claim that the world has already committed to warming in a range that could lead to "a major reduction of area and volume of Hindu-Kush-Himalaya-Tibetan (HKHT) glaciers, which provide the head-waters for most major river systems of Asia." And as land-based ice melts, the water flows into the oceans, contributing to potentially dangerous sea level rise.[3]

The 2007 report of the Intergovernmental Panel on Climate Change for the first time calculated the direct and indirect climate forcings of black carbon. It estimated the direct forcing from fossil fuel sources to be +0.2 watts per square meter (W/m^2) and from biomass sources to be +0.1 W/m^2. The report estimated the indirect forcing from black carbon's effect on snow and ice to be an additional +0.1 W/m^2, for a total forcing from black carbon of +0.4 W/m^2. More recent studies have estimated the figure is as high as +0.9 W/m^2, which is equal to about 55 percent of the forcing from CO_2. Although the net impact of black carbon seems squarely within a significant warming range, further study will be required to calculate its forcing more precisely.[4]

It is important to understand that when black carbon is emitted from incomplete combustion, other particles such as organic carbon, nitrates, and sulfates are emitted as well. Some of these other particles, such as sulfates, can have a cooling effect on the atmosphere. But because sulfates have been determined to harm public health, worldwide efforts are under way to reduce them. When this happens, further warming will be unmasked. Ramanathan and Feng claim that temperature-masking due to cooling aerosols is currently over 1 degree Celsius and that as aerosols continue to be reduced by local air pollution control measures, this warming will be felt—likely within the next 50 years. Thus it is all the more important to reduce black carbon to offset the increased warming that will result from the expected elimination of sulfates.[5]

A number of technologies for reducing black carbon emissions are available now and others are being developed. For diesel vehicles, highly effective diesel particulate filters (DPFs) can eliminate over 90 percent of the emissions, although DPFs require the use of ultra-low sulfur diesel fuel (ULSD), which is not yet widely available outside the United States, the European Union, or Japan. Increasing the availability of ULSD is an important step in reducing black carbon and other emissions from diesel vehicles worldwide. For vehicles without access to ULSD, other filters can be used. Newly developed flow-through, or partial, filters can eliminate 40–70 percent of emissions from vehicles using traditional diesel fuel. Programs for retrofitting diesel vehicles with the most efficient filters available will be essential, as older cars are often highly polluting.[6]

Ocean vessels generally use diesel engines as well and therefore emit substantial amounts of black carbon. Emissions from vessels in the northern hemisphere, especially those near the Arctic, can be especially harmful. Efficient DPFs for vessels are currently in development. In addition, the International Maritime Organization has been reducing the permissible sulfur content of bunker fuel due to the harmful effects of sulfur emissions. The possibility of also reducing black carbon emissions should lead to an acceleration of the low-sulfur fuels. Other means of reducing black carbon emissions from ocean vessels include reducing diesel fuel use by using shoreside electricity when at port and simply traveling more slowly at sea.[7]

Because of the higher ratio of cooling particles emitted with black carbon when biomass is burned, it is more complicated to get climate benefits from changing cooking and agricultural burning practices. But energy security and indoor air pollution concerns have already led to a variety of programs aimed at improving the efficiency of traditional cookstoves. A better understanding of the possible climate benefits of reducing black carbon provides an incentive to look closely at the potential to improve cooking techniques and agricultural burning practices. Some stoves not only reduce black carbon but also produce biochar, a stable form of carbon that can be stored permanently in soils and improve soil productivity.

Ensuring compliance and enforcement with existing national laws on black carbon can provide some relief from warming. This is being pursued by the International Network for Environmental Compliance and Enforcement. But new laws and regulations are needed at all levels for further and faster reductions. As a start, where ultra-low sulfur diesel fuel is available, diesel particulate filters should be required on both new vehicles and existing ones that may be used for many years. In addition, northern countries could restrict agricultural burning in the springtime melt season in order to reduce the impact of black carbon on snow and ice. Institutions such as the World Bank should make climate funding available for black carbon reduction programs.[8]

The faster-than-anticipated approach of dangerous tipping points is forcing society to consider the quickest way to mitigate climate change. Reducing black carbon may be the fastest means of all.

Women and Climate Change: Vulnerabilities and Adaptive Capacities

Lorena Aguilar

Although climate change will affect everyone worldwide, its impacts will be distributed differently between men and women as well as among regions, generations, age classes, income groups, and occupations. The poor, the majority of whom are women living in developing countries, will be disproportionately affected. Yet most of the debate on climate so far has been gender-blind.[1]

Gender inequality and climate change are inextricably linked. By exacerbating inequality overall, climate change slows progress toward gender equality and thus impedes efforts to achieve wider goals like poverty reduction and sustainable development. And women are powerful agents of change whose leadership on climate change is critical. Women can help or hinder strategies related to energy use, deforestation, population, economic growth, and science and technology, among other things.

Climate change can have disproportionate impacts on women's well-being. Through both direct and indirect risks, it can affect their livelihood opportunities, time availability, and overall life expectancy. (See Table.) An increase in climate-related disease outbreaks, for example, will have quite different impacts on women than on men. Each year, some 50 million women living in malaria-endemic countries become pregnant; half of them live in tropical areas of Africa with high transmission rates of the parasite that causes malaria. An estimated 10,000 of these women and 200,000 of their infants die as a result of malaria infection during pregnancy; severe malarial anemia is involved in more than half of these deaths.[2]

People's vulnerability to risks depends in large part on the assets they have available. Women, particularly poor women, face different vulnerabilities than men. Approximately 70 percent of the people who live on less than $1 a day are women. Many live in conditions of social exclusion. They may face constraints on their mobility or behavior that, for example, hinder their ability to relocate without a male relative's consent.[3]

In general, women tend to have more limited access to the assets—physical, financial, human, social, and natural capital—that would enhance their capacity to adapt to climate change, such as land, credit, decisionmaking bodies, agricultural inputs, technology, and extension and training services. Thus any climate adaptation strategy should include actions to build up women's assets. Interventions should pay special attention to the need to enhance women's capacity to manage risks with a view to reducing their vulnerability and maintaining or increasing their opportunities for development.[4]

Ways to reduce climate-related risks for women include improving their access to skills, education, and knowledge; strength-

***Lorena Aguilar** is the Global Senior Gender Adviser for the International Union for Conservation of Nature–IUCN.*

Direct and Indirect Risks of Climate Change and Their Potential Effects on Women

Potential Risks	Examples	Potential Effects on Women
Direct Risks		
Increased drought and water shortage	Morocco had 10 years of drought from 1984 to 2000; northern Kenya experienced four severe droughts between 1983 and 2001.	Women and girls in developing countries are often the primary collectors, users, and managers of water. Decreases in water availability will jeopardize their families' livelihoods and increase their workloads, putting their capacity to attend school at risk.
Increased extreme weather events	The intensity and quantity of cyclones, hurricanes, floods, and heat waves have increased.	An analysis of 141 countries in the period 1981 to 2002 found that natural disasters (and their subsequent impacts) on average killed more women than men in societies where women's economic and social rights are not protected, or they killed women at an younger age than men.
Indirect Risks		
Increased epidemics	Climate variability played a critical role in malaria epidemics in the East African highlands and accounted for an estimated 70 percent of variation in recent cholera series in Bangladesh.	Women have less access to medical services than men, and their workloads increase when they have to spend more time caring for the sick. Adopting new strategies for crop production or mobilizing livestock is harder for female-headed and infected households.
Loss of species	By 2050, climate change could result in species extinctions ranging from 18 to 35 percent.	Women often rely on crop diversity to accommodate climatic variability, but permanent temperature change will reduce agro-biodiversity and traditional medicine options.
Decreased crop production	In Africa, crop production is expected to decline 20–50 percent in response to extreme El Niño-like conditions.	Rural women in particular are responsible for half of the world's food production and produce 60–80 percent of the food in most developing countries. In Africa, the share of women affected by climate-related crop changes could range from 48 percent in Burkina Faso to 73 percent in the Congo.

Source: See endnote 2.

ening their ability to prepare for and manage disasters; supporting their political ability to demand access to risk-management instruments; and helping households gain greater access to credit, markets, and social security.

Despite the many challenges they face, women are already playing an important role in developing strategies to cope with climate change. They have always been leaders in community revitalization and natural resource management, and there are countless instances where their participation has been critical to community survival. In Honduras, for example, the village of La Masica was the only community to register no deaths in the wake of 1998's Hurricane Mitch. Six months earlier, a disaster agency had provided gender-sensitive community education on early warning systems and hazard management. Women took over the abandoned task of continuously monitoring the warning system. As a result, the municipality was able to evacuate the area promptly when the hurricane struck.[5]

Women also play a crucial role in forest preservation strategies and increasing carbon sinks through reforestation and afforestation. For example, since 2001 women in Guatemala, Nicaragua, El Salvador, Honduras, and Mexico have planted more than 400,000 Maya Nut trees as part of the Mayan Nuts Project supported by the Equilibrium Fund, which also increased food supplies for the communities. This shows how specific projects can improve the quality of life for women and at the same time be strategies for mitigation and adaptation to climate change.[6]

And in Kenya, the Green Belt Movement and the World Bank's Community Development Carbon Fund signed an emissions-reduction agreement in November 2006 to reforest two mountain areas. Women's groups are planting thousands of trees, an activity that will provide poor rural women with a small income and some economic independence as well as capture some 350,000 tons of carbon dioxide, restore eroded soils, and support regular rainfall essential to Kenya's farmers and hydroelectric plants.[7]

Women from indigenous communities often know a range of "coping strategies" traditionally used to manage climate variability and change. In Rwanda, women produce more than 600 varieties of beans, and in Peru Aguaruna women plant more than 60 varieties of manioc. These vast varieties, developed over centuries, allow them to adapt their crops to different biophysical parameters, including soil quality, temperature, slope, orientation, exposure, and disease tolerance.[8]

Despite their experience and knowledge, women have not been given an equal opportunity to participate in critical decisionmaking on climate change adaptation and mitigation. Any accurate examination of climate change must integrate social, economic, political, and cultural dimensions—including analysis of gender relations.

Courtesy Equilibrium Fund

This Guatemalan women's Maya Nut producer group has contracted to provide school lunches, using Maya Nut products, to three school districts.

The first important step is to promote international policy action on climate and gender. Negotiations on a post-2012 climate framework under the United Nations Framework Convention on Climate Change (UNFCCC) should incorporate the principles of gender equity and equality at all stages, from research and analysis to the design and implementation of mitigation and adaptation strategies. It is critical that the UNFCCC recognize the importance of gender in its meetings and take the necessary measures to abide by key human rights and gender frameworks, especially the Convention on the Elimination of All Forms of Discrimination against Women, known as CEDAW. The UNFCCC needs to develop a gender road map, to invest in specialized research on gender and climate change, and to guarantee the participation of women and gender experts at all meetings and in the preparation of reports. It should establish a system of gender-sensitive indicators for its national reports and for planning adaptation strategies or projects under the Clean Development Mechanism (CDM).

Second, governments need to take action at regional, national, and local levels, including translating international agreements into domestic policy. They can also develop strategies to improve and guarantee women's access to and control over resources, use women's specialized knowledge and skills in strategies for survival and adaptation to natural disasters, create opportunities to educate and train women on climate change, provide measures for capacity building and technology transfer, and assign specific resources to secure women's equal participation in the benefits and opportunities of mitigation and adaptation measures.

Third, all financial mechanisms and instruments associated with climate change should include the mainstreaming of a gender perspective and women's empowerment. For example, climate change adaptation funds could guarantee the incorporation of gender considerations and the implementation of initiatives that meet women's needs. Women could also be included in all levels of the design, implementation, and evaluation of afforestation, reforestation, and conservation projects that receive payments for environmental services, such as carbon sinks. And women should have access to commercial carbon funds, credits, and information that enable them to understand and decide which new resources and technologies meet their needs. Finally, the CDM should finance projects that bring renewable energy technologies within the reach of women to help meet their domestic needs.

Fourth, the many organizations, ministries, and departments that address women's issues, including UNIFEM, should play a more active role in climate change discussions and decisionmaking. Climate change cannot be considered an exclusively environmental problem; it needs to be understood within all its development dimensions.

One encouraging sign of progress on these issues was the establishment of the Global Gender and Climate Alliance (GGCA) in December 2007 at the Bali Conference of the Parties to the UNFCCC. This was set up by the U.N. Development Programme, the International Union for Conservation of Nature–IUCN, the U.N. Environment Programme, and the Women's Environment and Development Organization in response to the lack of attention to gender issues in existing climate change policymaking and initiatives.[9]

The GGCA plans to:

- integrate a gender perspective into global policymaking and decisionmaking in order to ensure that U.N. mandates on gender equality are fully implemented;
- ensure that U.N. financing mechanisms on mitigation and adaptation address the needs of poor women and men equitably;
- set standards and criteria for climate change mitigation and adaptation that incorporate gender equality and equity principles;
- build capacity at global, regional, and local levels to design and implement gender-responsive climate change policies, strategies, and programs; and
- bring women's voices into the climate change arena.

Climate change is a global security issue and a question of freedom and fundamental human rights. It represents a serious challenge to sustainable development, social justice, equity, and respect for human rights for

The Security Dimensions of Climate Change

Jennifer Wallace

Given its potential to cause a serious decline in the livability of different regions around the world, policymakers and others are beginning to identify climate change as a security threat. Although there is no consensus that this drives violent conflict, security concerns arise from its indirect impacts on local institutions in areas challenged by environmental degradation. Particularly in Europe, climate change is increasingly prominent in national security strategies and military policies, a reflection of the global reach of socioeconomic and political consequences. The fact that traditional security actors are involved in discussions on this issue confirms that state stability and security are no longer confined to the realms of territoriality and weapons-based threats. A broader understanding is needed of the threats to security posed by the direct and indirect impacts of climate change.

The direct impacts of climate change on human welfare are multiple and interlinked. The likely increase in the volatility of the water supply will threaten health and sanitation for the most vulnerable societies, for example. According to the Intergovernmental Panel on Climate Change (IPCC), 1.3 billion people today do not have adequate access to drinking water and 2 billion people lack access to sanitation. In Africa, anywhere from 75 million to 250 million people are projected to be exposed to increased water stress due to climate change by 2020. Yields from rainfed agriculture could be cut in half, adversely affecting the food supply and exacerbating malnutrition. Increased temperatures also have a direct effect on the spread of disease, adding to the potential for the disruption of social stability. The IPCC predicts more frequent temperature extremes, heat waves, and heavy precipitation events as well as more intense tropical cyclones, threatening the physical safety of people living in areas with limited capacity to adapt to these changes.[1]

The indirect impacts on states and communities are equally important. Migration, the collective impacts on human welfare, and the threat to livelihoods undermine political institutions in vulnerable states. They challenge the maintenance or establishment of political and socioeconomic stability—a worrying consequence since cooperative and legitimate governance is considered the key determinant in the peaceful management of scarce resources. The negative effect on governance structures is particularly relevant when an economy depends heavily on its resource base, which is the case in most developing countries.[2]

As centers of production shift to areas that remain viable during climate change, state and local institutions may be incapacitated. Loss of revenue, combined with the direct threats of climate change, bode ill for

***Jennifer Wallace** is a doctoral candidate in the Department of Government and Politics at the University of Maryland.*

institutions struggling to ease conflict, regardless of whether the tensions emerge over the division of scarce goods or other social, political, or economic divisions. The direct impacts and indirect institutional challenges linked to climate change can reinforce each other as security effects emerge at the state and transnational levels.

Recognizing these complex linkages, the U.S. Senate Committee on Foreign Relations held a hearing on the National Security Implications of Climate Change. In his opening statement Senator Richard Lugar acknowledged that "the problem is real and is exacerbated by man-made emissions of greenhouse gases. In the long run this could bring drought, famine, disease, and mass migration, all of which could lead to conflict."[3]

The military board of the CNA Corporation, a nonprofit research organization for public policy decisionmakers, notes that the National Security Strategy and National Defense Strategy of the United States should directly address the threat of climate change and prepare the military to respond to the consequences. So far this advice has not been implemented in security policy planning at the national level: the most recent National Security Policy of the United States did not mention anthropogenic climate change as an issue area of concern. In contrast, climate change is mentioned specifically as a security interest within the first few pages of the European Security Strategy.[4]

Researchers remain divided on the direct links between climate change and violent conflict. The models have been based on one of two scenarios: conflict over increasingly scarce resources such as water or arable land or migration as a trigger of conflict. Research in the early 1990s by Thomas Homer-Dixon on the resource scarcity–conflict relationship found limited evidence supporting a connection, but it did identify a causal link when resource competition was combined with other socioeconomic factors such as poor institutional capacity to govern the resource.[5]

One challenge in examining the relationship across a large number of cases was that both degradation and conflict data were only available at the national level, producing mixed results and masking the incidences of conflict within and between communities. A recent study by Clionadh Raleigh and Henrik Urdal used georeferenced data to look at the relationship of conflict occurrence to geographical boundaries rather than political ones. Although their analysis provided only moderate support for the effect of demographic and environmental variables on conflict, the authors called for further investigation into the links between physical processes and the political processes of rebellion.[6]

Migration is identified as the second primary climate-induced driver of conflict. In 2007 the *Stern Review* warned that "by the middle of the century, 200 million more people may become permanently displaced due to rising sea levels, heavier floods, and more intense droughts." Weak states are particularly vulnerable to climate-induced migration, since environmental impacts can be addressed by adaptation and mitigation or by leaving an affected area, but weak institutions are less capable of successfully implementing the former strategies.[7]

Resource competition can emerge when local and resettled populations are forced to share subsistence resources, which can serve to worsen preexisting ethnic or social tensions. Adrian Martin notes that in communities with resettled populations, "there is a growing concern that scarcity-induced insecurities can contribute to an amplification of the perceived significance of ethnic differences and inequalities, creating the conditions for unproductive conflict....In such cases, perceptions of resource use conflict and perceptions of inequity are mutually reinforcing." Nonetheless, some scholars emphasize that conflict in these cases is better explained by the migration of feuding parties or the weak institutional capacity of the receiving community.[8]

What the academic debate is unable to account for, based on historical incidences of conflict, is the threat to security and state stability posed by unprecedented levels of climate change due to human activities. The evidence from several areas indicates that climate change can act as a "risk multiplier," revealing a potential for unprecedented violent outcomes as climate conditions worsen.

In Sudan, for example, climate change is an additional stress in an area already unable to meet its resource demands. The U.N. Environment Programme (UNEP) reports that "desertification is clearly linked to conflict, as there are strong indications that the hardship caused to pastoralist societies by desertification is one of the causes of the current war in Darfur." The pastoralists were forced to move south to find arable land as the boundaries of the desert shifted southward due to declines in precipitation. In northern Darfur, the annual amount of rainfall has dropped by 30 percent over 80 years. As demand increases, in line with projected growth rates in the human and livestock populations, climate change is expected to aggravate conflicts in an area with an extensive history of local clashes over agricultural and grazing land.[9]

In one case reported by UNEP, the camel-herding Shanabla tribe had migrated southward into the Nuba mountains as a result of northern rangeland degradation, and the Nuba population "expressed concern over the widespread mutilation of trees due to heavy logging by the Shanabla to feed their camels, and warned of 'restarting the war' if this did not cease." While the primary drivers of the Darfur crisis include a range of social, political, and economic issues, episodes like this one demonstrate how declining resources can fuel an environment of competition and mistrust in regions plagued by conflict.[10]

Andrew Heavens

People frantically pull water from a well just filled by a tank truck with water from a nearby borehole, in the Oromiya region of Ethiopia during a severe drought, 2006.

Bangladesh is considered to be among the countries at highest risk from the effects of climate change, as floods, monsoons, tropical cyclones that increase in intensity, and sea level rise from melting glaciers threaten the population, particularly in coastal areas. Abnormally high destruction was already witnessed in the flood of 1998, when two thirds of the country was inundated. The flood led to more than 1,000 deaths, the loss of 10 percent of the country's rice crop, and 30,000 people being left homeless.[11]

Continued climate change may prevent future recovery in Bangladesh, since small islands in the Bay of Bengal are home to approximately 4 million people, many of whom will need to be relocated as the islands are rendered uninhabitable by rising sea levels. Conflict over territorial borders already plagues the region, and the resettlement of vulnerable populations threatens to add to these conflicts. The deteriorating socioeconomic and political situation in Bangladesh is already a security concern for other nations; following the U.S. invasion of Afghanistan in 2001, Taliban and Islamic extremists relocated to Bangladesh. Increasing extremism threatens to further destabilize the country as environmental stress combines with socioeconomic factors to weaken the government's ability to cope with multiple sources of instability.[12]

As the Darfur and Bangladesh cases demonstrate, the threat posed to security and stability at the global, state, and individual levels from environmental degradation is increasingly evident, despite academic criticism about the lack of precise evidence linking climate change to violent outcomes. Yet academic research suffers from improperly scaled national aggregate data, the challenge of capturing complex causal models, and the difficulty of accounting for the time-lagged effects of climate change. These constraints should not excuse policymakers who fail to address increasingly visible security challenges.

While preparing for the effects of climate change is receiving more attention through strategies of mitigation and adaptation, the developing world remains most at risk from the consequences of temperature rise—and it has the least access to financial, technical, and human resources to implement preventive measures. As threats to stability and security are increasingly seen to transcend political borders, climate change presents clear security challenges for industrial nations as well as for the most volatile or vulnerable regions of the world.

Climate Change's Pressures on Biodiversity

Thomas Lovejoy

The last 10,000 years has been a period of unusually stable climate very favorable to the development of human civilization. The world's ecosystems also adjusted to the stability. Now that is beginning to change. Nature is responding to the climate change that has taken place on a planet 0.75 degrees Celsius warmer than preindustrial levels. Some of the change is physical in nature, most notably the phase change between ice and water. Northern hemisphere lakes are freezing later in the year and ice is breaking up earlier in the spring.[1]

Glaciers are in retreat in most of the world, with those in the tropics due to disappear in 12–15 years. Some of these are important water sources for cities like La Paz and Quito. Others are important for the major rivers of China and the Ganges. This hydrological shift obviously has implications not only for human populations but also for the ecosystems that depend on them.[2]

The most dramatic changes are those taking place in the Arctic: the summer retreat of the sea ice of the Arctic Ocean has been accelerating, as might have been anticipated with more dark water exposed and absorbing heat from the sun. The biodiversity poster child for this situation is the Polar Bear, now listed as threatened under the U.S. Endangered Species Act because the critical habitat these huge animals need in order to survive is literally disappearing beneath them.[3]

Beyond the Arctic, the timing of the life cycles of many plant species is changing, plant and animal species ranges are shifting, and the rising temperature is having numerous other unexpected effects. (See Table.) These changes are no longer single examples. They constitute statistically robust documentation that nature is on the move all over the planet.[4]

Yet all these changes are relatively minor ripples in nature. The more important question is, What is in store? There is at least an additional 0.5 degrees Celsius of warming already implicit in the current atmospheric concentrations of greenhouse gases (GHGs) because of the lag time in the buildup of heat.[5]

Climate change is, of course, nothing new in the history of life on Earth. Clearly, glaciers came and went in the past hundreds of thousands of years without major biodiversity loss. But today's landscapes—highly modified by human activities—present obstacle courses to species as they attempt to follow the habitat conditions they need to survive. The obvious policy response to this is to restore natural connections in the landscapes in order to ease species dispersal.

A more difficult complication is that ecosystems and biological communities do not move as a unit. Rather, studies of past responses to climate change (for instance, after the retreat of the last glacier in Europe)

***Thomas Lovejoy** is President of The H. John Heinz III Center for Science, Economics, and the Environment in Washington, DC.*

Selected Examples of Climate Change's Effects on Biodiversity

Indicator	Changes to Date
Flowers	Blooming earlier at Kew Gardens in London
Tree swallows	Migrating, nesting, and laying eggs earlier
Butterflies	Ranges shifting for Edith's Checkerspot Butterfly in western United States and for many species in Europe
Golden Toad of Monteverde	Formerly found in Costa Rica, first species to become extinct due to climate change
Eel grass	Southern limit in U.S. Chesapeake Bay moving northward every year
American ash tree	Warmer and longer summers allowed Emerald Ash Borer to produce one more generation than before, producing massive tree death
Coral reefs	Algae expelled from reefs due to warming water, leading to coral "bleaching events"

Source: See endnote 4.

reveal that individual species each move in their own direction and in their own way. In essence, ecosystems will disassemble. After tracking their required conditions, the surviving species will assemble into novel ecosystems the likes of which are difficult to anticipate.[6]

Another complication is that change will not be linear and gradual. Much as there will be abrupt change in the climate system itself, there will also be abrupt change in ecosystems. In fact, such threshold changes are already being observed. In southern Alaska, British Columbia, the U.S. Northwest, and Colorado, as well as in Scandinavia and Germany, the longer and warmer summers and less severe winters are tipping the balance in favor of the native pine bark beetle, for example. With one more generation able to reproduce, there are now tens of millions of hectares where up to 70 percent of the trees are dead, creating an enormous timber and fire management problem and, through the loss of trees, adding yet more carbon dioxide (CO_2) to the atmosphere. It will be difficult to anticipate all such threshold changes; indeed, the world can expect a lot of surprises.[7]

In addition, change at an even greater scale—namely, system change—can be expected, such as with the hydrological cycle of the Amazon. It has been known for 30 years that the Amazon forest has the remarkable feature of generating an important fraction of its own rain. Windborne moisture from the Atlantic Ocean drops as rain, and most of it evaporates from the complex surfaces of the forest or is transpired by the trees, thus returning to the westward moving air mass to become rain and to then recycle again farther to the west. This is important in maintaining the forest and providing precipitation farther south on the continent.[8]

In 2005, the Amazon "rain machine" system failed, creating the greatest drought in recorded history in Amazonian Brazil. This was traced to changes in the Atlantic circulation—and believed to be a preview of what climate change could bring. Indeed, most of the major climate models, particularly that of Britain's Hadley Centre, predict "Amazon dieback" at somewhere around a temperature change of plus 2.5 degrees Celsius.[9]

An even more devastating system change is already taking place: the acidification of the oceans. This is change driven by the increased concentrations of carbon dioxide in the atmosphere. So much attention had been paid to the CO_2 absorbed by the oceans (which has kept climate change from being even greater) that the portion of the CO_2 converted to carbonic acid was largely overlooked. The oceans are today 0.1 pH unit more acid than in preindustrial times. That may seem inconsequential, but because pH is a logarithmic scale, this is equivalent to 30 percent more acid.[10]

Increasing acidity is a matter of great consequence because tens of thousands of marine species build shells and skeletons from calcium carbonate. They depend on a calcium carbonate equilibrium that is sensitive to both temperature and pH: the colder and more acid the water is, the harder it is for organisms to mobilize calcium carbonate. This includes species like corals or Giant Clams. It also includes tiny planktonic organisms that exist in huge profusion at the base of marine food chains. Change is already being detected at the base of food chains off Alaska and in the North Atlantic.[11]

What can be done to diminish as much additional climate change as possible and to simultaneously buffer natural systems against the change that is all but inevitable? The former involves producing a new energy basis for civilization. One component of that is reducing CO_2 emissions from the destruction of modern biomass, principally tropical forests. At the moment that accounts for roughly one fifth of the annual increase in GHG concentrations. This makes Indonesia and Brazil the third and four largest CO_2-emitting nations even though their fossil-fuel-derived emissions are relatively low.[12]

But forest carbon (other than reforestation and afforestation, and to a large extent plantation forests) was left out of the current arrangements for carbon trading through the Clean Development Mechanism established in the Kyoto Protocol. Its status is currently being explored in hopes of negotiating a system to reward countries for reducing emissions from deforestation and degradation. Finding a solution to lower and eliminate this source of CO_2 emissions is good for the forest, good for biodiversity, good for forest peoples, and good for the climate.

At the same time, a great deal of attention needs to be paid to "adaptation," to making biodiversity and ecosystems resilient in the face of climate change. Natural connections urgently need to be reestablished in landscapes to facilitate the dispersal of individual species as they follow the conditions they need to survive. Basically, the opposite of the current situation of patches of nature in human-dominated landscapes needs to be created, so that human needs and aspirations are embedded in a natural matrix.

A second obvious measure is to reduce other stresses on ecosystems so that they do not reinforce the changes taking place in a warming world—by reducing siltation, for instance, on coral reefs. Biodiversity and ecosystems basically integrate all environmental stress, so this aspect makes the existing conservation and environment agenda even more important.

A lot of the potential management and adaptation measures are hard to design using the extremely coarse scale of global climate models. Managers need a much more precise idea of what kind of change is likely to take place within a square kilometer and over the next few decades. This can be greatly facilitated by "downscaling," which can be done quickly and cheaply using laptops instead of supercomputers. There will undoubtedly be a succession of downscaled projections, each refining scientists' understanding of what is likely to happen in small units of landscape. One of the most useful things that can be done quickly is to produce a first set of downscaled projections that can highlight the challenges that managers need to address.

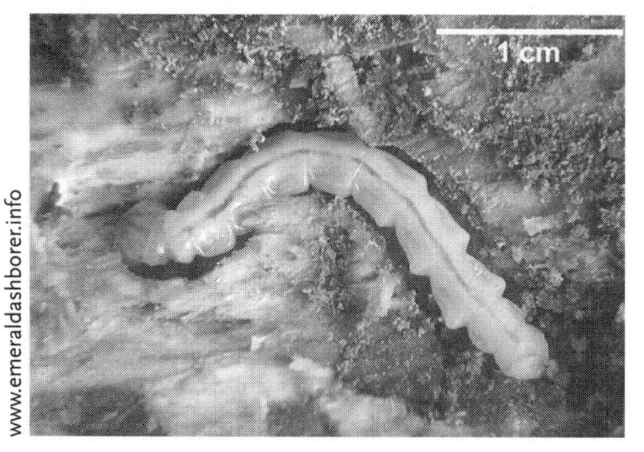

Emerald Ash Borer larva in fall

In particular situations, adaptation options can be illuminated through modeling projected range shifts of individual species, as has been done for some of the flowering plant species of the Cape Floral Kingdom in South Africa. And in very special situations management can actually "assist" migration. Unfortunately, the cold, hard reality is that the number of species in the world is far too large for these options to be used extensively.[13]

Adaptation to sea level rise, although virtually nonexistent for low-lying islands, is possible in coastal areas. The Nature Conservancy has a valuable experiment on this in Albemarle Sound in North Carolina that anticipates sea level rise and will facilitate the development of new freshwater wetlands as current ones become tidal. But such adaptation is useful only with gradual sea level rise, not the rapid rise likely to occur following major changes in the Greenland ice sheet, for example.[14]

The world's protected areas are certainly not being invalidated because the species for which many were established will move away. Indeed, they have a new conservation role: to be the safe havens from which species will move to new locations. Without the existing protected areas, there will be nowhere for the new biogeographical pattern to emerge from. Species will of course need safe havens once they have moved and will need natural connections in the landscape to facilitate the movement to those areas.

The most important conclusion is that ecosystems and biodiversity are extremely sensitive to climate change and represent one of the most urgent reasons to limit additional change. The warming inherent in current GHG levels will bring the planet's average temperature increase to 1.3 degrees Celsius. This is just a bit more than half a degree short of the level at which many conservation organizations anticipate ecosystems will be in serious trouble. Yet since there are already threshold changes in ecosystems and ocean acidification at current levels of climate change and greenhouse gas concentrations, dangerous change is likely to appear before 2.0 degrees Celsius. In essence, biodiversity is indicating climate change needs to be treated with unprecedented immediacy and urgency.[15]

Small Island Developing States at the Forefront of Global Climate Change

Edward Cameron

The world's small island developing states (SIDS) are often cited as the most vulnerable countries to climate impacts and the first nations on Earth to face critical climate change thresholds. Yet they have contributed least to the growing concentrations of greenhouse gases in the atmosphere and so have the least responsibility for the crisis the world now faces. They are least likely to be heard at the negotiating table, as they lack the political weight of the major emitters. As a result, their vulnerability goes unnoticed and their voices go unheard. They are also least likely to be the beneficiaries of climate funds, most of which get spent on mitigation (particularly energy projects) rather than adaptation. And when action is taken they are least likely to be involved in the consultations.

The Caribbean states provide a good example of the vulnerability of small islands states. According to the New Economics Foundation, the increased strength of storms and hurricanes and the surge in their destructive forces have affected hundreds of thousands of victims and led to multimillion-dollar damages. In 2004 Grenada, an island considered to be outside the hurricane belt, was devastated when Hurricane Ivan struck, destroying over 90 percent of the country's infrastructure and housing stock and causing over $800 million in damages, the equivalent of 200 percent of Grenada's gross domestic product. The increase in frequency and intensity of these storms expected due to climate change could well place further strain on political, social, and economic systems and act as an additional constraint on development in the region.[1]

These islands depend on fragile ecosystems such as coral reefs. Globally, coral reefs provide critical habitat for more than 25 percent of marine species and contribute more than $30 billion in annual net economic benefit. Recent studies estimate that a third of the world's reef-building coral species are facing extinction. Climate change, coastal development, overfishing, and pollution are the major threats. A new analysis shows that before 1998, only 13 of the 704 coral species assessed would have been classified as threatened. Now the number in that category is 231.[2]

The Caribbean has the largest proportion of corals in high extinction risk categories, but reefs in the Indian Ocean and the Pacific are also likely to be decimated. Sea level rises, flooding, and storm surges are a particular concern for the atoll states in the Pacific and Indian Oceans. If the projections of the Intergovernmental Panel on Climate Change prove correct, these island nations will effectively disappear by the end of this century.[3]

SIDS also suffer from a lack of natural resources, often have limited freshwater supplies, and are constrained by poor trans-

***Edward Cameron** is a Washington-based climate change specialist who has worked extensively with small island states.*

port and communication infrastructure. This means they are particularly susceptible to even small changes in the global climate. Furthermore, the chronic lack of adaptive capacity, including financial, technical, and institutional resources, means they are ill prepared to deal with these multiple threats.

Today small island states are striving to achieve long-term sustainable development and implement the Millennium Development Goals (MDGs). Climate change impacts are already undermining their efforts, however.

The first MDG—to eradicate extreme poverty and hunger—is being affected by changing patterns of food production and the gradual undermining of livelihoods. Many of these islands depend heavily on tourism and natural resources for their economic livelihood. They also depend on local staples and species for the bulk of their food. Threats to biodiversity and coral reef systems will reduce these livelihood assets, undermine economic performance, and threaten regional food security.

The second goal—to achieve universal primary education—is being compromised by extreme weather events that create a cycle of destruction and reconstruction and that reduce the amount of investment flowing into long-term development. Tropical cyclones destroy schools and hospitals, damage public utilities and infrastructure (including energy, water, and transport connections), and so reduce access to education, health care, and other public services. Loss of national revenue from associated impacts may also undermine public spending on education.

The third MDG—to promote gender equality and empower women—is jeopardized, as women living in poverty are often the most threatened by the dangers that stem from climate change. Cultural norms can mean that women do not have the appropriate skill sets to deal with myriad impacts. The statistics indicating fatalities from extreme weather events are revealing in this regard. Moreover, as resources become scarcer, women and young girls spend more time collecting food and water and less time caring for their health and education.

Three of the MDGs deal with health and aim to reduce child mortality, improve maternal health, and combat HIV/AIDS, malaria, and other diseases. The World Health Organization and leading health providers are anticipating an increase in waterborne and vector-borne diseases, in diarrheal diseases, and in malnutrition as a result of associated climate impacts. This could lead to increases in child mortality, a reduction in maternal health, and the undermining of nutritional health needed to combat HIV/ AIDS.[4]

In the Maldives, a small islands nation in the Southern Indian Ocean, the human drama of climate change is a daily reality for 300,000 residents. In 1987 the President of the Maldives, Maumoon Abdul Gayoom, became the first world leader to draw attention to the threat of climate change. In a landmark speech to the United Nations General Assembly, he warned that this would result in the death of his nation and others like it. Twenty years on and the effects of climate change are already evident: storm surges and coastal erosion destroy homes, pose dangers to infrastructure and utilities, and divert limited resources from strategic development.[5]

In the medium term, rising ocean temperatures, coupled with growing acidification, threaten the survival of coral reefs in the Maldives—the very lifeblood of the economy. The island's two principal industries, tourism and fisheries, are entirely dependent upon the reefs. They account for 40 percent of the national economic output and more than 40 percent of the jobs. Together, these industries have fueled the sustained and enviable economic development that has enabled the Maldives to grow from being one of the poorest countries in South Asia in the 1970s to the richest country per capita in the region today.[6]

In the long term, it is not economic development but the country's very survival that is threatened. With most of the islands lying less than one meter above sea level, this generation—the most fortunate one to have ever lived on these islands—may be the last one to live in the Maldives.

Nevit Dilmen

Part of the Maldive tourist infrastructure at risk of sea level rise

Since some degree of climate change is already inevitable as the effects of current concentrations of greenhouse gases in the atmosphere continue to be felt for the next few decades, the government of the Maldives has developed a comprehensive program of domestic adaptation. Work has concentrated on reinforcing vital infrastructure, particularly related to transport and communications. Public services ranging from water supply and electricity generation to the provision of health care and education are being strengthened against climate threats. Flood defenses have been constructed, and measures are being taken to minimize coastal erosion.[7]

Perhaps the most innovative adaptation measure is the development of the "safe island" concept. This initiative is designed to minimize climate vulnerability by resettling communities from smaller islands that are more vulnerable onto larger, better-protected ones. This lets the government concentrate limited resources on protecting the more viable islands. It also allows for public services to be strengthened and economic opportunities to be developed.[8]

Domestic adaptation in the Maldives and throughout other vulnerable societies will involve significant engineering projects and large financial investments. It will also require large-scale capacity building to strengthen institutional capacity, to enhance knowledge, human, and financial resources, and to encourage an awareness-raising program to prepare people for the inevitable changes.

Adaptation without mitigation will result in little more than a temporary respite, postponing catastrophic climate change to a later date. Urgent and ambitious action must be taken to reduce greenhouse gas emissions. Small island states have been active in attempts to find a global consensus on climate action from the very beginning. Indeed, the momentum to create the United Nations Framework Convention on Climate Change and the Kyoto Protocol was in part a result of moral and ethical arguments advanced by members of the Alliance of Small Island States (AOSIS), earning the organization the title of "Conscience of the Convention."

Today AOSIS members are participating actively in the Bali process, which seeks to find an appropriate global climate regime to succeed the Kyoto Protocol's first commitment period, which expires in 2012. The AOSIS negotiating position for the Bali process is entitled *No Island Left Behind*. It outlines three long-term strategic objectives:

- An ambitious long-term goal for reducing greenhouse gas emissions should be the organizing point for all other processes within the Bali process. This implies deep and aggressive cuts in emissions to levels that keep long-term temperature increases as far below 2 degrees Celsius above preindustrial levels as possible.
- More funding for adaptation is needed, with priority access given to SIDS on an expedited basis based on their specific vulnerabilities and lack of capacity.
- SIDS need support and technical assistance to build capacity and gain access to technologies to respond and adapt to climate change across a wide range of socioeconomic sectors.[9]

AOSIS favors an expanded and broadened Kyoto protocol, with clear opportunities for developing countries that may wish to enter into full Kyoto commitments. The overall outcome should use impacts on SIDS as a benchmark for effectiveness and success. Although AOSIS has had a legitimate and important voice in the climate change process, the organization has often suffered from its own capacity constraints and from division among its members.

Many countries have become frustrated at the lack of urgency and ambition in international negotiations and believe that the time has come to change the dynamic by introducing new approaches to solving the climate crisis. In March 2008, the government of the Maldives, working closely with a number of other island nations and drawing on the support of more than 70 countries, introduced a resolution on climate change and human rights at the United Nations Human Rights Council in Geneva. It called on the Office of the High Commissioner for Human Rights to conduct an analytical study exploring the interface between human rights and climate change. This groundbreaking and innovative initiative seeks to import the rhetorical, normative, and operational force of international human rights law into the climate change discourse.[10]

A rights-based approach to climate change holds a great deal of promise for small island states as they seek to inject urgency and ambition into mitigation policy while simultaneously lobbying for increased financial flows to support mitigation. First, a rights-based approach could help improve analysis of the human impacts of climate change by linking it to realizing more than 50 international human rights laws, such as the right to life, health, and an adequate standard of living. Second, a rights-based approach replaces policy preferences with legal obligations and turns the communities most vulnerable to climate change from passive observers of climate negotiations into rights holders. This will give voice to the vulnerable and compel the major emitters to act on climate change before the clock runs out on small island states.

The Role of Cities in Climate Change

David Satterthwaite and David Dodman

Cities are often blamed for contributing disproportionately to global climate change. Numerous sources state that cities are responsible for 75–80 percent of all human-caused greenhouse gases (GHGs)—although the scientific basis for these figures is unclear. One detailed analysis concluded that the number is more like 40 percent.[1]

In fact, many cities combine a good quality of life with relatively low levels of greenhouse gas emissions per person. There is no inherent conflict between an increasingly urbanized world and reduced global GHG emissions. Focusing on cities as "the problem" often means that too much attention is paid to the reduction of greenhouse gas emissions, especially in low-income nations, and too little to minimizing climate change's damaging impacts. Certainly, the planning, management, and governance of cities should have a central role in reducing GHG emissions due to human activities worldwide. But this should also have a central role in the often neglected activities of protecting people in cities from the floods, storms, heat waves, and other likely impacts of climate change.

The main sources of greenhouse gas emissions in cities are the use of energy in industrial production, transportation, and buildings (heating or cooling, lighting, and appliances) and waste decomposition. Transport is an important contributor to GHG emissions in almost all cities, although its relative contribution varies a lot—from around 11 percent in Shanghai and Beijing (in 1998) to around 20 percent in London and New York and as much as 30–35 percent in Rio de Janeiro and Toronto.[2]

GHG inventories show more than a tenfold difference in average per capita emissions between cities—with São Paulo responsible for 1.5 tons of carbon dioxide-equivalent per person compared with 19.7 tons per person in Washington, DC. If figures were available for cities in low-income nations, the differences could well be more than 100-fold. In most cities in low-income nations, GHG emissions per person cannot be high because there is scant use of fossil fuels and little else to generate other greenhouse gases. There is little industry, very low levels of private automobile use, and limited ownership and use of electrical equipment in homes and businesses.[3]

Thus perhaps it is not cities in general that are the main source of greenhouse gas emissions but only cities in high-income nations. Yet an increasing number of studies of particular cities in Europe and North America show that they have much lower levels of greenhouse gas emissions than their national averages. New York and London, for example, have much lower emis-

***David Satterthwaite** is a Senior Fellow in the Human Settlements Group at the International Institute for Environment and Development (IIED) in London. **David Dodman** is a researcher in IIED's Human Settlements and Climate Change Groups.*

sions levels per person than the average for the United States and the United Kingdom.[4]

Of course, it is not cities (or small urban centers or rural areas) that are responsible for greenhouse gas emissions, but particular activities. Figuring out how to allocate emissions to different locations is not a simple exercise. For instance, locations with large coal-fired power stations are marked as very high GHG emitters, even though most of the electricity they generate may be used elsewhere. That is why it is common in GHG inventories for cities to be assigned the emissions generated in providing the electricity used within their boundaries, and it explains why some cities have surprisingly low per capita emissions: much of the electricity they import comes from hydro, nuclear power, or wind and solar, not from fossil-fueled power stations.

There are other difficulties too. For instance, do the emissions from gasoline used by car-driving commuters get attributed to the city where they work or the suburb or rural area where they live? Which locations get assigned emissions from air travel? Total carbon emissions from any city with an international airport are greatly influenced by whether or not the city is assigned the fuel loaded onto the aircraft—even if most of the fuel is used in the air, outside the city.

A more fundamental question is whether greenhouse gas emissions used in producing goods or services are allocated to production or consumption. If they are assigned to the location of the final consumer, much of the greenhouse gas emissions from agriculture and deforestation would go on the tally sheet of the cities where wood products are used and food is consumed. If instead they are assigned to where goods are produced, then a city that was a major producer of windmills, photovoltaic cells, and hydrogen-fueled buses could have high greenhouse gas emissions per person even though its products help keep emissions down wherever they are used.

These questions over how to assign GHG emissions have enormous significance for allocating responsibility for reducing emissions between nations and, within nations, between cities and other settlements. If China's major manufacturing cities are assigned all the GHG emissions related to exported goods (including the coal-fired electricity that helped produce them), this implies a much larger responsibility for moderating and eventually reversing emissions than if the nations or cities where goods are used are held accountable. Thus, seeing cities as "the problem" misses the fact that the driver of most GHG emissions is the consumption patterns of middle- and upper-income groups in wealthier nations, including those who live outside cities.

This attitude also misses the extent to which well-planned and well-governed cities can provide high living standards without high greenhouse gas emissions. Consider the large differences among wealthy cities in gasoline use per person. People in most U.S. cities use three to five times as much gasoline as people in most European cities because of much higher private automobile use. But even within U.S. cities people can have a relatively small carbon footprint. On a per capita basis, for example, New Yorkers emit just 30 percent as much greenhouse gas as the national average, in part because of smaller and more concentrated houses and apartments and the greater use of public transportation. Many of the most desirable (and expensive) residential areas in the world's wealthiest cities have high densities and buildings that minimize the need for space heating and cooling—in distinct contrast to houses in suburban or rural areas. Most European cities have high-density centers where walking and bicycling are common. High-quality public transport can keep down private car use.[5]

Cities also concentrate so much of what contributes to a high quality of life that does not involve high material consumption (and thus high GHG emissions): theater, music, museums, libraries, the visual arts, dance, and the enjoyment of historic buildings and districts. They have long been places of social, economic, and political innovation—something already evident regarding climate change. In many high-income nations, city politicians like Mayor Michael Bloomberg of New York and Ken Livingston (when he was mayor of London) are more committed to reducing greenhouse gas emissions than national politicians are.[6]

How a city is planned, managed, and governed also has important implications for how it will cope with the impacts of climate change. Most cities in low-income nations in Africa and Asia have very low emissions per person. Yet they house hundreds of millions of people who are at risk from the increased frequency or intensity of floods, storms, and heat waves and from the water supply constraints that climate change is likely to bring. Discussions of climate change priorities so often forget this. And these risks are not easily addressed, especially by international aid agencies that show little interest in tackling the reasons so many people are at risk, such as the lack of provision for urban infrastructure (such as drains) and the high proportion of people living in poor-quality homes in informal settlements. A great deal of urban expansion increases risks from climate change, because the only sites that low-income groups can find for their houses are on floodplains or other dangerous sites.[7]

Albert Cesario

Traffic on Avenida Atlântica, Copacabana Beach, Rio de Janeiro, with a pall of pollution on the horizon

But there are some good precedents to show what can be done. Manizales in Colombia, for example—long an innovator in environmental policies—has shown how to reduce risks for vulnerable populations, as people living on dangerous hillsides were provided with alternative sites and the hillsides were turned into locally managed eco-parks. Yet the good examples will remain isolated and unusual unless national governments and international agencies learn how to support these kinds of local innovations on a much larger scale.[8]

Climate Change and Health Vulnerabilities

Juan Almendares and Paul R. Epstein

Climate change has multiple direct and indirect consequences for human health—all of which are important. Climate change also threatens to disrupt Earth's life-support systems that underlie health and well-being. After all, human health and well-being basically depend on the health of crop systems, forests, other animals, and marine life. Health is the final common pathway for environmental and social conditions. Thus, the well-documented threats that climate change holds for societies and for ecosystems—for coral reefs, forests, and agriculture—ultimately pose the greatest long-term threats to health, nutrition, and well-being.[1]

One of the first direct and most obvious results of climate change—an outcome clearly tied to rising average temperatures—is heat waves. These are expected to take an increasing toll in all nations. The disproportionate increase in nighttime temperatures since 1970 and the rising humidity that stems from warming oceans and a heated atmosphere increase the health threats from heat waves.[2]

Extreme weather events, especially heavy downpours, can create conditions conducive to "clusters" of diseases carried by mosquitoes, rodents, and water. In addition, more-intense hurricanes, droughts, and sea level rise are all projected to increase substantially the number of refugees and internally displaced persons across the globe—conditions that will squeeze resources (like water and food) and raise the risk of epidemics of communicable diseases.[3]

Intense storms have other, less obvious effects on health. When Hurricane Mitch hit Central America in October 1998, it deposited six feet of rain in three days, causing flooding, landslides, and mudslides, and it dislodged pesticide-laden soils from banana, sugarcane, and African palm plantations as well as sediments from ancient Mayan ruins. Areas surrounding gold mines became heavily contaminated with toxic chemicals and heavy metals, and surveys of the local population have shown a dramatic rise in skin and eye diseases. Along with causing 11,000 deaths, Mitch brought epidemics of malaria, dengue fever, cholera, and leptospirosis, and the damages continue to affect development in Honduras today.[4]

Changes in the availability of water due to climate change are another area of concern. Droughts and disappearing glaciers are projected to take an increasing toll on health, agricultural yields, and hydropower. Drought, for example, is associated with epidemics of bacterial meningitis across the African Sahel. And water shortages contribute to waterborne disease outbreaks.[5]

Warming also expands the potential range

__Dr. Juan Almendares__ is Physiology Professor at the Medical School, Universidad Nacional Autonoma de Honduras. __Dr. Paul R. Epstein__ is the Associate Director of the Center for Health and the Global Environment at Harvard Medical School.

Sustainable Harvest International

A Honduran boy excited about a carrot from his parents' organic garden—established with technical assistance and support from Sustainable Harvest International

of infectious diseases and disease carriers. In southern Honduras, the warming has been so great that malaria no longer circulates. But people have also found the temperatures inhospitable and have moved north into forested areas ripe with malaria; so the indirect impact of climate change is that more people are exposed to health threats. Insect pests can affect not just humans but also forests and crops, as well as livestock and wildlife.[6]

To deal with these escalating problems, health care systems must be supported and public health services strengthened. Needed environmental measures include ecologically sound control of vector-borne diseases, such as malaria and dengue fever. Community research on the prevention of malaria has demonstrated that integrated control can be achieved without using DDT. The measures needed include community participation and training, treatment of infected populations, and larval control of anopheline mosquitoes.[7]

Such solutions require organizing communities and mobilizing international forces to address these vital problems. Education is an essential component of all solutions. The development of schools that are ecologically sustainable in Colombia, El Salvador, Honduras, and Guatemala by Friends of the Earth International has helped in the search for appropriate solutions to climate change and health problems. Organic, locally grown agriculture promotes health and nutrition—the basis of resistance to disease.[8]

Health ministries must have convening power and support to work with ministries of agriculture, planning, education, and finance on climate change protection, preparedness, and prevention. At the international level, the World Health Organization can provide guidelines for all nations to prepare for climate change–related ills and "natural" catastrophes. Financial support for this initiative is sorely needed.

Energy poverty is standing in the way of achieving the Millennium Development Goals. The bottom line is that health must again take center stage and—as it did in the nineteenth and early twentieth centuries, when vast improvements in basic water and sanitation were made—become the cornerstone of clean, healthy, and sustainable development.[9]

India Starts to Take on Climate Change

Malini Mehra

In 2009, the eyes of the world will be on China, India, and the United States. The threat of climate change is now so great, and dealing with it effectively is so central to the future of national economies, that new scripts are being called for. The role of the United States as the world's single largest polluter in per capita terms remains pivotal. But China and India are now assuming an importance they did not have in 1997, when the world came together in Kyoto to do a deal on climate change.

This is a moment of decision for India. How can a country with one sixth of the global population, and more billionaires than Japan, not play a leadership role on the climate agenda? As the world's third largest economy (in purchasing power parity terms), and the fourth largest emitter of greenhouse gases (GHGs), India's positive engagement will be crucial to constructing a "global deal" on climate at Copenhagen, the next pivotal meeting of governments that are party to the Framework Convention on Climate Change.[1]

For India, the stakes are too high to continue with politics as usual. Many studies have underscored the nation's vulnerability to climate change. The impacts are already being seen in unprecedented heat waves, floods, cyclones, and other extreme weather events. With its long coastline, India is experiencing sea surges and salinization, affecting infrastructure, agriculture, fisheries, livelihoods, and human health. Food security is being compromised through reduced crop yields, and water security is under threat everywhere with declining water tables, conflicts over rivers and basins, and the prospect of severely diminished freshwater resources due to glacier retreat in the Himalayas.[2]

The government claims it is already spending over 2 percent of gross domestic product (GDP) on measures to adapt to the impacts of the changing climate. The Carbon Disclosure Project estimates that climate change could result in a loss of 9–13 percent in the country's GDP in real terms by 2100.[3]

Given India's deeply stratified society, the hardest hit will be the poor and the marginalized. India is home to one third of the world's poor and a still growing, predominantly youthful population. By 2045 India will have overtaken China as the most populous nation, with an estimated population of 1.501 billion compared with China's 1.496 billion.[4]

Although India has not been an emitter historically, the past is no predictor of the future. As the economy grows and consumption patterns change, there is little doubt that emissions will rise and the country's carbon footprint will increase dramatically. The International Energy Agency projects that India will become the third-largest emitter by 2015, precisely when global GHG emissions need to peak if the world is to avoid the severest impacts of climate change.[5]

***Malini Mehra** is founder and CEO of the Centre for Social Markets, based in London and Kolkata.*

India's problem is its energy economy. The country has an extremely high dependence on fossil fuels—in particular on imported oil and dirty coal, which it has in abundance. Fossils fuels are responsible for 83 percent of India's carbon dioxide emissions; coal alone accounts for 51 percent. Addressing climate change effectively therefore will require a transformation of India's energy economy.[6]

The government's rhetoric on this topic remains tinged with fear. While it recognizes that "global warming will affect us seriously," the government concludes that "the process of adaptation to climate change must have priority" and that "the most important adaptation measure is development itself."On mitigation, the government is unequivocal: "With a share of just 4 percent of global emissions, any amount of mitigation by India will not affect climate change." Instead the government calls for action by industrial countries and a burden-sharing formula based on historical culpability, common but differentiated responsibilities, differences in respective capabilities, and the per capita emissions principle. The Prime Minister has pledged, however, that India's per capita emissions (presently 1.2 tons annually) will never exceed those of industrial countries.[7]

Yet if the Intergovernmental Panel on Climate Change's figures are to be believed, India will experience "the greatest increase in energy and greenhouse gas emissions in the world if it sustains eight percent annual economic growth or more as its primary energy demand will then multiply at least three to four times its present levels." A change of direction therefore is very much needed.[8]

The government has recognized that business as usual is no longer tenable. For example, the Eleventh Five Year Plan (2007–12) commits the country to reducing energy intensity per unit of GHG by 20 percent in the period 2007 to 2017. Further, it seeks to boost access to cleaner and renewable energy by "exploiting existing resources (e.g., hydropower and wind power), developing nuclear power, and also supporting research in newer areas such as biofuels from agro-waste, solar energy, etc."[9]

In June 2008, the Prime Minister released the much-anticipated National Action Plan on Climate Change. It focuses on eight areas intended to deliver maximum benefits for climate change mitigation and adaptation in the broader context of promoting sustainable development: solar energy, energy efficiency, sustainable habitat, water, sustaining the Himalayan ecosystem, green India, sustainable agriculture, and sustainable knowledge for climate change.[10]

The plan was launched with much fanfare, but the detailed action plans for each area are yet to be worked out, and the document contains virtually no targets or timetables. The Climate Challenge India coalition concluded that while the Action Plan is a more coherent approach to sustainable development across government departments, it is not a new agenda based on ensuring climate security or a strategy for a low-carbon pathway for India. The group gave the report a B+ for effort and a D for vision.[11]

Although the Action Plan may have been a missed opportunity for leadership by the government, it did contain some innovations such as a domestic cap-and-trade system as an incentive for emissions reductions in nine energy-intensive sectors. It also stands full-square behind market tools such as the Clean Development Mechanism (CDM), which the government has used as a lever to accelerate take-up of clean technology by Indian firms and to encourage participation in the global $30-billion carbon market.

India now accounts for more than one third of all CDM projects registered worldwide.[12]

The government can claim some credit for a few achievements that provide a good foundation for further improvements. For example, India was the first country to establish a ministry for non-conventional energy sources and has the world's fourth largest installed wind power capacity. Since 2004, India has managed to decouple economic growth from energy use, with the economy growing annually at a rate of over 9 percent but energy growing at less than 4 percent. The country has had an energy labeling program for appliances since 2006, with almost all fluorescent tube lights and about two thirds of refrigerators and air conditioners now covered by the scheme.[13]

Yet it will take more than a smattering of good examples to make the changes needed. Instead of following the example of industrial countries, India needs to opt for smart, low-carbon growth and make sustainability the organizing principle of its economy and modernization agenda. For a country with an advanced nuclear program and space exploration ambitions, leapfrogging from a high-carbon to a low-carbon energy economy is timely and possible.

The good news is that change is coming. And India's business community appears to be setting the pace. Dismayed at the lack of government leadership, many people in the business community are beginning to tackle climate change head-on as a business issue. In a recent survey of Indian business leaders, 83 percent of those questioned said they had a fair to good understanding of climate change, 65 percent said that India should be leading the way, and almost half said that climate change is a crucial and urgent issue that should be near the top of India Inc.'s agenda.[14]

Many Indian businesses are beginning to look to the future and invest in clean energy, energy conservation and efficiency, smart buildings, and green products. They realize the market is changing and the time to act is now. Industrialist Anand Mahindra relishes his "eco-warrior" tag, for instance, and views climate change as an emerging consumer and competitiveness issue. He wants his group to be at the forefront of addressing it and is redesigning his automotive portfolio accordingly.[15]

ITC's headquarters in Gurgaon is LEED Platinum-rated by the U.S. Green Building Council. Infosys, another sector leader, has embraced carbon neutrality, and Bangalore's Reva car is now the world's biggest selling electric vehicle. Cleantech and renewable energy investments are soaring in India, and the domestic wind power giant Suzlon is now the largest in Asia and fifth largest globally. Solar energy is undergoing a renaissance, with companies such as Tata BT Solar betting on it meeting the majority of India's energy needs by 2100.[16]

These examples show that India Inc. is prepared to move and doing so voluntarily in many respects. None of this should surprise anyone familiar with the country's deep-rooted enterprise culture. Where there is a market opportunity, Indian business will find it. With a supportive policy environment—in particular, a carbon tax to level the playing field—fiscal incentives, improved infrastructure, and clear standards and guidance, Indian business can do much more. What is needed is a strategic partnership between government and the private sector for a low-carbon development path. Establishing low-carbon innovation zones to incubate and promote such initiatives, as is being piloted in China, could be one imaginative way forward.[17]

Cities and municipalities are also tackling energy and environmental challenges, particularly in transportation—India's fastest-growing user of energy. Bangalore is leading the way with a state-of-the-art low-emissions mass transit system, and Delhi is subsidizing the purchase of Reva electric cars. A new breed of eco-developer is focusing on housing, seeking to capitalize on a projected $4-billion market for green buildings by 2012 and pushing existing building codes on energy efficiency.[18]

Adam Cohn, www.adamcohn.com

A charcoal vendor in Mysore, India

Civil society groups are mobilizing, and initiatives such as Climate Challenge India are leading the way with new thinking and optimism. India's "generation next" is coming together in networks such as the Indian Youth Climate Network. Media leadership is emerging, with national papers and magazines dedicating themselves to climate coverage. Madhya Pradesh, one of India's largest states, has broken new ground by establishing a committee on climate change. So all across India there is a palpable sense that the country has awoken and is on the move on climate change.[19]

These are small beginnings, but they represent a huge opportunity. The year 2009 is very different from those before it. With elections in both India and the United States, and with domestic electorates more alive to the need for action and leadership, it is a game-changing moment. Both India and the United States need a new narrative that looks forward, not backward. One where the politics of blame is replaced by the recognition of a shared dilemma and the value of collective action.

The shaky global economy provides a stark backdrop to why cooperation in an interdependent world is the only way forward. To succeed, climate change must be reframed as an agenda of hope, growth, innovation, and opportunity. It must be used to mobilize a new sense of national purpose and imbue people with optimism. India has a billion good reasons for leadership on climate change. Addressing this could be the best way for the country to secure prosperity and development. If India truly aspires to greatness, no other issue is more timely or compelling.

A Chinese Perspective on Climate and Energy

Yingling Liu

As the world considers what to do about climate change, attention has turned to China, the most populous and fastest-developing country, for its current and potential emissions of greenhouse gases (GHGs). China's carbon dioxide emissions are now estimated to be about 24 percent of the global total, surpassing the U.S. contribution of 21 percent, although China's per capita emissions are still far below those in industrial countries. But China's rapid economic growth shows no sign of leveling off, making it more than likely that the country's energy consumption and GHG emissions will continue to grow. The energy path China follows is going to determine not only its own development course but also global environmental well-being.[1]

The dominance of coal in China's energy portfolio is responsible for much of its GHG emissions, accounting for 85 percent of the total. The country relied heavily on this dirty conventional energy source throughout three decades of economic boom, and it has the world's third largest remaining proven recoverable coal reserves. The share of coal in total primary energy consumption has come down only slightly, from 72 percent in 1980 to 69 percent in 2006.[2]

Another source of concern is that China's energy consumption has shot up drastically since the beginning of this century, after almost two decades of low and stable growth, mainly due to the country's skewed industrial structure. Industry uses 70 percent of China's total energy, and energy-intensive industries such as steel, nonferrous metals, petrochemicals, and construction materials account for almost half of national energy use.[3]

Emphasis on these energy-intensive industries has been driven by demands from an expanding Chinese urban population and from overseas markets as a result of economic globalization. The number of city dwellers increased from 370 million in 1997 to 594 million by 2007. Thus 224 million people—roughly as many as live in all U.S. cities—were added to China's cities in just one decade. Urbanites normally use three to four times as much energy as rural residents. And the need to accommodate, move around, and entertain this expanding urban population has driven up energy-intensive sectors such as power generation, steel and cement production, and the manufacture of cars, appliances, and machinery.[4]

China's increased emissions have also been tied to economic globalization. This encourages the global flow of capital and resources, driving businesses to wherever they can maximize profits. This has meant the massive relocation of energy-intensive industries to places with good investment environments and lower costs—and China is a major destination. Its entry into the World Trade Organization in late 2001 fundamentally integrated the country into the

Yingling Liu is China Program Manager at Worldwatch Institute.

world economy.

While exports bring wealth to the country, continuous global demands for energy-intensive products have also accelerated heavy industrialization. China is the third largest trading nation, and industrial products accounted for more than 90 percent of total exports. A September 2008 study suggested that about one third of China's emissions were embedded in exports in 2005, a figure that was just 21 percent as recently as 2002.[5]

Since 2005, as the unbridled development of China's energy-intensive industries pressured the country's energy supply and environment, policymakers have hastened their efforts to move the country in a more sustainable direction. Although many of the energy policy changes have been for economic, health, and security reasons, they are also bringing immediate benefits to the climate.

The government is experimenting with a mix of state-led regulatory and policy tools to restructure energy-intensive industries. The most notable has been the ambitious national target for increasing energy efficiency. In early 2006 the government announced a plan to cut energy consumption per unit of gross domestic product by 20 percent by 2010.[6]

This announcement followed the 2005 launch of 10 national energy-saving projects that targeted major energy-intensive sectors, closing down and phasing out inefficient power and industrial plants and improving energy efficiency through technological innovations, financial support, and pilot projects. In September 2006, the government made eight energy-intensive industries pay more for electricity. It has also adjusted export rebates and tariffs to discourage energy-intensive exports. Since 2004, China has changed the export tariff on steel products more than 10 times, not only scrapping export rebates of 15 percent but also levying an export tariff of 25 percent. And in June 2007 the government abolished export rebates for 533 energy- and resource-intensive and polluting commodities and imposed export tariffs on 142 of these items.[7]

At the same time that it is weaning energy guzzlers from cheap electricity and export rebates, the government has increased financial support for energy conservation projects. In mid-2007 it added 10 billion yuan to the existing 6.3 billion yuan of state bonds and an earlier input of 5 billion, making a total of 21.3 billion yuan ($2.9 billion) dedicated to energy-saving and emission reduction projects. Some 9 billion yuan was set aside for the national energy-saving projects, 13 times as much as in 2006. The government requires financial institutions to increase credit support for such projects and encourages enterprises to raise funds through markets. Local governments have gradually followed suit and set up special funds for energy conservation, although these are too limited to make visible changes.[8]

These state-led measures need local government support to make real changes on the ground. The top leaders have taken steps to get local officials to cooperate. The macroeconomic planning body forced local governments to abandon preferential policies on land, taxes, and electricity prices for energy-intensive industries. In June 2007, the State Council (China's cabinet) made it clear for the first time that performance in meeting energy-saving and emission reduction targets could be the decisive "one-vote veto" in assessing local leaders' political performance. In other words, local officials risk their political careers if they fail to save energy.[9]

The legal framework for reducing emissions has improved gradually. In 2007, China revised its decade-old Energy Conservation Law, defining energy saving as a basic state policy. The law requires energy conservation to be integrated in all development plans and eases the way for enforcing and reporting through the institutional system. The law also has specific stipulations targeting industries, setting up a system to evaluate and assess energy efficiency in fixed capital investment projects, and providing more severe punishments for enterprises that fail to reach energy efficiency goals.[10]

Unfortunately, the state-led efforts have not sparked much enthusiasm from local governments and industries because of a lack of incentives. And the annual energy efficiency targets are rather arbitrary, with few considerations of the time frame needed by industries for such changes. As a result, the targets were not reached in 2006 or 2007. The policies and regulations do, however, indicate the central government's political will, and they have cleared many obstacles for optimal market functioning. This has opened up a potential business realm for energy efficiency technologies and services.[11]

Yet energy saving alone cannot solve China's emissions problem. Despite some improvements, energy consumption will still rise. The country's urban population will continue to swell, with 45 percent of Chinese already living in urban areas by 2007 (compared with about 70 percent in industrial countries). Domestic demand for heavy industrial products will thus keep on growing and is far from being saturated. The country will remain a major exporter as well. Even if China meets the 2020 emission cutting target proposed in its national climate change assessment report, its emissions could more than double by then.[12]

Thus China urgently needs to introduce clean energy technologies. The country has the necessary industrial base and potential vast market for new clean energy options and poses as a potential world leader in renewable energy. This sector has seen breathtaking development over the last three years, driven by a mix of domestic and international factors. Its rapid evolution shows how state policies can encourage the development of industries for a new market niche and how market forces can inject vitality in the private sector and achieve policy goals at a much faster pace. The reinforcement of policies and markets will likely provide the most lasting and profound force in pushing China onto a new energy path.

Aiming to diversify the country's energy portfolio, China enacted a landmark renewable energy law at the start of 2006. It requires the government to formulate development targets, strategic plans, and financial-guarantee measures for renewable energy. It also establishes a framework for sharing the extra costs of renewable energy among users and requires power utilities to buy more renewable power. In addition, the law establishes fixed premium prices and pricing mechanisms for biomass and wind power.[13]

As a result of this law and related implementation regulations, China's renewable energy sector has taken off, with wind and solar being the two leading stars. Meanwhile, a surging demand in the global market for renewable energy products, especially for photovoltaic (PV) systems in Europe and the United States, has encouraged a world-class solar PV manufacturing base to spring up in China literally from scratch.

China is quickly becoming a global leader in wind and solar power, in addition to its leading position in the production and installation of solar water heaters and in hydro and biogas development. Wind power is the fastest-growing renewable energy sector. New installed capacity grew by over 60

Courtesy of Li Junfeng

Yang Ba Jing grid-connected solar PV station, Tibet

percent in 2005, and it more than doubled in both 2006 and 2007. By the end of 2007, cumulative wind power capacity had reached roughly 6 gigawatts (GW), up from 0.8 GW in 2004—making China fifth in global wind installations. The cumulative wind installations in 2007 exceeded the target that had been set for 2010 only one year earlier. And the target for 2020 of 30 GW is now likely to be reached by 2012—eight years ahead of schedule.[14]

China's PV manufacturing has witnessed phenomenal development in recent years as well. Total solar cell production has jumped from less than 100 megawatts (MW) in 2005 to 1,088 MW in 2007, making China the world's top solar cell producer. Chinese experts and business leaders believe that solar cell production will exceed 5 GW by 2010, accounting for one third of the world total, and 10 GW by 2015, or two thirds of the world total by then. The country is already turning into a major solar PV base, with the lion's share of production being for export.[15]

In just a few years, renewable energy has become a strategic industry in China. There are more than 50 domestic wind turbine manufacturers, over 15 major solar cell manufacturers, and roughly 50 companies constructing, expanding, or planning polysilicon production lines, the key components for solar PV systems. Those two sectors employ some 80,000 people. Together with the 266,000 working in biomass and 600,000 in the solar thermal sector, renewable energy industries employ some 946,000 people in a new market niche independent of conventional energy industries.[16]

China currently gets 8 percent of its primary energy from renewable sources, with large hydro being dominant. The country aims to expand that share to 15 percent by 2020. Developments in the marketplace show that this target could well be exceeded. Policy tools and market forces can together push China toward a less carbon-intensive energy path. And there is considerable room for international cooperation and business initiatives to accelerate the process.[17]

Trade, Climate Change, and Sustainability

Tao Wang and Jim Watson

International trade has continued to increase in the last few decades as a result of deepening globalization. Relocation of production in the pursuit of comparative advantages has brought economic growth to many regions. Some developing countries have benefited from this trend due to abundant resources or labor supply or both. But the environmental consequences of international trade have been increasingly highlighted. Within discussions about the international targets for and mechanisms to achieve large-scale reductions in carbon emissions, the emissions embodied in traded goods have often been highlighted as a particular challenge.[1]

Emissions embodied in internationally traded goods are currently attributed to the producing nation under the U.N. Framework Convention on Climate Change's definition. Many of these goods are manufactured in developing countries, such as China, that lack binding carbon emissions targets. Exports now account for more than a third of China's total economic output, much higher than most economies of similar size. In 2006, some 58 percent of China's exports were from multinational ventures and around 70 percent of foreign direct investment went to manufacturing. Given the treaty's definition of the source of emissions, it is not surprising that China is now the world's largest emitter of carbon dioxide.[2]

Thus the industrial world is becoming ever more reliant on importing goods from China and at the same time "exporting carbon" to this nation to meet carbon reduction targets. A recent Tyndall Centre for Climate Change Research assessment of the carbon emissions embodied in China's international trade found that net exports accounted for 23 percent of China's carbon emissions in 2004. This is partly because of China's large trade surplus but also because of China's higher carbon intensity due to an inefficient, coal-dominated energy system. The carbon embodied in China's exports was comparable to Japan's total carbon emissions in 2004 and more than double the emissions from the United Kingdom. (See Table.) And several studies show a clear trend of increasing embodied carbon during the last decade as well as an increasing share of total carbon emissions over time.[3]

These findings highlight the importance of an issue that has been underplayed in climate policy. They show that consumers in industrial countries are indirectly responsible for a significant proportion of China's carbon emissions. This evidence adds weight to the view that industrial countries should help developing ones reduce their carbon emissions through technical assistance and finance.

__Tao Wang__ is a Research Fellow at the Sussex Energy Group and the Tyndall Centre for Climate Change Research in England. __Jim Watson__ is Deputy Director of the Sussex Energy Group and Deputy Leader of Tyndall's Climate Change and Energy Programme.

Carbon Dioxide Emissions from China's Net Exports and Total Emissions from Selected Countries, 2004

Country	CO_2 emissions
	(million tons)
United States	5,800
China	4,732
China, from net exports	1,109
Japan	1,215
Germany	849
United Kingdom	537
Australia	354

Source: Wang and Watson.

The scale of this "carbon leakage" to developing countries through international trade is so significant that it needs to be taken into account in the next round of international climate agreements. Some observers have called for a radical change from production-based to consumption-based national emissions accounts so that emissions embodied in traded goods are included within the consuming country's targets. But this would be impractical due to data uncertainties and the large amount of political capital that has already been invested in the current accounting system. Measurement of consumption-based carbon emissions could, however, be used as a "shadow indicator" in negotiations and could complement official nationally based emissions inventories.[4]

This shadow indicator could help inform a range of policies for the mitigation of emissions. Some of these are known as "sectoral agreements." These are designed to deal with sectors that are not only exposed to high levels of international competition but are also carbon-intensive. If they are sufficiently binding, sectoral approaches could help reduce emissions while helping companies in developing countries improve their technological capacity.

Another approach to including traded goods might be to impose border tax adjustments on goods brought into countries or regions with emissions caps. Senior policymakers in both the European Union and United States are considering such an approach since it would internalize the embodied carbon cost of imports and would "level the playing field" with goods produced domestically. These proposals have inevitably been criticized as being "protectionist," however, by some developing countries and may be subject to challenge within the World Trade Organization. Again, if this policy were implemented carefully, with compensatory financial and technological assistance to developing country producers, it might be seen more favorably.[5]

It is important not to overstate the impacts of carbon leakage on international competitiveness. Contrary to the arguments of some industrial lobbyists, emissions caps in the United States or the European Union will only significantly affect the competitiveness of a few energy-intensive industries, such as steel and cement. The products of these industries make a relatively small contribution to China's exports—and to the emissions embodied in them. China's exports are instead dominated by consumer goods such as textiles, footwear, and electronics, which are not as carbon-intensive or sensitive to carbon taxes.[6]

Whichever way forward is followed, the solution will require trust, not suspicion. Collaboration rather than confrontation in bilateral or multilateral relationships is required, as no country can deal with climate change alone in a globalized trade network. International trade policy could play a significant role in the future climate regime as well as in sustainable development. It is important to make sure trade is more ethical and more environmentally friendly and that the costs and benefits are more fairly distributed.

Adaptation in Locally Managed Marine Areas in Fiji

Alifereti Tawake and Juan Hoffmaister

On small islands, adaptive management and planning is the cornerstone for survival. Climate change is putting additional stress on islands' limited resources, thereby threatening livelihoods. The need for local solutions that are flexible and responsive to the local context is paramount.

The Republic of Fiji has recently gained valuable experience with local adaptive management. In an archipelago of more than 300 islands in the South Pacific Ocean, Fijians have learned to coexist with the ocean for centuries and to make a living through the management of limited resources. As the sea level rises and severe storms become more frequent and damaging due to the changing climate, this management is being put to the test.

Over the past decade, more than 300 communities in Fiji have adopted a management model tailored to the needs of the community: the locally managed marine area (LMMA). The LMMA Network was launched in August 2000 as a learning network after a series of workshops to provide guidance on community-based management of marine areas. Promoting models of adaptive governance and knowledge-sharing networks, the network now has members in Indonesia, Palau, Papua New Guinea, the Philippines, Micronesia, the Solomon Islands, and Fiji.[1]

As an alternative to conventional centralized resource management or typical government approaches, the LMMA approach is a local community-driven effort to design, manage, and monitor marine resources through co-management by community members, with the support of traditional leaders, government agencies or ministries, nongovernmental organizations (NGOs), and educational institutions, in collaboration with other stakeholders such as businesses. LMMAs do not necessarily exclude government or other institutions; they engage them as partners rather than as commanders. It is an inclusive governance model for resource management, integrating different stakeholders in the decisionmaking process.

Communities are empowered to decide how to best use their resources in light of the predicted effects of climate change. The strategies created by the communities have multiple benefits, including community-based risk reduction, protection of endangered resources that are critical for food security, community engagement and capacity building, and enhanced disaster risk management. In Fiji, improvement in the integrity of the marine ecosystem is measured by monitoring reefs with the help of NGOs, the University of the South Pacific, and trained members of the community.

Many communities using the LMMA

__Alifereti Tawake__ is the national coordinator for the Fij LMMA Network. __Juan Hoffmaister__ is a Watson Fellow 2007–08 researching community-based adaptation.

model have found practical solutions to emerging problems by reviving traditional knowledge, which can then be combined with modern tools. To decide the best combination, communities use an adaptive management approach, which the LMMA Network defines as "the integration of design, management, and monitoring of a project to systematically test assumptions in order to adapt, learn, and improve the results of their efforts." With the help of the Network, practitioners increase the effectiveness and efficiency of local strategies over time.[2]

The community engagement process involves initial awareness-raising about marine issues and dialogue with stakeholders to engage them in the goals of the LMMA concept and ensure that the community is in harmony with the process of developing their local plans. This is followed by a workshop in which the community develops a marine resource management plan, which might include:

- declaration of a *tabu* (no-catch area) area and other traditional management practices,
- reduction in the number of fishing licenses,
- banning the use of *duva* (fish poisoning) and destructive fishing measures,
- restoration of economically important species such as clams,
- reduction of marine pollution,
- replanting of mangroves and coastal trees to reduce coastal and riverside erosion,
- marine awareness raising, and
- alternative livelihood options.

LMMAs and *tabu* areas are set up not just for conservation but to improve the yield of marine resources that people use for subsistence and trade. The *tabu* is implemented by the communities, led by a local headman whom the community trusts to decide on implementation. A community with a demonstrated sustainable and secure source of food and livelihood is better prepared to address and adapt to climate change. A common Fijian saying is that "a hungry community will be handicapped in making and acting on good decisions."

Communities are trained to do biological and socioeconomic monitoring to monitor the effects of their management actions. Meetings are held regularly to review progress and see if changes in the action plan are needed.

Drama is an important component of community education and awareness in Fiji, as many elders who are key decisionmakers in villages find reading to be very challenging. Drama provides an interactive and innovative means of translating complex technical concepts such as climate change, and it can also paint a picture of its likely impacts.

More than 200 different localities in Fiji are using the LMMA model. The results vary, but most groups have found that effective implementation of this strategy can help recover marine resources through improved habitat quality (coral cover, seagrass, and mangroves) and increased fish populations. By being engaged in the LMMA, communities are also better prepared to implement practical solutions to emerging external threats such as climate change

The key element of the LMMA work in Fiji is that the communities are in control. Information on management options is provided by co-managers to help make decisions, but the community members make all decisions, such as location of the *tabus*. Thus the goal of informed decisionmaking on resource management is as important as the actual resource improvement. This will be increasingly valuable in a warming world. For while LMMAs initially focused on food security issues and resource depletion, Fijian communities are learning important lessons about managing the impacts of climate change.

Building Resilience to Drought and Climate Change in Sudan

Balgis Osman-Elasha

Drought, population pressures, and conflict are degrading lands and undermining the resilience of ecosystems. Severe drought across many parts of Sudan is affecting several million people, many of whom are at acute risk of food insecurity. Low and sporadic rainfall has severely affected water resources and agricultural production, particularly in the traditional rainfed sector.[1]

Population and economic pressures have driven people to intensify cultivation of drylands, extend cultivation into more marginal areas, overgraze rangelands, and overharvest vegetation. These factors have degraded lands, reduced the availability of water, and depressed the production of food, fodder, and livestock. Competition for these lifelines has been a source of conflict and has contributed to the tragic violence that has engulfed parts of Sudan.[2]

Climate change is an additional source of uncertainty and risk. Sudan has experienced more than 20 years of below average rainfall during which there have been many localized droughts, as well as a severe and widespread drought from 1980 to 1984. These conditions can be expected to worsen during future climate change, with the country's climate becoming even drier.[3]

Balgis Osman Elasha is a Senior Researcher at the Higher Council for Environment and Natural Resources in Sudan and a lead author of the Fourth Assessment Report of the Intergovernmental Panel on Climate Change.

Sudan's rural communities are adapting in order to reduce their risks in a harsh, variable, and changing environment. While the adaptations are not necessarily driven by climate change, they are nonetheless building resilience to it. The measures being adopted include using water harvesting and special irrigation methods, expanding food storage facilities, managing rangelands to prevent overgrazing, replacing goats (which are heavy grazers and are sold at a lower value) with sheep (which have less impact on grassland and are sold at higher value), planting and maintaining shelterbelts, planting backyard farms or *jubraka* to supplement family food supplies and incomes, supplying microcredit and educating people about its use, and forming and training community groups to implement and maintain these various measures.[4]

Environmental Strategies for Increasing Human Resilience in Sudan is a regional assessment that was undertaken by the Sudan Higher Council for Environment and Natural Resources in collaboration with the Stockholm Environment Institute–Boston Center and with the participation of researchers from the University of Khartoum and experts from local nongovernmental organizations (NGOs). The assessment examined three cases of community efforts to improve livelihoods and manage natural resources that succeeded in increasing the overall resilience of the communities. All three projects were prompted by the adverse consequences of Sahel-wide droughts in the early 1980s. The vulnerability and adaptation of these communities were assessed in terms of their financial, physical, human, social, and natural capital, taking into account productivity, equity, and sustainability as well as risk factors. This involved the collection and analysis of "resilience indicators" data and an analysis of national and local institutions.[5]

The first assessment looked at 17 villages of Bara province in North Khordofan. Located in the Sahel zone, which has undergone a general decline of rainfall since the late 1960s, the area is marked by high rainfall variability. The severe 1980–84 Sahel-wide droughts deeply affected family and tribal structures among pastoralists and agro-pastoralists, deepening poverty, marginalization, and food insecurity. Thousands of people ended up in refugee camps surrounding towns and cities.[6]

The project in Bara aimed to sequester carbon by setting up a resource management system that prevents degradation and that rehabilitates or improves rangelands to reduce the risks of production failure so as to limit outmigration and stabilize population. The measures undertaken included the following:

- establishment of local institutions such as Village Community Development Committees to coordinate natural resource management and community development activities;
- development of land use master plans to guide future resource use and implementation of sustainable rotational grazing systems;
- establishment of community mobilization teams to conduct outreach and training;
- rangeland rehabilitation, including land management, livestock improvement, agroforestry, and sand dune fixation to prevent overexploitation and restore rangeland productivity;
- water harvesting and management, rural energy management, and diversification of local production systems and income-generating opportunities to reduce pressure on rangelands; and
- creation of water management subcommittees to better manage wells.[7]

Original expectations for the project were more than met: the achievements to date include revegetation and stabilization of 5 kilometers of sand dunes to halt desert encroachment; construction of 195 kilometers of windbreaks to protect 30 farms from soil erosion, restocking of livestock by replacing goat herds with sheep, establishment of 17 women's gardens to produce vegetables for household consumption and to sell at local markets, and establishment of five pastoral women's groups to support supplemental income-generating activities such as sheep fattening, handicrafts, and milk marketing.[8]

The second project, carried out by the British NGO SOS Sahel, focused on Khor Arba'at in Red Sea state, home to Beja pastoralist and agro-pastoralist tribal groups. It aimed to improve the livelihoods of tribal groups in farming, local water resources, and other natural resources. Rainfall is highly variable, with averages recorded between 1900 and 1980 ranging between 26 and 64 millimeters per year. The rocky and compact nature of soils, steep slopes, heavy downpours, and poor vegetation cover all contribute to the high rates of runoff. A traditional pattern of natural short-term recovery was shattered after the long drought and famine of the 1980s.[9]

Because the Khor Arba'at delta had been neglected by the government in recent years, local communities were eager to participate in a broad-ranging program that promised to improve their livelihoods. The main objectives of the project were to rehabilitate the Khor Arba'at delta, to realize the full agricultural potential of the area for the benefit of tribal groups, to tailor the sustainable management of natural resources so as to meet local needs, to ensure the sustainability of food security, to set up an equitable water harvesting scheme, and to enhance grassroots participation in the overall development of the community. It was also hoped that success in this project could be replicated elsewhere.[10]

The third case study involved communities in North Darfur that implemented various measures to cope better with variable water resources and land productivity. This is one of the most drought-affected regions of the Sudan. The target group in this project was the most vulnerable households practicing subsistence farming and raising livestock in the area. Following the drought years of 1983–85, many people left their homes due to increasing poverty, famine, desertification, and land degradation. This was accompanied by tribal conflicts, the growth of shantytowns, and changes in the pattern of livestock raising and agricultural production. Most people lost more than half their cattle as well as large numbers of sheep, goats, and camels.[11]

Unlike the other two cases, these adaptation measures were initiated by the local community and only later were supported by external funding. The key measures undertaken included adoption of better water harvesting techniques and construction of terraces that helped farmers grow vegetables that can be harvested up to five months after the rainy season, restocking of gum trees (*Acacia Senegal*) and retention of part of the tree cover in agricultural fields with alluvial soils for the provision of fuelwood, and cultivation of clay soil, easing pressure on the sandy soil.[12]

Indicators on the sustainability of livelihood assets and adaptation measures showed improvement in all three projects as a result of the efficiency of the local Community Development Committees and the efforts of the Sudanese Environment Conservation Society–Kordofan Branch, which has been very active in supporting the continuity of measures and the dissemination of information on rainfall, new production inputs and technologies, and prices. Another key aspect of the success has been high loan repayment rates to the community revolving funds.

A number of key conclusions emerge from these three cases. First, successful strategies emphasize livelihoods and are embedded in community development efforts rather than being implemented in isolation. Typically suites of measures are implemented that provide the means to improve and diversify livelihood opportunities, advance sustainable management of natural resources, and hedge against risks of variable incomes and variable access to food, water, and other resources. Successful strategies that have added to human and social capital include training farmers in techniques to diversify their production activities and improve resource management, involving women in home gardening of vegetables, and aiding traditional farmers, fruit growers, and vegetable growers to form unions to help harvest and market products.[13]

Second, adaptation requires effective involvement of local institutions, tribal leaders, community-level committees, and NGOs. Such a participatory, bottom-up approach is essential to successfully engage at-risk groups in decisionmaking processes. Farmers, herders, women, and minorities gain a better understanding of their vulnerability, priorities, and adaptation needs.

It also facilitates cooperation within the community and the mobilization of local resources and indigenous knowledge. Local institutions can ensure continuity of development and adaptation activities after externally funded projects end.

Third, the sustainability of adaptation measures depends on enhancing the sense of responsibility among communities. To ensure proper implementation of policies, work should focus on improving communities' knowledge and capacity to manage their own natural resources. Regulations and policies that are based on real knowledge of communities and a sense of responsibility lead to positive results and improved performance. Central to the success of the interventions are efforts to strengthen institutions with training and resources, form new community institutions, empower local institutions with skills and information to plan and implement project activities, and promote the participation of community members in different sustainable livelihood activities and decisionmaking processes.

A final lesson is that adaptation falls short of what is needed. Existing efforts to cope and adapt are insufficient in the face of present-day risks. Drought already exacts an unacceptably high toll on the people of Sudan, and the suffering is likely to grow further with climate change. The adaptive responses that have been applied and shown to be successful in building resilience need to be replicated and expanded even as new approaches are explored and tried.

Children learn how to use an ox-plow in Twic County, southern Sudan. In this remote province, the use of animal traction in agricultural work is not yet common.

Geoengineering to Shade Earth

Ken Caldeira

In June 1991, Mount Pinatubo in the Philippines erupted explosively—the biggest eruption of the twentieth century. The volcano created a column of ash and debris extending upward 40 kilometers (about 25 miles). The eruption ejected around 20 million tons of sulfur dioxide into the stratosphere, where it oxidized to form sulfate dust particles. The stratosphere is the part of the atmosphere that is higher than where jets normally fly.[1]

As a result, about 2 percent of the sunlight passing down through the stratosphere was deflected upward and back into space. The dust particles were big enough to scatter sunlight away from Earth but small enough to allow Earth's radiant heat energy to escape into space. Earth cooled about half a degree Celsius (almost 1 degree Fahrenheit) the following year, despite the continued increase in greenhouse gas concentrations. This raises an obvious question: Could we similarly put dust into the stratosphere to offset climate change?[2]

Earth is heated by sunlight and cooled by the escape of radiant heat into space. Earth's atmosphere is relatively transparent in the wavelengths that make up sunlight but somewhat opaque in the wavelengths that make up escaping radiant heat energy. As greenhouse gases accumulate, the atmosphere becomes more opaque to outgoing radiant heat. With greater amounts of radiant heat trapped in the lower atmosphere, Earth's surface warms.[3]

The most obvious approach to keeping Earth cool is to reduce greenhouse gas concentrations in the atmosphere, so that heat energy can escape more easily into space. But another strategy involves reducing the amount of sunlight absorbed by Earth. If greenhouse gases accumulating in the atmosphere are like closing the windows of a greenhouse and trapping heat inside, then "geoengineering" approaches seek to keep Earth cool by putting the greenhouse partially in the shade. They try to reverse warming by preventing sunlight from being absorbed by Earth.[4]

A number of modeling and theoretical studies have looked into such climate engineering schemes. The consensus appears to be that these will not perfectly reverse the climate effects of increased greenhouse gases but that it might be technically feasible to use geoengineering to reduce the overall amount of climate change. Obviously, however, these schemes would not reverse the chemical effects of increased carbon dioxide (CO_2) in the environment, such as ocean acidification or the CO_2-fertilization of land plants.[5]

Several approaches have been suggested for deflecting sunlight away from Earth. The most science-fiction scheme would be to place sunlight-blocking satellites between Earth and the sun. But in order to compen-

***Ken Caldeira** is a climate scientist at the Department of Global Ecology at the Carnegie Institution for Science in Stanford, California.*

sate for the current rate of increases of greenhouse gases in the atmosphere, governments would need to build and put in place more than a square mile (about 3 square kilometers) of satellite every hour. Most people would probably agree that such an enormous effort would be better applied to reducing greenhouse gas emissions.[6]

The placement of sulfur dust particles in the stratosphere appears to be the leading candidate for most easily engineering Earth's climate. (Numerous other approaches have been suggested, including some designed to increase the whiteness of clouds over the ocean with sea salt particles formed by spraying seawater in the lower atmosphere.) Tiny particles have a lot of surface area, so a lot of sunlight can be scattered with a relatively small amount of dust. The full amount of sulfur from Mount Pinatubo, if it had remained in the stratosphere for a long time, would have been more than enough to offset the warming (at least, on a global average) from a doubling of atmospheric carbon dioxide content. The actual short-lived cooling from the Mount Pinatubo eruption turned out to be much less because the oceans helped keep Earth warm despite the reduction in the amount of absorbed sunlight.[7]

The sulfur from Mount Pinatubo remained in the stratosphere only for a year or two. To maintain a dust shield in the stratosphere for the long term would require continual dust injection. It is thought that a small fleet of planes, or perhaps a single fire hose to the sky suspended by balloons, would be enough to keep the dust shield in place. Costs are uncertain, but it might total less than a few billion dollars a year. The amount of sulfur required would be a few percent of what is currently emitted from power plants and so would contribute somewhat to the acid raid problem.[8]

Why might policymakers want to deploy

Karin Jackson, U.S. Air Force

Mount Pinatubo erupting on June 12, 1991, as seen from Clark Air Force base eight miles away

climate engineering systems? The main reason is to reduce climate damage and the risk of further damage from greenhouse gases. Some commentators deny the reality of human-caused greenhouse warming but think it worth developing climate engineering systems as an insurance policy—just in case events prove them wrong. Others accept human-induced climate change but think reducing emissions will be either too costly or too difficult to achieve, so they favor climate engineering as an alternative approach. Some people fear that a climate crisis may be imminent or already unfolding and that these systems are needed right away to reduce negative climate impacts such as the loss of Arctic ecosystems while the world works to reduce greenhouse gas emissions in the longer term. Still others think climate engineering is needed as an emergency response system in case an unexpected climate emergency occurs while greenhouse gases are being reduced.[9]

There are also many reasons not to develop climate engineering, some of them having to do with climate science and some having to do with social systems. These schemes will not work perfectly, for example, and there is some chance that unanticipated consequences will prove even more environmentally damaging than the problems they are designed to solve. Concerns include possible effects on the ozone layer or patterns of precipitation and evaporation. Climate engineering would not solve the ocean acidification problem, although it would not directly make it worse either.[10]

Some observers fear that the mere perception that there is an engineering fix to the climate problem will reduce the amount of effort placed on emissions reduction. Climate engineering could lull people into complacency and produce even greater emissions and ultimately greater climate damage. (On the other hand, such schemes also could frighten people into redoubling efforts to reduce greenhouse gas emissions.) And it might work well at first, with negative consequences manifesting themselves strongly only as greenhouse gas concentrations and the offsetting climate engineering effort both continued to grow.[11]

Climate engineering will affect everyone on the planet, but there is no clear way to develop an international consensus on whether it should be attempted and, if so, how and when. It would likely produce winners and losers and therefore has the potential to generate both political friction and legal liability. Conflict over deployment could produce political strife and social turmoil. (On the other hand, any success at reducing climate damage could lessen strife and turmoil.)

From the perspective of physical science and technology, it appears that climate engineering schemes have the potential to lower but not eliminate the risk of climate damage from greenhouse gas emissions, yet unanticipated effects and difficult-to-predict political and social responses could mean increased risk. Thus the bottom line is that climate engineering schemes have the potential to make things better, but they could also make things worse.

Carbon Capture and Storage

Peter Viebahn, Manfred Fischedick, and Daniel Vallentin

Rising oil and gas prices, insecure energy supplies, and increased energy consumption in transition economies have boosted the use of coal—the most abundant fossil fuel and one that many countries have considerable reserves of. The United States, China, and some other countries are highly dependent on coal. In the United States, coal-powered plants generate more than half the electricity, and some observers expect that expanding the use of coal will help reduce U.S. reliance on foreign oil.

But coal is the most carbon-intensive fossil fuel. Thus a new technology called carbon capture and storage (CCS) has recently gained considerable attention. CCS aims to capture carbon dioxide (CO_2) from any large point source, liquefy it, and store it underground. Because of its high costs and complex infrastructure, CCS is by necessity suited primarily for centralized, large-scale power stations or big industrial facilities like cement plants and steelworks.

With today's technologies, there are three ways to capture CO_2. Post-combustion capture, which involves capturing CO_2 from flue gases in conventional power stations, is basically available today, but it has not yet been demonstrated at a commercial power station scale. In the longer term, this technology is unlikely to become widely established unless its energy consumption can be reduced significantly.

A more efficient method is pre-combustion capture of CO_2 in coal-fired power stations with integrated gasification combined cycle technology. These plants use heat to gasify coal that is then burned to generate electricity. During the gasification step, CO_2 can be removed relatively easily. Apart from its higher efficiency levels, the prime advantage of this method lies in its flexibility in terms of both fuel (coal, biomass, and substitute fuels) and product (electricity, hydrogen, synthetic gas, and liquid fuel). Pre-combustion capture of CO_2 has not yet been demonstrated on a large scale.

The so-called oxyfuel process currently offers the best prospects for CO_2 capture in terms of achievable overall process efficiency as well as costs because it is largely based on conventional power station components and technology. Combustion takes place in 95 percent pure oxygen rather than air, enabling efficient CO_2 capture due to the concentrated flue gas. This process is still near the beginning of its demonstration phase. It is expected to capture 99.5 percent of the emissions directly at the stack, while the post-combustion and pre-combustion methods would reduce CO_2 by 88–90 percent on average.[1]

Peter Viebahn *is a Project Co-ordinator and research fellow, and* ***Daniel Vallentin*** *is a PhD student, in the Research Group on Future Energy and Mobility Structures at the Wuppertal Institute for Climate, Environment and Energy in Germany.* ***Manfred Fischedick*** *is head of the Research Group and Vice-President of the Institute.*

Once CO_2 has been captured from industrial sources and pressurized into a quasi-liquid form, it can be pumped into geological formations such as deep saline aquifers more than 2,000 feet underground, depleted oil and gas fields, and deep and non-exploitable coal seams. It can also be deposited deep in the ocean. Furthermore, the productivity of oil and gas fields in their final stages of exploitation can be increased by injecting CO_2 into them, something the oil and gas industry has been doing for years. A mineralization process for binding CO_2 to silicates is also under discussion as a way to sequester and store the gas, along with a method for fixing CO_2 using algae to produce biomass that can be turned into animal fodder, biodiesel, or construction materials.[2]

Along with the overriding motivation of climate protection, questions of security of energy supply, technological aspects, and in some cases immediate economic considerations have increased interest in carbon capture and storage. Technology that can facilitate progress in international climate protection negotiations is of particular importance. Some of the strongest supporters of CCS are governments that have so far rejected the international climate protection process or adopted a wait-and-see stance, such as the United States.[3]

Yet several constraints make it questionable that a global rollout of CCS will consist of more than demonstration plants and some initial commercial plants. The first concern is the time frame. CO_2 capture technologies are more likely to become available in the medium than the short term. Most experts anticipate large-scale applications between 2020 and 2030. But the rush to build new coal-fired power plants will likely take place within the next 10 years—too soon to take advantage of CCS technologies. And decisions on new power plants made today will influence the energy mix 40–50 years from now, when greenhouse gas emissions need to be substantially lower than today. For plants built before CCS is mature, only retrofitting of CCS technology, usually with the low-efficiency post-combustion method, would be an option. And retrofitting power stations would cost more and be less efficient than newly built plants fitted with CCS from day one.[4]

The number and location of safe reservoirs is a second concern. To be able to store billions of tons of CO_2 "safely and cheaply, on a global scale, both in the West and in the developing world," one observer notes, advanced methods other than "simple" enhanced oil and gas recovery will be required. For various reasons, storage possibilities for CO_2 are restricted at both national and global levels. Gas fields are believed to have the largest potential, followed by coal seams, oil fields, and aquifers. From a purely technical perspective, there appears to be enough capacity to store global CO_2 emissions for many decades. Yet there is a great degree of uncertainty about the fundamental suitability of the various storage options. Ultimately a case-by-case analysis will be required to obtain practical and relevant results for each storage site considered. Another important question is that of liability. Undoubtedly, similar questions will arise as in discussions of nuclear energy waste disposal.[5]

High energy penalties and environmental impacts are a third constraint. Capturing CO_2 requires additional fuel consumption of 20–44 percent to generate the same amount of useful energy, which in turn leads to more CO_2 and other harmful emissions. But only the CO_2 emitted directly at the stack can be captured, in contrast to the CO_2 and other emissions of upstream and downstream

© Lyle Rosbotham

Coal trains near Gillette, Wyoming

processes. For instance, methane emissions during coal mining or natural gas pipeline transport cannot be reduced by CCS. Yet according to the Kyoto Protocol, greenhouse gas (GHG) emissions as a whole—not just CO_2 emissions—must be reduced. Recent life-cycle assessments show that assuming a CO_2 capture rate of 88 percent, GHG emissions along the whole value chain can be reduced by only 67–78 percent, depending on the fuels or power station technologies used. Furthermore, other environmental impacts like photooxidant formation, eutrophication, or particle emissions will increase with CCS, while acidification will decrease slightly.[6]

One of the most pressing environmental issues could be water use: it is expected that CCS will require 90 percent more fresh water. The increase of hazardous waste production due to the chemical reaction of the scrubbing agents is also important. Last but not least, CCS would only worsen many major local environmental problems tied to the extraction and transport of coal, such as habitat destruction, damage to waterways, and air pollution.[7]

The fact that alternatives to CCS have already entered the market could reduce interest in this technology. The GHG emissions associated with electricity generated from solar thermal power or wind power are just 2–3 percent of the amounts for fossil-fueled CCS plants. And the GHG emissions of electricity generated by advanced natural-gas-fired combined heat and power stations are roughly the same as those for power stations using CCS. Thus there are even fossil fuel technologies commercially available that are already as "green" as CCS power stations aim to be in 2020. Expanding use of these alternatives will of course require significant structural changes in the overall energy system.[8]

Cost is another constraint. CO_2 capture requires high investment costs in addition to the costs resulting from the energy penalty. Different sources put CO_2 capture costs at between 35 and 50 euros per ton of CO_2 in 2020, translating to a 50-percent increase in electricity generation costs (assuming no increase in fossil fuel prices). This assumes that significant learning processes will have occurred by then. Yet just when the first CCS power stations might be coming on stream, some individual renewable technologies (such as offshore wind and solar thermal power plants) could already be offering cheaper electricity. In the longer term, renewables can be expected to have considerable cost advantages due to their independence from fuel price fluctuations.[9]

A final constraint is infrastructure requirements. Suitable storage sites will not usually be located in the immediate neighborhood of the power stations, which means that large investments in a completely new pipeline infrastructure will be necessary. In the United States, deploying a national CCS system at the scale needed would require "no less fundamental a transformation of the country's energy infrastructure than would a huge-scale adoption of wind energy," noted the World Resources Institute. If the storage locations are 500 kilometers or more away from the big emitters, CO_2 transport will likely not pay off. A possible solution to this problem would be to place new power stations directly at potential storage sites and to transport the electricity instead of the carbon dioxide.[10]

It is possible that technological developments might be able to offset some of the significant constraints on CCS. In the future, for example, the combination of CCS with biomass-fired power plants could be an interesting option due to the negative carbon balance of such a system. CO_2 is first captured from the atmosphere during biomass growth and then could again be captured from the power plant's flue gases and sequestered afterwards. If storage works, this process could help achieve drastic CO_2 emission reductions. On the other hand, processes using biomass could meet only a part of the energy demand due to limited acreage.[11]

In general, several national and global energy scenarios show that even ambitious greenhouse gas emission targets can be met by a three-step strategy without assuming any appreciable use of CCS within the next few decades: increased energy efficiency, more-efficient use of primary energy by using combined heat and power plants, and ambitious development of renewable energy.[12]

Even if CCS is supposed to just be a bridging technology, significant research and development efforts are needed. Furthermore, if this technology can be demonstrated successfully, additional financing instruments will be needed to help spread the use of CCS. Including CCS-CO_2 in the carbon market, as planned by the European Union, would mean that deployment of CCS will strongly depend on the price development of CO_2 certificates. If CCS is included as an avoidance measure in the Kyoto Protocol, these projects could also be handled via flexible instruments such as the Clean Development Mechanism and Joint Implementation. Another incentive under discussion is government subsidies to make the technology competitive. Yet all these instruments raise fears that financing CCS could take funds away from renewable energy or energy efficiency measures, which would be counterproductive as these are the most robust climate protection strategies.[13]

In the end, a lot of open questions about CCS remain to be solved—technical as well as legal and socioeconomic ones. Today it cannot be foreseen if, how much, where, and when CCS will play a significant role as a strategic climate protection option. If it proves to be both commercially available and competitive, the question of suitable and safe storage places may become the tipping point for extensive use. What is clear is that there will not be a large-scale deployment of CCS in the next 10–15 years. If this time is used for ambitious development and diffusion of renewable sources, the argument for CCS as a "bridge" to renewable energy will lose its force.

Using the Market to Address Climate Change

Robert K. Kaufmann

There is near-universal agreement among scientists and economists that climate is changing in ways that reduce economic well-being due in part to emissions of carbon dioxide (CO_2) during the combustion of fossil fuels. The price of these fuels does not reflect these effects, and this omission is a classic example of what economists call an externality. By definition, externalities are not corrected by the market—government intervention is required.[1]

Policymakers are discussing two forms of intervention to abate CO_2 emissions: carbon taxes and a cap and trade system. Both systems seek to reduce emissions toward their optimal level—the point at which the marginal benefit of burning fossil fuels equals the damage caused by its combustion. But this optimum is unknown and probably will never be known with a high degree of certainty. This uncertainty creates advantages and disadvantages for the two mitigation strategies that arise from the ways in which they reduce emissions.

Carbon taxes raise the price of fuels based on the amount of carbon they emit. Ideally, the tax rate is set by the marginal damage done by a unit of carbon emitted. Because this quantity is unknown, a carbon tax starts with a political decision about a tax rate per unit of carbon emitted. Emission rates vary for the different fossil fuels. One thousand BTUs of coal emit 26 grams of carbon, a thousand BTUs of oil emit 21.4 grams, and a thousand BTUs of natural gas emit 14.5 grams. Nonconventional fossil fuels, such as oil shale, emit even more carbon per BTU.[2]

Using these emission rates to tax fossil fuels changes their relative prices. Based on prices to U.S. electric utilities in 2007, a $100 tax on a ton of carbon emissions would raise coal prices by 14.6 percent, oil prices by 2.5 percent, and natural gas prices by 2.0 percent. (Surprisingly, even though natural gas emits less carbon than oil, the percentage rise in the prices of the two fuels is very similar. This is because natural gas is generally less expensive than oil, so a smaller nominal increase in price resulting from the tax can result in a larger percentage increase.)[3]

These price increases can lower energy use and carbon emissions in two ways. Consumers can reduce activities that use energy and emit carbon. Examples include driving fewer miles or turning down the thermostat for home heating (or raising it for air conditioners). Because this strategy lowers emissions by reducing activities that consumers enjoy, these actions often are considered a reduction in well-being.

A second strategy allows consumers to lower emissions while maintaining activity levels. This involves two forms of substitution. In one, consumers purchase more-efficient machinery or use more labor, which allows them to use less energy and hence

Robert K. Kaufmann is Director of the Center for Energy and Environmental Studies at Boston University in Massachusetts.

emit less carbon. For example, buying a more-efficient car or installing insulation lets people maintain their driving habits or their comfort level while reducing emissions.

A second form of substitution, interfuel substitution, changes the mix of fuels away from those that emit large amounts of carbon per heat unit. Because a carbon tax raises the price of coal relative to oil or natural gas, consumers can reduce their energy costs and carbon emissions by substituting natural gas or oil for coal. Many industrial boilers, for instance, can switch among coal, oil, and natural gas relatively quickly.

The other major form of market intervention is a cap and trade system. As implied by its name, policymakers choose a cap or upper limit for the amount of carbon that can be emitted in a given period, most often a year. This cap is represented by an equal number of allowances. Individual allowances entitle the holder to emit a specified quantity of carbon. Allowances are then allocated to those included in the cap and trade system.

Allowances can be allocated using two general approaches. "Grandfathering" allocates allowances to emitters based on their emissions during some earlier reference period. Alternatively, allowances can be auctioned and sold to the highest bidder. These two mechanisms are not mutually exclusive—the U.S. cap and trade system for sulfur emissions made initial allocations using a combination of grandfathering and auctions. Once they are allocated, allowances can be bought or sold by participants.[4]

By issuing fewer allowances than the amount of current emissions, a cap and trade system forces individuals to make a simple decision: Should they reduce emissions and sell some allowances? Or should they buy allowances to cover all their emissions? The answer varies because the cost of reducing a unit of carbon emissions (by eliminating activities, using more energy-efficient machinery, or substituting among fuels) differs among individuals. If someone's cost of reducing emissions is less than the price of an allowance, the person will reduce emissions, sell a corresponding number of allowances, and make money by lowering emissions. If the cost of reducing emissions is greater than the price of an allowance, the person will buy allowances. Despite these purchases, the cost of compliance is typically less than a system in which all individuals are forced to lower emissions by a fixed amount or adopt a prescribed technology (known as a command and control strategy).

The price of allowances is determined by the market balance between buyers and sellers. When the market for permits is in equilibrium (the number offered for sale equals the number wanted by buyers), the market-clearing price will represent the marginal cost of abatement. In theory, the price for permits will be the same as the carbon tax rate that is required to reduce emissions by the amount specified by the cap. As such, carbon taxes are functionally equivalent to a cap and trade system, although the costs imposed by the latter are less visible, so a cap and trade system may be more politically palatable than a carbon tax.

But carbon taxes are not equivalent to a cap and trade system when it comes to real-world practicalities. One important difference concerns uncertainty. Because the price increase associated with carbon taxes is determined by the price per unit of carbon emitted, the increase is known at inception. What is not known is the quantity of emissions abated. This number is uncertain because economists have not quantified precisely how increases in energy prices affect energy demand—both directly via price elasticities and indirectly by altering rates of economic activity. These uncertainties are reversed by a cap and trade system. By printing a limited number of allowances, the quantity of abatement is known. Yet uncertainty about the cost of abatement and the effect of higher energy prices on economic activity makes it difficult to forecast the price of the allowances.

Recently policymakers have talked about reducing this uncertainty with a hybrid system. Under this scheme, cap and trade programs would have a "safety valve." When the market price for permits reaches a preestablished threshold, the government would sell an unlimited number of allowances at that price. This would in effect establish a backstop carbon tax—the price for allowances could not go beyond this maximum. But this hybrid approach would ease constraints on emissions and thereby introduce uncertainty about abatement.

The two systems also have different implications for inclusivity and costs. The number of consumers who will pay a carbon tax is likely to be greater than the number of consumers who buy and sell allowances in a cap and trade system. The efficiency of cap and trade depends in part on the technical sophistication of the participants and the cost of enforcement. Participants must be able to make economically rational decisions to buy or sell allowances and must have access to capital that would support economically rational investments to reduce emissions. These requirements imply that a cap and trade system is most likely to include only large energy consumers, such as manufacturers and electricity generators. A small subset of large emitters also reduces the costs of enforcement.

A cap and trade system is likely to have higher overhead costs. While carbon taxes can be implemented and collected through the existing infrastructure, cap and trade will require a monitoring system to ensure that those who emit carbon dioxide have the requisite number of allowances. To be efficient, there must be a market in which allowances can be bought and sold. Furthermore, firms will have to establish organizational structures that will allow them to participate in the program.

Photodisc

Will these emissions be "grandfathered"?

This last set of costs may increase the effectiveness of a cap and trade system. Organizational structures for participation in such a system will have a mandate to quantify emissions and reduce them where economically effective. A carbon tax is less likely to generate such structures, and most firms do not have structures specifically aimed at evaluating options for reducing energy use. The lack of such structures may be one reason why the academic literature describes vast opportunities to save energy and money even though many of those opportunities go unrealized in the real world.

Finally, the way in which allowances are introduced highlights equity issues. If allowances are distributed free to existing emitters, these individuals or entities retain a property right, the right to emit carbon, that becomes valuable once the cap and trade system is implemented. As such, those receiving the allowances benefit. That property right and economic gain is confiscated by the government if emitters are forced to purchase their initial allocation. In this case, the government captures the economic value of this new property right. The capture of this property right lies behind the debate about whether the additional revenues raised by a carbon tax should be returned to taxpayers by lowering some other tax.

Even if permits are given away for free, equity issues arise regarding how they are distributed. If allowances are distributed based on previous levels of emissions, existing patterns are merely reinforced. That is, heavily polluting industries and affluent consumers have an advantage over newer industries or poorer individuals. Conversely, heavily polluting industries and affluent consumers bear the brunt of efforts to reduce emissions if allowances are distributed more evenly.

The relative magnitude of these advantages and disadvantages differs at various geographical scales. Equity issues associated with the initial distribution of allowances probably preclude a global cap and trade system. In an auction, consumers from industrial nations would largely outbid consumers from developing nations, and this would make it difficult for developing nations to expand their energy use, which is closely correlated with economic development. Conversely, a distribution scheme based on population would force significant reductions in existing emissions (or wealth transfer) from industrial nations, which would reduce their economic well-being.

For individual nations, a cap and trade system may be the most effective means for meeting emission targets. Equity issues are reduced because participants are drawn from a single nation, which minimizes differences in income, technical sophistication, or governance. In many nations, the population of large emitters is sufficient to justify the fixed costs of trading allowances.

A carbon tax probably would be most effective at the subnational scale, where the market may be too thin to establish a cap and trade system. Instead, the low overhead and wide incidence of a carbon tax would be consistent with the popularity of efforts that would be needed to generate climate change policy at a local level.

Technology Transfer for Climate Change

K. Madhava Sarma and Durwood Zaelke

Climate governance at all levels, from local to international, must be designed to promote technology innovation and to ensure fast and fair dispersion of new technologies. The governments that signed on to the United Nations Framework Convention on Climate Change (UNFCCC) and the Kyoto Protocol recognize the vital importance of technology. They also recognize that industrial countries need to provide funding for climate-friendly technologies, which ultimately are the key to avoiding dangerous levels of climate change due to human activities.[1]

A technology needs assessment done by the UNFCCC identified many advanced technologies that can simultaneously reduce greenhouse gases (GHGs), increase profits, and create jobs. The benefits of these are particularly important for the developing world. Many of these technologies already exist and simply need to be deployed effectively. They include technologies:

- to use renewable energy sources;
- to improve energy efficiency in key sectors (cement, aluminum, steel, and other industries, along with transport, building, and consumer sectors);
- to recover or prevent methane emissions, fluorine gases, and other greenhouse gases beyond carbon dioxide (CO_2);
- to improve forest and soil management; and
- to adapt to climate change.

In addition, numerous technologies are currently in development, ranging from those in the research phase to those ready for demonstration and accelerated commercialization.[2]

Technology transfer has three key components: capital goods and equipment, skill and know-how for operating and maintaining equipment, and knowledge and expertise for generating and managing technological change. Since 1991 various institutions and funding mechanisms have successfully promoted technology transfer to developing countries—including the Global Environment Facility, the Special Climate Change Fund and the Least Developed Countries Fund of the UNFCCC, the Multilateral Fund (MLF) of the Montreal Protocol on Substances that Deplete the Ozone Layer, the World Bank, regional development banks, international partnerships, national development assistance programs, and nongovernmental organizations. The Clean Development Mechanism (CDM) of the Kyoto Protocol has also played a part.[3]

The Montreal Protocol's MLF is widely acclaimed as having been particularly successful at helping developing countries meet scheduled reduction targets for the 97 ozone-depleting substances that had to be phased out. These substances are also powerful greenhouse gases, so phasing them out will mitigate climate change by 11 billion

***K. Madhava Sarma** is the former Executive Secretary, Montreal Protocol Secretariat. **Durwood Zaelke** is President of the Institute for Governance & Sustainable Development.*

tons of CO_2-equivalent a year between 1990 and 2010. Because ozone-depleting substances are covered under the Montreal Protocol, they were not addressed in the Kyoto Protocol.[4]

Despite ongoing efforts since 1995, however, governments have been unable to agree on a technology transfer mechanism to address climate-specific emissions. In December 2007, at a Conference of the Parties to the UNFCCC, they laid out guidelines for revising the Kyoto Protocol in the Bali Action Plan, with the goal of finalizing agreement at a meeting in Copenhagen in December 2009. Several primary points of disagreement remain to be negotiated. (See Box.)[5]

Intellectual property rights (IPR) remain a contentious topic. For many years industrial countries have viewed IPR as essential for promoting innovation, while developing countries have viewed them as a hindrance to the transfer of critical technologies. In the last 20 years, however, many developing countries have achieved impressive economic progress and have been attracting foreign investment by creating a stable and enabling economic environment at the domestic level. These countries increasingly have access to technologies from anywhere in the globe. In fact, globalization favors developing-country production of many of the most advanced technology products, such as photovoltaic cells.

The 193 Parties to the Montreal Protocol found that most of the technologies needed to phase out the use of ozone-depleting substances were already in the public domain. Of the few technologies covered by IPR, most were owned by private businesses operating in a competitive market and eager to sell those rights on reasonable terms. In only two cases out of more than 4,000 projects were technologies held by companies that insisted on unreasonable conditions. In both cases developing-country enterprises came up with their own processes in order to avoid paying licensing fees or meeting conditions they considered unacceptable. The lesson repeated over and over again is that IPR did not present a barrier to immediate action under that regime.[6]

Nonetheless, cases may arise where useful IPR-protected technologies are owned by only a few companies, operating in a less competitive market, that refuse to sell the technology unless monopoly profits are paid. Other cases where a premium may be demanded might include sales to buyers in developing countries deemed to have extensive internal problems or in markets that are too small for substantial investment. In such cases, industrial countries may need to apply political pressure on their domestic industries to share the technologies. They may also need to relax their IPR regime, as they have already done for essential medicines under the Trade-Related Intellectual Property Rights regime. But the bottom line is that IPR may not be a difficult problem in the vast majority of cases, nor a barrier to immediate significant action.

Funding is another area where a few key changes could overcome the current stalemate. The various climate funds for technology transfer are all voluntary, which makes funding highly unpredictable. This is not acceptable for near-term mitigation programs needed to avoid passing tipping points for abrupt climate changes, nor is it acceptable for the medium-term or long-term and larger-scale programs required. The voluntary nature also gives effective control to the donors, whatever the governing structure.

Committed contributions are needed to create trust among developing countries. A formula for the percentage to be paid by each donor should be agreed upon in advance. The ratio of state contributions to the United

Nations is a time-tested model.

Funding needs to be committed for as long as it takes to reduce emissions and enhance sinks to avoid dangerous anthropogenic interference with the climate syestem. Because it often takes practical experience to gain an understanding of actual costs, funding commitments need to be reassessed every few years, as is done in the Montreal Protocol's MLF, which is replenished in three-year cycles. Similarly, governing bodies need to include equitable representation from all countries, both industrial and developing. A double majority, consisting of a majority of both donor and recipient countries, works well for the MLF and ensures transparency.[7]

The UNFCCC Secretariat estimates that more than $200 billion will be needed annually to bring about a 25-pecent reduction in global greenhouse gas emissions by 2030. Developing countries will need 35–40 percent of this total, or up to $80 billion a year. In addition, adapting to climate change may cost $35–60 billion a year for developing countries. While this amount may seem large, it represents only about 1.5 percent of total global investment projected for the year 2030.[8]

The actual amount needed for funding on climate change cannot be calculated, even in the short term, until it is clear which items will be paid for as grants and which will be concessional loans, loan guarantees, market borrowings, the stock market, the CDM, and so on. Once this is known, it will be necessary to develop an indicative list of agreed incremental costs, as was done by the MLF. When the final numbers are uncertain, it is prudent to start funding and then to later replenish with additional amounts needed to achieve environment goals once recipient institutions prove that investments are cost-effective.

Funds must not only be predictable; they must be adequate. What is adequate

Government Positions on Key Issues in Revision of Kyoto Protocol

On commitments by developing countries to reduce GHG emissions

- Developing countries prefer the status quo and do not want to make additional commitments beyond those that apply to all parties under Article 4.
- Industrial countries maintain that "major and emerging economies" should commit to goals.

On technologies

- Developing countries want industrial countries to assure transfer of technologies (and meet incremental transfer costs through a financial mechanism) and to relax intellectual property rights on privately owned technology if needed.
- Industrial countries want intellectual property protection to continue under the current Trade-Related Intellectual Property Rights regime. They maintain that an enabling environment in developing countries will attract technologies via markets.

On a funding mechanism

- Developing countries call for a new mechanism modeled on the Montreal Protocol's highly successful MLF (committed funding by industrial countries to meet all agreed incremental costs, including cost of technologies, capacity building, and conversion of facilities).
- Industrial countries prefer existing mechanisms, with voluntary funding to be enhanced.

On the governance structure

- Developing countries want assurance of equitable representation in governance.

depends on the climate goals of the beneficiaries and the specific national actions they intend to achieve them. Options include overall emissions-reduction goals, sector-specific goals, or mitigation activity goals. They can be long-term, medium-term, or short-term. The goals will have to be worked out by each country. Most developing countries are already taking up mitigation activities on their own because of the many strong co-benefits. However, they are reluctant to commit themselves—especially to "stretch goals"—unless they are sure that they will get the technologies and funding needed.

In addition, goals can be mandatory or voluntary. They can be a part of an amended Kyoto Protocol or can be included in a new agreement with a new funding mechanism. In the Montreal Protocol, many developing countries agreed to phase out ozone-depleting substances in advance of the prescribed schedule in return for access to additional MLF funding.

Country goals are essential for all developing nations, not just the major or emerging economies. Climate change cannot be solved without the participation of all countries, industrial and developing. While the "biggest bang for the buck" can be achieved if the focus is on major or emerging economies, the other developing countries are also investing in infrastructure for economic progress. If they are ignored, their emissions will reduce the achievements of others. Also, some energy-inefficient industries might migrate to countries that do not have tough regulations or other programs to mitigate climate change. Including every country in the mitigation activities is the best way to ensure a decrease in GHG emissions—and to guard against an increase.

But there may be some latitude for differentiation among developing-country groups. Industrial countries proposed that wealthier developing countries not receive assistance from any climate fund, even where they agree to accept mandatory targets. This approach was argued during the Montreal Protocol negotiations as well. That treaty ended up prescribing that developing countries with per capita consumption of ozone-depleting substances greater than the prescribed limits be treated on a par with industrial countries. A similar arrangement may be appropriate for addressing climate change.[9]

An upside of all these considerations for implementing technology transfer is that developing countries can start with modest initial investment goals and then strengthen these over time. With more confidence in the system, recipient countries will come forward with stiffer goals and donors may in turn provide more funds. Rather than let the quest for the perfect become the enemy of the effective, it will be better to implement a system capable of progress in the near term and of being strengthened over time. This has been the Montreal Protocol's successful approach: "start and strengthen." Hence the UNFCCC should agree immediately on goals and committed funding and then start negotiating on the few remaining contentious details so that governments can arrive at a functional solution in 2009. Overall, the path forward should aim to start now where consensus exists and then strengthen investments in technology transfer as further agreement is reached.

Electric Vehicles and Renewable Energy Potential

Jeffrey Harti

Skyrocketing energy prices and concerns about energy security and climate change have sparked interest in alternatives to transport systems run on fossil fuels. In recent years, electric vehicles have emerged as the preferred alternative thanks to the environmental benefits of zero tailpipe emissions and the vehicles' ability to take power directly from the power grid. (See Table.)

In the same way that a cell phone or computer runs on its battery when someone is on the go and is then plugged in when the person is at work or at home, an electric vehicle will run on battery power when someone is driving and can be plugged in at night or while the owner is at work to recharge, drawing about as much power from the grid as a dishwasher would. The main change will be no more trips to the gas station and extra money in drivers' pockets. At current U.S. electricity prices, running an electric vehicle would cost the equivalent of 75¢ a gallon (20¢ a liter).[1]

This ability to "fuel" electric vehicles directly from the existing power grid raises questions about how much additional generation capacity would be required to meet increased demand for power if there were widespread use of electric vehicles. In particular, people promoting the transition to an energy system based on renewable sources are concerned about whether such forms of energy can meet growing power as well as transportation needs.

Fortunately, there is more than enough renewable energy to do both. For instance, concentrating solar power plants built on less than 0.3 percent of the deserts of North Africa and the Middle East could generate sufficient energy to meet the local needs of these regions as well as the electricity needs of the entire European Union. One effort to tap this potential is the DESERTEC Concept, which envisions 100,000 megawatts (MW) of concentrating solar power plants being developed in North Africa and the Middle East by 2050 and transported via underwater cables across the Mediterranean to Europe. Algeria already has plans to build a 3,000-kilometer cable to Germany, allowing it to export 6,000 MW of solar thermal power by 2020 and providing a perfect complement to Germany's significant wind energy capacity.[2]

The renewable energy potential in the United States is similarly vast, and only a small fraction of it would need to be harnessed to electrify the current fleet of cars and light trucks (which account for more than a third of the total world vehicle fleet). If half of the light vehicle fleet were plug-in hybrid electrics, a transformation that is likely to require several decades, an increase in U.S. wind energy capacity of roughly 105,000 MW would be needed to run the vehicles. This is equivalent to total

Jeffrey Harti is an MSc candidate in the Lund University International Masters of Environmental Studies and Sustainability Science program in Sweden.

global wind capacity today and could be supplied by the addition of 63,000 5-MW wind turbines. (These vehicles would still rely on gasoline for longer trips, depending on individual driving and charging behavior.) A 2008 U.S. Department of Energy report concluded that wind power could supply 20 percent of U.S. electricity by 2030, or more than 300,000 MW—which is almost three times as much electricity as is used to run half the country's light vehicles today. And—unlike oil—the sun, wind, water, and biomass needed to produce renewable energy are available throughout the world.[3]

As an additional benefit, existing manufacturing facilities (some of which are now idle) could be converted to produce wind turbines, creating thousands of green-collar jobs. China provides an example of a booming wind manufacturing industry, which is set to produce 11,000 MW of turbine capacity (or more than 7,000 1.5-MW wind turbines) in 2008. And Chinese companies are beginning to export their turbines as well.[4]

Are there benefits to using renewable energy to power electric vehicles in addition to the cleaner air and reduced acid rain–forming emissions that would result? The government of Denmark thinks so and is now looking to lead in the move toward electric vehicles. Wind accounted for 21 percent of Danish power production in 2007. Since wind speeds are higher at night, when electricity demand is lower, Denmark has looked for ways to offload its excess generation. Electric vehicles could help here, as most of them will be charged at night and will be able to make use of excess wind energy. As Torben Holm of Danish Oil and Gas has noted, "by charging the cars at night Denmark will be able to use wind energy that otherwise would have to be exported to neighboring countries, typically at relatively low prices. Moreover...that energy can be sent back to the grid during peak hours."[5]

Working with the Danish government to make this vision a reality is an Israeli-American start-up company called Better Place. It plans to provide electric vehicles and countrywide electric recharge grids in both Israel and Denmark beginning in 2009 and expects to have 100,000 vehicles in place in each country by the end of 2010. The electric recharge grid would include recharging points for drivers but also battery exchange stations where they could swap depleted lithium-ion batteries for fully charged ones for trips of longer than 120 miles (200 kilometers), the projected range of these vehicles, thus eliminating concerns about the limited range of full electric vehicles. The cars in Israel and Denmark will rely on solar and wind power respectively, providing truly clean transportation. Better Place is currently discussing similar projects with 25 countries.[6]

Key to the implementation of this plan is the development of vehicle-to-grid technology. Switching electric vehicles to this system would create a large, flexible, distributed power generation, storage, and transmission network, eliminating the need for other means of energy storage as well as reducing the need for new transmission and distribution infrastructure.[7]

Google is already putting this approach into practice at its California headquarters, using a fleet of six plug-in hybrid electric vehicles and a 1.6-MW solar array that currently feeds power into the California grid. Google plans to eventually include 100 plug-in hybrid electric vehicles in its fleet and to install solar charging stations so that these vehicles can be charged from Google's solar array. The company has conducted a successful demonstration of vehicle-to-grid technology in cooperation with the utility company Pacific Gas & Electric. These vehi-

Electric Vehicle Technologies	
Technology	Description
Hybrid electrics	Have an electric motor as well as an internal combustion engine; not able to plug in to recharge
Full electrics	Solely battery-powered; recharged by plugging into main electricity source
Plug-in hybrids	Combine batteries, which allow them to run in electric mode, with smaller internal combustion engines (which use gasoline or other liquid fuels) to power the vehicle on longer trips when battery power has been depleted; recharged by plugging into main electricity source
Vehicle-to-grid	A form of smart grid technology that allows utilities to communicate with vehicles as they would larger power plants, with energy able to flow in both directions and the vehicle owner debited or credited for energy taken from or provided to the grid

cles will be able to store energy produced from Google's solar array for later use or to supply additional power to Google or the grid during peak demand hours. After their first year on the road, the cars in the Google fleet are getting an average of 93.5 miles per gallon of gasoline.[8]

Building on this interest in electric vehicles, major automakers such as General Motors and Nissan-Renault have plans to market full electric vehicles or plug-in hybrid electrics within the next two years. Smaller automakers such as REVA of India and BYD Auto of China are also getting in on the game, with plans to have electric vehicles on the market both at home and in Europe and North America in the next two years as well. The electric range of these vehicles should be 40–100 miles (about 65–160 kilometers), with prices that are competitive with today's mid-range vehicles.[9]

Plug-in hybrid electrics provide an alternative that eliminates concerns related to the limited range of fully electric vehicles. This flexibility does come with increased emissions, however, although they are still far below those produced by cars run just on gasoline. The relatively lower emissions are due to the fact that 71 percent of cars in the United States are driven for at most 40 miles a day in weekday travel, a distance that could be driven entirely in electric mode in most plug-in hybrid electrics (assuming that the vehicles are recharged every night for the next day's driving), with gasoline only being used by people driving beyond the vehicle's electric range each day.[10]

In many countries, powering vehicles with the existing electricity mix would mean that much of the energy would come from coal. But even in the worst-case scenario of using 100 percent coal-fired electricity, the carbon dioxide emissions associated with electric vehicles are expected to decline—up to 50 percent by some estimates—relative to those of gasoline and diesel-fueled vehicles because electric motors are three to four times more efficient.[11]

The chief barrier to wider use of electric vehicles is the lack of political will. To date, the private sector has driven the transition, but governments can greatly accelerate it through preferential taxation of electric vehicles, higher taxes on gasoline and diesel, and public incentives and research to advance technologies and infrastructure.

Another barrier to deployment is current battery technology. The nickel–metal hydride

Better Place

The Better Place electric car prototype

batteries installed in most current hybrid-electric vehicles are limited in both power and energy storage density. But the development of lithium-ion batteries promises significant improvements. These are becoming the industry standard due to their higher energy and power densities and lower cost, and most automakers are banking on them to provide the necessary extension in driving range for the next generation of electric cars and plug-in hybrids.[12]

An infrastructure needs to be created to ensure that drivers can easily charge their vehicles wherever they happen to be. Public and private investment is also required to develop the "smart" grid technologies needed to ensure that all new plug-in hybrid electric and fully electric vehicles are vehicle-to-grid capable and that electricity grids can handle the demands of these vehicles. As hundreds of thousands of electric vehicles hit the roads, their charging patterns will have to be actively managed to reconcile the needs of drivers and grid operators.[13]

Electric vehicles have a great potential to help reduce the impact of transportation systems on the environment. But it is important to remember that their adoption must be seen as a part of a larger process of moving toward a more sustainable transportation system. A holistic approach to future transport must include not only electric vehicles but also the promotion of better urban design, walking, cycling, and carpooling as well as increased investment in public transit and rail—programs that will lead to less driving and a more efficient transit system. And that will help the world meet sustainable mobility goals with renewably generated energy sooner.

Employment in a Low-Carbon World

Michael Renner, Sean Sweeney, and Jill Kubit

As climate action grows urgent, some observers warn that economies will suffer as a result. But economic prosperity and employment depend in fundamental ways on a stable climate and healthy ecosystems. Without timely action, many jobs could be lost due to resource depletion, biodiversity loss, the impacts of increasing natural disasters, and other disruptions. Meanwhile, employment that actually contributes to protecting the environment and reducing humanity's carbon footprint offers people a tangible stake in a green economy.

The pursuit of so-called green jobs will be a key economic driver as the world steps into the uncharted territory of building a low-carbon global economy. "Climate-proofing" the economy will involve large-scale investments in new technologies, equipment, buildings, and infrastructure, which will provide a major stimulus for much-needed new employment and an opportunity for retaining and transforming existing jobs.[1]

The number of green jobs is already on the rise. Most visible are those in the renewable energy sector, which has seen rapid expansion in recent years. Current employment in renewables and supplier industries stands at a conservatively estimated 2.3 million worldwide. The wind power industry employs some 300,000 people; the solar photovoltaics (PV) sector, an estimated 170,000; and the solar thermal industry, more than 600,000 (this relatively high figure is due to low labor productivity in China, the leading producer of solar thermal systems). More than 1 million jobs are found in the biofuels industry—growing and processing a variety of feedstocks into ethanol and biodiesel.[2]

Some industrial regions that have become Rust Belts, such as parts of the U.S. Midwest or Germany's Ruhr Valley, are gaining new vitality from wind and solar development. Rural communities receive additional income when farmers place wind turbines on their land. Installing, operating, and servicing renewable energy systems provides additional jobs; local by definition, these are resistant to outsourcing. In Bangladesh, the spread of solar home systems—which might reach 1 million by 2015—could eventually create some 100,000 jobs.[3]

Wind and solar are poised for continued rapid expansion. Under favorable investment projections, wind power employment worldwide could reach 2.1 million in 2030, and the solar PV industry might employ as many as 6.3 million people by then. Although renewables are more labor-intensive than the fossil fuels they replace, the energy sector does not account for a very large portion of employment. Many more green jobs will eventually be created through the pursuit of more-efficient machinery and appliances. Energy performance services are already a

***Michael Renner** is a Senior Researcher at the Worldwatch Institute. **Sean Sweeney** and **Jill Kubit** are Director and Assistant Director, respectively, of Cornell University's Global Labor Institute in New York.*

growing phenomenon.[4]

Construction jobs can be greened by ensuring that new buildings meet high performance standards. This is particularly important in Asia, which is undergoing a construction boom. And retrofitting commercial and residential buildings to make them more energy-efficient has huge job potential for construction workers, architects, energy auditors, engineers, and others. For instance, the weatherization of some 200,000 apartments in Germany created 25,000 new jobs and saved 116,000 existing jobs in 2002–04 at a time when the construction industry faced recession. Providing decent and efficient housing in the developing world's urban agglomerations and slums presents an unparalleled job creation opportunity—if the necessary resources are mobilized.[5]

The transportation industry is a cornerstone of modern economies, but it also has the fastest-rising carbon emissions of any sector. Incorporating the very best in fuel efficiency technology would dramatically lessen the environmental footprint of motor vehicles. An assessment of the most-efficient cars currently available suggests that relatively green auto manufacturing jobs may today number no more than about 250,000 out of roughly 8 million direct jobs in the auto sector worldwide. But a concerted push toward much greater efficiency and carbon-free propulsion systems is needed. Likewise, retrofitting highly polluting two-stroke engines that are ubiquitous, especially in Asia, to cut their fuel consumption and emissions would create many jobs.[6]

Overall, the reliance on cars and trucks needs to be reduced. Railways offer an alternative, yet many jobs have been lost over the last few decades as rail has been sidelined. In Europe, railway manufacturing and operating employment is down to about 1 million. Even in China and India, rail jobs fell from 5.1 million to 3.3 million from 1992 to 2002. A recommitment to rail, as well as to urban public transit, could create many millions of jobs. In growing numbers of cities, good jobs are being generated by the emergence of bus rapid transit systems. There are also substantial green employment opportunities in retrofitting old diesel buses to reduce air pollutants and in replacing old equipment with cleaner compressed natural gas (CNG) or hybrid-electric buses. In New Delhi, the introduction of 6,100 CNG buses by 2009 is expected to create 18,000 new jobs. More-affordable and nonpolluting transportation networks also give poor people in developing-country cities better access to job opportunities.[7]

Basic industries like steel, aluminum, cement, and paper may never be truly "green," as they are highly energy-intensive and polluting. But increasing scrap use, greater energy efficiency, and reliance on alternative energy sources may at least render them a pale shade of green. Secondary scrap-based steel production requires up to 75 percent less energy than primary production. Worldwide, 42 percent of steel output was based on scrap in 2006, possibly employing more than 200,000 people. Likewise, secondary aluminium production uses only 5–10 percent as much energy as primary production. About one quarter of global aluminum production is scrap-based. No global employment numbers exist for this, but in the United States, Japan, and Europe, it involves at least 30,000 jobs. The cement and the paper and pulp industries have similar greening potential, but like the aluminum industry they are relatively small employers.[8]

The number of recycling and remanufacturing jobs worldwide is another unknown. In the United States, these are estimated at more than 1 million. With higher rates of recycling, Western Europe and Japan can be assumed to have greater employment in this sector. In developing countries, paper recycling is often done by an informal network of scrap collectors, sometimes organized into cooperatives in order to improve pay and working conditions. Jobs and livelihoods in informal communal recycling efforts are difficult to document; in Cairo, the Zabbaleen have received considerable international attention. Believed to number some 70,000, they recycle an estimated 85 percent of the materials they collect. Brazil is thought to have some 500,000 recycling jobs. China, with estimates as high as 10 million jobs, trumps all other countries in this area.[9]

For many developing countries, a key concern is the future of agriculture and forestry, which often still account for the bulk of employment and livelihoods. Small farms are more labor- and knowledge-intensive than agroindustrial farms, and they use less energy and chemical inputs. But relatively sustainable forms of smallholder agriculture are being squeezed hard by energy- and pesticide-intensive plantation and specialized crops, trade liberalization, and the power of global supply and retail chains. Organic farming is still limited, although expanding. More labor-intensive than industrialized agriculture, this can be a source of additional green employment in the future. A study in the United Kingdom and Ireland showed that organic farms employed one third more full-time equivalent workers than conventional farms do.[10]

Afforestation and reforestation efforts, as well as generally better stewardship of critical ecosystems, could support livelihoods among the more than 1 billion people who depend on forests, often through non-timber forest products. Planting trees creates large numbers of jobs, although these are often seasonal and low paid. Agroforestry—which combines tree planting with traditional farming—offers significant environmental benefits in degraded areas, including carbon sequestration. It has been shown to provide food and fuel security and to create employment and supplementary income for small farmers. Some 1.2 billion people already depend on agroforestry to some extent.[11]

NREL

Installing solar panels on a roof in Atlanta, Georgia

There is additional job potential in dealing with the accumulated environmental ills of the past and improving the ability to cope with the climate change that is already inevitable. The building of much-needed adaptation infrastructure, such as flood barriers, to protect communities from extreme weather events has barely started but presumably would employ large numbers of people, even if only temporarily. Activities such as terracing land or rehabilitating wetlands and coastal forests are labor-intensive. Efforts to protect croplands from environmental degradation and to adapt farming to climate change by raising water efficiency, preventing erosion, planting trees, using conservation tillage, and rehabilitating degraded crop and pastureland can also support rural livelihoods.

So the potential for green jobs is immense. But much of it will not materialize without massive and sustained investments in the public and private sectors. Research and development programs need to shift decisively toward clean technology, energy and materials efficiency, and sustainable workplace practices, as well as toward environmental restoration and climate adaptation.

Governments need to establish a firm and predictable policy framework for greening all aspects of the economy, with the help of targets and mandates, business incentives, and reformed tax and subsidy policies. It will also be critical to develop innovative forms of technology transfer to spread green methods around the world at the scale and speed required to avoid full-fledged climate change. Cooperative technology development and technology-sharing programs could help expedite the process of replicating best practices.

To provide as many workers as possible with the qualifications they will increasingly need, an expansion of green education, training, and skill-building programs in a broad range of occupations is crucial. Some jobs involve sophisticated scientific and technical skills. But green job development also needs to offer opportunities for the broad mass of workers, including those who have too often found themselves in underprivileged situations.

The transition to a low-carbon future will involve major shifts in employment patterns and skill profiles. Resource extraction and energy-intensive industries are likely to feel the greatest impact, and regions and communities highly dependent on them may face serious consequences. They will need proactive assistance in diversifying their economic base, creating alternative jobs and livelihoods, and acquiring new skills. Today, such a "just transition" remains a theoretical notion.

Green jobs need to be decent jobs—offering good wages and income security, safe working conditions, dignity at work, and adequate workers' rights. Sadly, this is not always the case today. Recycling work is sometimes precarious, involving serious occupational health hazards and often generating less than living wages and incomes, as is the case for 700,000 workers in electronics recycling in China. Growing crops for biofuels at sugarcane and palm oil plantations in countries like Brazil, Colombia, Malaysia, and Indonesia often involves excessive workloads, poor pay, exposure to pesticides, and oppression of workers. Also, the expansion of biofuels plantations has driven people off their land in some cases, thus undermining rural livelihoods.[12]

These cautionary aspects highlight the need for sustainable employment to be good not only for the environment but also for the people holding the jobs. Still, an economy that reconciles human aspirations with the planet's limits is eminently possible.

Climate Justice Movements Gather Strength

Ambika Chawla

In Operation Climate Change, members of a Nigerian indigenous peoples' rights movement attempt to shut down oil flow stations in the Niger Delta. Ecuadorian environmental activists risk their lives to protest construction of an oil pipeline through the Amazonian forest that is home to the native Quichua, Shuar, and Achuar people. More than 200 farmers from 20 countries march in Bali, Indonesia, during a meeting on the U.N. climate convention to demand that food sovereignty be addressed by negotiators.[1]

These are just three of the many grassroots initiatives worldwide that are seeking to mobilize governments and the public to tackle climate change. In doing so, most groups argue that human rights include people's rights to a clean environment and access to critical natural resources. Many of them advocate the participation of marginalized communities in the U.N. climate negotiation process. Although their specific agendas differ, these groups are part of an emerging global movement for climate justice—an intricate web of grassroots initiatives from diverse regions calling for attention to the inequities inherent in climate change and the need to consider human rights when addressing this pressing global issue.

These grassroots groups tend to be self-organized and oriented toward visible action and advocacy rather than research. They typically define themselves as "economically marginalized" peoples, "disadvantaged," or "poor." The livelihoods of many members depend extensively on climate-sensitive sectors for their survival, such as farming, forestry, and fisheries. Others are union members seeking alternative employment opportunities within a growing green economy or young people concerned about their future.

Local struggles for climate justice connect at the international level with a shared understanding that in addition to accelerating environmental degradation and species loss, global climate change will jeopardize human rights and exacerbate socioeconomic inequities. According to a recent report by the U.N. Development Programme, climate change is "intensifying the risks and vulnerabilities facing poor people, placing further stress on already over-stretched coping mechanisms."[2]

The first-ever Climate Justice Summit took place in The Hague, Netherlands, in November 2000 at the same time as the Sixth Conference of the Parties to the U.N. Framework Convention on Climate Change (UNFCCC). More than 500 grassroots leaders from Asia, Africa, Latin America, and North America gathered to build bridges across borders and thematic issues. Regional and international networks quickly merged, building the foundation of a global grassroots movement to tackle climate change.[3]

Members of international coalitions such as the Indigenous Environmental Network,

Ambika Chawla *is the 2009 State of the World Fellow and Project Coordinator for this year's book.*

the World Rainforest Movement, Oilwatch International, and Friends of the Earth International joined to craft the climate justice movement's initial guiding principles and to organize parallel events to the official meeting, such as cultural activities and mass mobilizations. "We affirm that climate change is a rights issue. It affects our livelihoods, our health, our children and our natural resources. We will build alliances across states and borders to oppose climate change inducing patterns and advocate for and practice sustainable development," proclaimed the summit's action statement.[4]

Orin Langelle

Representatives of indigenous peoples protest their exclusion from the UNFCCC meeting in Bali.

In addition, individual international coalitions presented statements on specific issues of concern. Indigenous groups had collaborated on the Declaration of the First International Forum of Indigenous Peoples on Climate Change, calling for the creation of an adaptation fund with financing allocated for indigenous groups and the inclusion of indigenous peoples in all levels of decision-making in the UNFCCC process. The World Rainforest Movement presented the Mount Tamalpais Declaration, demanding deep greenhouse gas emissions cuts and an end to the inclusion of tree plantations as "carbon sinks" within the Clean Development Mechanism established in the Kyoto Protocol. According to the declaration, "licensing the burning of fossil fuels by financing tree plantations to 'absorb' carbon dioxide would expand the ecological and social footprint of the rich, making existing social inequalities worse." As an alternative strategy, the declaration recommended that local communities manage forest ecosystems.[5]

The global climate justice movement has since grown and evolved as a widening circle of civil society groups worldwide integrate climate protection objectives into their strategic agendas. At the Thirteenth Conference of the Parties to the UNFCCC in Bali in December 2007, a diverse spectrum of social movement groups held street demonstrations, press conferences, and educational side events. An Asian Young Leaders Climate Forum brought together 35 young people from 14 nations who developed a regional climate action plan that was later presented to the official conference.[6]

Farmers from around the world filled the streets with bright and colorful banners, calling for small-scale, sustainable agriculture as an alternative to industrial farming. "Sustainable agriculture will cool the earth!" they cried. Oilwatch International activists demanded the redirection of financing from fossil fuels to emissions mitigation and clean renewable energy technologies. Numerous groups at the meeting decided to form the Climate Justice Now! coalition, demanding that industrial nations implement drastic emissions, increase financing to support adaptation programs in the developing world, and support rights-based conservation programs that promote community control over energy, forests, and water.[7]

Assessing the impact of climate justice movements on domestic and international climate governance can be a challenge, as these movements tend to participate outside of climate conventions, have no voting authority within official negotiations, and often use international conferences as an opportunity to strengthen their agendas through networking and alliance building. The movement's overarching principles—climate equity, inclusive participation, and human rights—play a limited role in the arena of global policymaking on climate change. The UNFCCC, for example, makes no mention of human rights.[8]

But those key principles are beginning to emerge in the activities of nongovernmental organizations (NGOs), some branches of the United Nations, and some intergovernmental organizations. International humanitarian organizations such as ActionAid, Christian Aid, Oxfam, and Tearfund have developed climate campaigns based on equity and human rights, often acting as a bridge between underrepresented communities and official policymakers. The United Nations is increasingly integrating the concerns of marginalized groups such as indigenous peoples into its program work on climate change. According to a recent U.N. report, "the proposals of indigenous communities to integrate their social, political, cultural, and economic rights and their aspirations into future development strategies must be considered so that the challenges they are facing are fully addressed, respect for their rights and cultures is ensured, and their survival and well-being is protected." And in mid-2008 the World Bank initiated a program on human rights and climate change, with a focus on developing policies and procedures that build resilience to climate change and reduce vulnerability by using a rights-based approach.[9]

Moreover, climate justice movements collaborate closely with NGOs that in turn incorporate the movement's principles into proposals they submit to the UNFCCC Secretariat. Tearfund, for example, has submitted a proposal on disaster risk reduction that focuses on community participation within the context of adaptation planning. The Global Forest Coalition has presented a proposal addressing the need to involve indigenous peoples in policymaking programs to reduce emissions from deforestation in developing countries. And the Climate Action Network has put in a proposal calling for governments to initiate a collaborative dialogue on how adaptation of the most vulnerable groups of a population can be effectively supported. Thus although members of marginalized communities may not join in international climate negotiations at the official level, they have succeeded in making their voices heard by influencing more-established NGOs that work closely with the negotiators.[10]

Shifting Values in Response to Climate Change

Tim Kasser

Tamsen Butler was living the busy life of a mother of two, a college student, and a freelance writer when her 15-month-old son could no longer breathe properly. She carried him into the ambulance, clutching "my son in one arm while I used my other arm to balance my laptop bag." After a couple of nights in a Nebraska hospital tending to her son and staying up late trying to meet writing deadlines, she had an epiphany: "My son was in the hospital and I was a fool." Rather than working while her son slept, Butler realized she should have been resting. Rather than "clutching my son with only one arm I should have had both arms wrapped around him." When her son recovered, Butler and her family began spending less time rushing from here to there and reorganized their lives around their health and their time together. They also gave to charity the many extra toys and clothes they had accumulated.[1]

J. Eva Nagel awoke one autumn night in upstate New York to find her house was burning. The fire moved slowly enough that she got her children, pets, and photo albums out, but she watched as her clothes, her books, and her dissertation notes were destroyed. Nagel eventually came to see the fire not as a tragedy but rather as "a wake-up call." Now, she writes, "Our priorities...are as clear as a crisp autumn evening": her family, her health, the pursuit of joy, and giving back to the community.[2]

These are true stories, but they are also metaphors for the situation facing humanity. The world is full of busy people whose lives are jam-packed with appointments and possessions. The Earth is ill and, although not on fire, it is warming at a dangerous rate. As these problems worsen, humanity is faced with a choice: Continue with life as usual, like Butler first did during her son's hospitalization, or "wake up," realize that only "fools" persist in a damaging lifestyle, and use the environmental threats humanity faces as an opportunity to shift priorities and values.

The scientific evidence is quite clear about the environmental dangers of continuing to focus on the values and goals so prominent in today's hyperkinetic, consumeristic, profit-driven culture. A growing body of research shows that the more people value money, image, status, and personal achievement, the less they care about other living species and the less likely they are to recycle, to turn off lights in unused rooms, and to walk or bicycle to work. One study of 400 American adults showed that the more people pursue these extrinsic, materialistic goals, the higher were their "ecological footprints." And when researchers have asked people to pretend to run a timber company and bid to harvest trees from a state forest, those who care more about money, image,

Tim Kasser, Ph.D., is Associate Professor of Psychology at Knox College in Illinois and a member of the Environmental Protection and Justice Action Committee of Psychologists for Social Responsibility.

and status act more greedily and cut trees down at less sustainable rates.[3]

In psychological parlance, life challenges spurred Tamsen Butler and Eva Nagel to care less about such materialistic aims and instead focus more on "intrinsic" values and goals. Intrinsic goals are those focused on self-acceptance (personal growth and pursuing an individual's own interests), affiliation (close relationships with family and friends), physical health (fitness), and community feeling (contributing to the broader world).[4]

Such shifts toward intrinsic values after people experience a very stressful life event are well documented in the psychological literature on "post-traumatic growth." Sometimes traumatic events (including brushes with death) jar people loose from their typical ways of living and the standard goals they thought were important. As they struggle to understand and assimilate these traumatic events, many people reject materialistic, self-enhancing values and goals and instead express a newfound appreciation for family and friends, for helping others, and for personal growth.[5]

Two recent sets of experiments have even documented that "virtual" death experiences can help people shift, at least temporarily, away from extrinsic and toward intrinsic values. In one study, people scoring high in materialism who were asked to deeply imagine their own death and reflect on the meaning it held for their life later behaved in a more generous, less greedy fashion than did materialistic individuals who thought about neutral topics. In another experiment, sustained reflection on their own death over six days helped intrinsically oriented people maintain intrinsic values, while daily reminders of death helped more materialistic people become more intrinsically oriented.[6]

It is crucial not to underestimate the importance of this shift toward intrinsic values as a way of helping humans avert ecological catastrophe. For just as scientific research has documented that materialistic, self-enhancing values contribute to climate change, the pursuit of intrinsic values has been empirically associated with more sustainable and climate-friendly ecological activities. What's more, to ensure that ecological damage is not borne primarily by the most vulnerable (whether that be poor people, other species, or future generations), a shift toward intrinsic values will again be beneficial, as such aims promote more empathy and higher levels of pro-social and cooperative behavior. And, in a happy convergence, a shift toward intrinsic values may also benefit humanity's well-being: whereas dozens of studies show that materialistic, self-enhancing goals are associated with lower life satisfaction and happiness, as well as higher depression and anxiety, intrinsic values and goals promote greater personal well-being.[7]

But here is the rub. While Butler and Nagel were both able to use their life challenges to reorient their lives, and while some people do grow out of traumas, this does not always occur. Stress, trauma, and fear often lead people to treat themselves, others, and the environment in more damaging ways. Experiments show that when people are led to think only superficially (instead of deeply) about their own death, they become more defensive, more focused on consumption and acquisition, more greedy, and more negative in their attitudes toward wilderness. Similarly, studies show that economically difficult times often increase people's levels of materialism and decrease their concern for the environment and for other people.[8]

Thus, there are both potentially very scary and very hopeful outcomes of the looming climate crisis. On the one hand, humanity might respond in a defensive fashion, becoming increasingly fearful and insecure as the climate changes, as species go extinct, and as Earth's resources become scarcer. If this happens, psychological—and, indeed, international—forces are likely to perpetuate the very materialistic values that have contributed to current environmental and social challenges. On the other hand, the present climate crisis could be the "wake-up call" necessary to help humans realize how foolish they have been to fixate on material progress and personal achievement to the detriment of Earth, civil society, and human well-being.

Caroline Hoos

Extrinsic house of worship? Shopping mall in Hamburg, Germany

Butler and Nagel had little time to prepare for the crises they faced, but scientists have given humanity substantial forewarning about the ecological challenges ahead. Fortunately, there is still a window of opportunity to change lifestyles and societal practices so as to lessen the coming damage. There is much that can be done right now to promote such a shift in values.

First, it is important to recognize that advertising messages almost always activate and encourage the materialistic values known to undermine environmental sustainability. Rather than allowing young children to be exposed to such messages and encouraged to develop such values, some Scandinavian nations have banned advertising to children, thereby helping lessen their materialistic values. Other countries need to follow this precedent. And rather than allowing corporations to deduct the costs of marketing and advertising, the government could tax the tens of billions of dollars spent each year inculcating materialistic values and use that revenue to promote intrinsic values.[9]

American families also need help to reorient their lives away from the pursuit of material affluence and toward the pursuit of "time affluence." Research shows that people who work fewer hours per week are more likely to be pursuing intrinsic goals, are happier, and are living in more sustainable ways. What's more, a recent cross-national study concluded that "If, by 2050, the world works as many hours as do Americans it could consume 15–30 percent more energy than it would following Europe. The additional carbon emissions could result in 1 to 2 degrees Celsius in extra global warming." Rather than maintaining practices and policies that promote time poverty, time affluence can be enhanced by passing laws mandating that Americans be given paid vacations and family leave (which is the case in most every other nation in the world, rich or poor). And the number of holidays per year can be increased while the number of hours worked per week can be decreased so that people commute less and have more time to live in sustainable ways.[10]

It is also important to recognize that seeking economic growth above all else is just another way that materialistic values dominate intrinsic pursuits. Rather than allowing a flawed measure like gross domestic product to direct national policy, new indicators such as the Happy Planet Index can be used that incorporate values like people's well-being and environmental sustainability. And rather than focusing on green consumption and the business case for sustainability, environmental organizations can stop capitulating to materialistic values and instead argue for the reduced levels of consumption that most know are necessary to avoid massive climate change.[11]

And if these and other efforts are too little or too late to avert climate change, leaders from every arena will need to help people experience and interpret the coming ecological challenges in ways that maximize the likelihood that humanity will grow from them rather than succumb to and worsen them. This will be an enormous challenge, for facilitating growth in the face of trauma entails a tricky balance of helping people to acknowledge, process, and accept the disturbing realities around them while at the same time seeing these realities as opportunities to create a new and better life. It will require leaders who can help people develop a new set of beliefs and meanings, a fundamentally new narrative of what it means to be a civilized human. Ultimately, this will have to be a narrative that promotes growing as people, loving each other, and transcending self-interest to benefit the poor in flooding countries, the species on the verge of extinction, and future generations of humans rather than the current dominant narrative that is obsessed with acquisition, self-enhancement, and profit.[12]

Not Too Late to Act

Betsy Taylor

Scientific reports on how quickly climate change is proceeding have divided climate experts on the issue of how far the delicate global ecological balance has tipped. While virtually all experts agree that the situation is precarious, a growing number of influential leaders are quietly whispering that we are already too late to avoid cataclysmic change. Don't believe them.

When you think things seem impossible, consider migrating birds. By some miraculous combination of genetic coding and sheer determination, warblers, waterfowl, hummingbirds, and hawks travel thousands of miles each year, despite increasingly scarce habitat and food, to mate and nest. Some birds weigh less than an ounce yet travel at high altitudes from hillsides in Canada to mountains in the Dominican Republic. *Homo sapiens* can be equally amazing. We consciously sacrifice ourselves to protect others. We do extraordinary things to ensure that life flourishes. The drive for a safe future should never be underestimated.

Earth is resilient, and humans have a remarkable capacity to overcome tough odds in the quest for survival, freedom, and justice. We have our stories—Nelson Mandela, Vaclav Havel, Wangari Maathai, Mahatma Gandhi, Mother Teresa, the Berlin Wall, and the journey to the moon among the most inspiring. Now a worldwide movement inspired by the prospect of climate change is refusing to accept the traditional pattern of incremental change. Bold actors in a variety of fields and professions are rapidly emerging on every continent and in every sector of human society.

Author and academic Michael Pollan, for example, envisions a radical restructuring of the global food system that rapidly eliminates the need for fossil-fuel fertilizers and minimizes global transport of most foods. Stephen Heinz of Rockefeller Brothers Fund, Jules Kortenhorst of the European Climate Foundation, and Uday Harsh Khemka of the Nand & Jeet Khemka Foundation together are harnessing the power of philanthropy on a global scale to support energy innovations in China, India, Europe, and the United States. Local officials like Mayor Marcelo Ebrard of Mexico City, former school teacher and now California legislator Fran Pavley, and Mayor Michael Bloomberg of New York City are changing transportation, land use, and energy policies to reduce carbon emissions.[1]

The case for bold action is indisputable. The roadmap into the future is becoming increasingly clear.

Imagine it is 2025 and you are speaking with a young person, describing the role you played in helping your school, community, or country turn away from fossil fuel consumption and catastrophic climate change

***Betsy Taylor** is a consultant on climate solutions and founder of the 1Sky Campaign and the Center for a New American Dream based in Takoma Park, Maryland.*

and toward a more just, secure, sustainable path. Rather than a story of mass migrations, rising sea levels, and collapsing ecosystems, it is a story of heroism, unexpected breakthroughs, and climate solutions. Instead of feeling hopeless about a global economic machine built on coal and oil, you are feeling excited about the new ways we generate power, grow food, and live together. Let's look ahead and imagine that together with millions of other concerned citizens we defied the doomsday prophets and created the foundation for transformational change.

By 2025 global carbon emissions are rapidly decreasing. The world has embarked on a green economic recovery program that stimulates new jobs, businesses, and sustainable growth. Millions of new jobs have been created on every continent to hasten the transition to a zero-carbon economy. In the United States alone, 2 million new jobs were created by 2010. Today roofers, electricians, civil engineers, assembly-line workers, lawyers, loan officers, and urban planners are building zero-carbon schools, health clinics, and homes. They manufacture and market wind turbines, solar panels, and hybrid vehicles. Many plant and harvest community gardens and small farms, while others design and construct mass transit systems, a smart grid, and new solar-powered irrigation systems. Work is valued, and green jobs in New Orleans as well as in Haiti, Zimbabwe, and Liberia lift people out of poverty.[2]

The world has experienced a revolution in energy efficiency. We stop wasting energy and realize that conserving energy means saving money. The extra money we initially spend on efficient cars, appliances, and buildings is quickly paid back in reduced energy and gas bills. By 2012 incandescent light bulbs were banned everywhere, and our homes, offices, and buildings have LED lights that have displaced at least 700 coal-fired plants. We have exciting and climate-friendly new technologies to use, such as solar-powered electric bicycles that go up to 20 miles per hour and cars that get the equivalent of 200 miles per gallon of gasoline—but that use no gas at all. Zero-carbon mass transit systems operate in nearly every major metropolitan area, funded by dramatic reductions in global military spending.[3]

Offices in New York, Beijing, and Bombay have task lighting, occupancy sensors, high-efficiency windows, and white or pastel roofs that deflect rather than absorb heat. Old-fashioned energy conservation is popular. Millions of people in industrial countries dry their clothes in the sun and turn off unnecessary lights and electronics. Women-owned enterprises in India, Honduras, and Ethiopia produce naturally dyed clotheslines, and inner-city youth in Detroit employed by the Conservation Jobs Corps ensure that homes are properly insulated.

Nations have saved billions of dollars and hundreds of gigawatts of electricity by establishing aggressive codes and standards for buildings, vehicles, appliances, and power plants. These standards unleash market-driven innovations in lighting, heating and cooling, building materials, insulation, vehicles, industrial processes, power generation, and appliances. Looking back, it is clear that 2009 and 2010 were pivotal years—a time when forward-thinking nations recharged the global economy with a program to retrofit half of the world's buildings with energy-efficient technologies and restart auto assembly lines to produce affordable plug-in hybrid vehicles.[4]

Electricity from renewable sources has displaced coal and traditional fossil fuel power plants. Renewable sources of energy such as wind, solar, geothermal, and biomass have displaced conventional power plants in the

United States and in many other parts of the world. In northern Africa, parabolic solar troughs span the equivalent of nearly 45 football fields in the Sahara. Along with smaller-scale community-owned photovoltaic and hot water systems, solar power from this region alone annually supplies the equivalent of half of the Middle East's annual oil production. Community-owned solar installations in developing nations provide power for water pumping and drip irrigation, health clinics, schools, homes, streetlights, and wireless Internet.[5]

Wind power is a source of rural economic development in China, the United States, Spain, Tunisia, Egypt, and elsewhere. China exceeded its goals and had more than 150,000 megawatts of wind power by 2020. Communities and buildings look different: Rooftop solar panels and flat-profile residential windmills feed electricity back into the grid. Pedestrians generate electricity just by walking on energy-generating sidewalks, while health clubs produce electricity through treadmills and aerobics classes.

Terri Heisele

Part of our solar-powered future

Solar-powered rickshaws in Milan, Calcutta, and Jakarta offer mobile coffee bars. In the United States, all coal-fired power plants have been shut down, replaced by renewable energy and the short-term use of natural gas. Former coal workers in Wyoming and Huainan now build small-scale, state-of-the-art underground storage facilities for zero-carbon food preservation.[6]

The world is greener as we look around in 2025. By the end of 2009, more than 7 billion new trees had been planted. The success of this initial campaign led to vast tree planting around the world. Early successes with reforestation and tree planting in tropical zones were pivotal in demonstrating a global spirit and commitment to action. With leadership from Costa Rica, Ethiopia, Kenya, and Papua New Guinea, and financing from the industrial world, each of the billions of trees on average absorb 6 tons of carbon dioxide a year.[7]

Developing nations are on the path to greater prosperity, and industrial countries are on the path to sufficiency without excess. Population is declining because men and women in the developing world have access to family planning and to food and shelter. We joined together in 2009 and 2010 and said NO to business as usual and YES to a fundamental change in direction. We are preoccupied with creating convergence between those who do not have nearly enough and those who have more than their share. We do this joyfully, for we recognize that climate change can no longer be denied and that social justice and global cooperation at all levels are prerequisites for a safe future. Our economies are more local than global and as a result, small businesses and farms thrive.

We have begun to see that life can be safer, slower, and less driven by anxiety. More of us have a sense of sufficiency and security. Global capitalism driven by ever-rising consumption has been fundamentally altered by a new system of rules that unleash innovation while protecting freedom and all life on the planet. We have transcended powerful special interests and centuries-old debates about economic systems to build a dynamic and promising world. Young people feel hopeful.

We will not reach this world of 2025 until we rise above the narrow confines of our individual concerns to embrace the concerns of all humanity. We must also act with a sense of overwhelming urgency. The incremental baby steps of the past will not be enough. Millions of students, business leaders, engineers, community activists, and local elected officials are taking bold action and we must join with them. So don't believe those who whisper that we are acting too late. The future will be determined by those who defy the odds to imagine a very different tomorrow and then get serious about the kind of transformational change needed to get there.[8]

CHAPTER 4

An Enduring Energy Future

Janet L. Sawin and William R. Moomaw

In 1992, Güssing was a dying town not far from the rusting remains of the Iron Curtain and the capital of one of Austria's poorest districts. Just nine years later, Güssing was energy self-sufficient, producing biodiesel from local rapeseed and used cooking oil, as well as heat and power from the sun, and had a new biomass-steam gasification plant that sold surplus electricity to the national grid. New industries and more than 1,000 jobs flocked to the town. Today, not only do Güssing residents enjoy much higher living standards, they have cut their carbon emissions by more than 90 percent. And Güssing is not an isolated case. The Danish island of Samsø and several other communities have achieved similar transformations using various combinations of innovations.[1]

A growing number of towns are rapidly transitioning to low-carbon renewable energy, and larger cities are attempting to follow their lead. But most of the world remains wedded to polluting, carbon-intensive fossil fuels, despite rising economic costs and threats to human health, national security, and the environment. Until recently, fossil fuels were cheap and abundant; as a result, they have been used very inefficiently. The tenfold rise in the price of oil in the past decade and recent increases in natural gas and coal prices mean fossil fuels are no longer cheap, and their volatile prices have devastated many economies. Readily accessible conventional fuels are in increasingly shorter supply as discoveries fail to keep up with demand, and extraction requires developing ever more remote resources and using increasingly drastic measures—from removing mountain tops to heating tar sands. Competition for fossil fuels is heightening international tensions, a trend likely to intensify over time. The urgent need to reduce the release of carbon dioxide (CO_2) and methane in order to avoid catastrophic climate change has finally focused the

William R. Moomaw is Director of the Center for International Environment and Resource Policy at The Fletcher School at Tufts University.

world's attention on the need for a rapid shift in how energy services are provided.[2]

Energy scenarios offer a wide range of estimates of how much renewable sources can contribute and how fast. The International Energy Agency (IEA) recently projected that the share of primary world energy from renewables will remain at 13 percent between 2005 and 2030. But if national policies now under consideration are implemented, that share could rise to 17 percent, and renewables could be generating 29 percent of global electricity by then. The Intergovernmental Panel on Climate Change (IPCC) projects that, with a CO_2-equivalent price of up to $50 per ton, renewables could generate 30–35 percent of electricity by 2030. As a 2007 review of global energy scenarios noted, the "energy future we ultimately experience is the result of choice; it is not fate."[3]

The transition away from fossil fuels involves a dual strategy: reducing the amount of energy required through energy efficiency and then meeting most of the remaining needs with renewable sources. The IEA estimates that $45 trillion in investment, or an average 1 percent of annual global economic output, will be needed between now and 2050 in order to wean the world off oil and cut CO_2 emissions in half. It is imperative that the vast majority of these investments be in efficiency improvements and renewable energy.[4]

Renewables already provide a significant share of the world's energy. In 2007 renewable energy, including large hydro, generated more than 18 percent of global electricity. At least 50 million households use the sun to heat water. Renewable resources are universally distributed, as are the technologies. While much of the current capacity is in the industrial world, developing countries account for about 40 percent of renewable power capacity and 70 percent of existing solar water heating.[5]

As this chapter describes, a range of renewable technologies are used to produce electricity and meet heating and cooling needs. They are available now and ready for rapid scale-up. Most of them are experiencing annual growth rates in the double digits, with several in the 20–50 percent range. Once these technologies are in place, the fuel for most of them is forever available and forever free. The current technical potential of renewable resources is enormous—many times current global energy use. (See Figure 4–1.)[6]

Some observers propose that coal with carbon capture and storage or nuclear power may be needed to address climate change while meeting rising energy demand. But renewable energy combined with energy efficiency can do the job, and renewables are the only technologies available right now that can achieve the emissions reductions needed in the near term. Efficiently delivered energy services that use natural energy flows will protect the global climate, strengthen the economy, create millions of new jobs, help developing countries reduce poverty, increase personal and societal security in all countries, reduce international tensions over resources, and improve the health of people and ecosystems alike. Although this chapter focuses on industrial countries and rapidly developing emerging markets, it is important not to forget the needs of people in the poorest economies.

Making Every Building a Power Plant

Buildings use about 40 percent of global energy and account for a comparable share of heat-trapping emissions. About half of this demand is for direct space heating and hot water needs, and the rest is associated with the production of electricity for lighting, space cooling, appliances, and office equipment.[7]

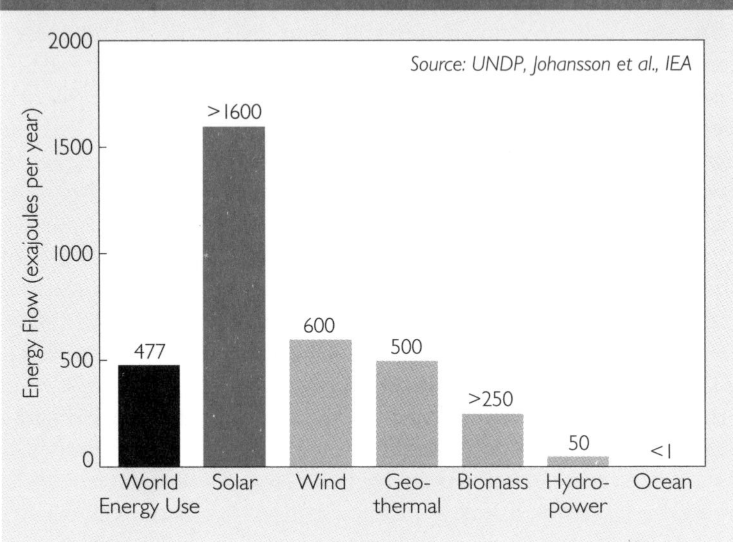

Figure 4–1. World Energy Use in 2005 and Annual Renewable Energy Potential with Current Technologies

The advent of cheap and readily available energy enabled modern buildings to work in spite of nature rather than with it. But it is possible to reduce demand in existing buildings by insulating them properly, controlling unwanted air infiltration, and improving performance for space and water heating, lighting, ventilation, and air conditioning. For new construction, an integrated design with multiple energy efficiency measures can reduce energy use to at least half of a conventional building, and gains of greater than 80 percent have been achieved. The use of information technology to manage multiple functions can also help make the best use of energy.[8]

The potential savings could be great. The fragmented nature of building and lighting codes in the United States, for example, has meant the continuing construction of inefficient buildings and the unavailability of technologies that are common in Europe and Canada. India has no mandatory efficiency codes for commercial buildings, and most contractors do not know how to install insulation. But greener buildings are on the way in India as well. One of the largest green commercial developments in the world is under construction near Delhi; it is expected to exceed international energy performance standards.[9]

As energy efficiency improves, each energy unit is cheaper, so consumers might choose to use more energy or to spend their savings on other goods that require energy. This is known as the rebound effect or leakage, and it is measured by the difference between projected and actual energy savings that result from an increase in efficiency. Evidence suggests that in developing countries the rebound effect can be 100 percent or greater, meaning that efficiency improvements have at best no impact on energy use. In mature or wealthier markets, however, efficiency improvements in electrical equipment result in energy savings that are 60–100 percent of projected levels. Case studies in the United States have concluded that energy savings in commercial buildings—from schools to office towers—have frequently been greater than projected. Perhaps most promising are more-efficient, or even zero-net-energy, buildings.[10]

A zero-energy, zero-carbon building is one that produces all its own energy on site with renewable energy and emits no CO_2. Most buildings will need an energy supply from outside to meet peak demands at particular times of day. Still, such buildings can be zero-net-energy if they produce as much energy as they consume in a year. And if they import renewable, or zero-carbon, energy from elsewhere, they can also be zero-carbon buildings. The United Kingdom has mandated that all new homes built after 2016 and all commercial buildings constructed after 2019 must be zero-carbon.[11]

Once buildings are as efficient as possible, remaining energy demands can be met with renewable technologies. Passive solar heating and thermal storage in buildings can significantly lower the need for additional heating, and judicious placement of windows and roof shading can reduce cooling needs. In many locations, solar thermal panels can cost-effectively provide water and space heating. Solar photovoltaics (PVs) can be integrated into rooftops and even building facades, where they are often cheaper than traditional sidings. In some locations PVs are already cost-competitive with conventional power at peak demand times as they compete with the price customers pay for power rather than utility wholesale rates. They are also often cheaper than extending the electric grid or using diesel generators.[12]

The Passivhaus Institute in Germany has built more than 6,000 dwelling units that consume about one tenth the energy of standard German homes. These low demand levels are achieved through passive solar orientation for heating and daylighting, efficient lighting and appliances, super insulation and ultra-tight air barriers on doors and windows, and heat recovery ventilators. As peak loads decline for lighting, heating, ventilation, and cooling, so does the required size of fans, boilers, and other equipment, providing greater savings.[13]

There are multiple benefits to the on-site generation of energy from a variety of distributed production sites or nodes. New production units can be brought on line in small increments that conform to the new demand in a timely manner without requiring additional transmission lines and often without additional distribution lines. Since power is produced and consumed on site, transmission and distribution (T&D) of electricity from central plants are lowered, and losses are reduced—so less energy must be generated to meet the same demand. Local thermal power systems allow the capture and use of waste heat along with the production of electricity, providing heat to adjacent buildings and thereby reducing energy use and associated emissions; using renewable energy sources reduces emissions even more.

A multi-nodal system of distributed energy is more resilient and more reliable, especially where the power grid is subject to frequent interruptions from accidents or other failures. Having more but smaller power production units reduces a system's vulnerability to major disruptions. An analysis following a major 2003 blackout in the U.S. Northeast found that a few hundred megawatts (MW) of distributed PV, strategically placed around the region, would have reduced the risk of power outages dramatically. Wind turbines might play a similar role, placed along transmission corridors, highways, or train tracks, as they are in parts of Denmark.[14]

Other options for distributed power include fuel cells and thermal power systems fueled by solid biomass, biogas or liquid biofuels, or conventional natural gas turbines; all provide heat as well. The U.K. government projects that distributed generation could provide electricity for 40 percent of Britain's homes by 2050.[15]

The full benefits of interconnected systems could be achieved by transforming the outdated central-power-plant-dominated T&D system into a dynamic "hybrid" network that relies on diverse and multiple production nodes, much like the Internet. This network would consist of locally sited renewable power and combined heating, cooling, and power units, some large-scale centralized power plants, and electricity storage systems. The development of "smart grids," which use information technology to manage supply and demand, will also be critical to achieving the full potential of renewables and multiple distributed storage devices. (See Box 4–1.)[16]

Many technical issues of interconnection for distributed systems have been addressed in Europe, where homes, farms, and businesses produce significant shares of electricity for the grid. Most remaining problems are due to regulatory rules. In many places, electric utilities have monopolistic control over generation; in others, distributed generators must pay retail rates for electricity from the grid but are then paid only wholesale rates—or, in some cases, nothing—for power

Box 4–1. Building a Smarter Grid

In today's "dumb" electricity grid, communication is only one-way, from consumers to utilities, which attempt to adjust to changes in demand by ramping production up and down. When utilities cannot respond as needed, system problems such as blackouts can occur. Kurt Yeager, former president of the U.S.-based Electric Power Research Institute, compares the current electromechanically controlled grid to "a railroad on which it takes 10 days to open or close a switch."

Yet the digital age, which has increased demand for electricity and highly reliable power systems, is now allowing the transition to a faster, smarter grid that can provide better-quality power with two-way communication, balancing supply and demand in real time, smoothing out demand peaks, and making customers active participants in the production as well as consumption of electricity. The smart grid allows more-efficient use of existing power capacity and of T&D infrastructure by reducing line losses through the use of more local, distributed generation. As the share of generation from variable renewable resources increases, a smart grid can better handle fluctuations in power when the wind ebbs or clouds hide the sun. It will also allow electric vehicles to store power for transport use or to sell back to the grid when needed.

Smart technologies—including smart meters, automated controls systems, and digital sensors—will provide consumers with real-time pricing and enable them to save money and power by setting appliances, entire building heating and cooling systems, or industrial loads to shut off at specific times or when electricity prices exceed a certain level or there is a drop in generation from large wind plants. They can help shift loads to low-demand periods, when line losses are lowest and the dirtiest, least efficient plants are not operating. And they allow grid controllers to anticipate and instantly respond to troubles in the grid. Pilot programs have demonstrated significant consumer savings and demand reductions.

Full development of smart grids is 10–30 years away, depending on the policies enacted. But many countries and regions are well on their way. Pacific Gas and Electric in California, for example, is in the process of installing 9 million smart meters for its customers, while the Netherlands aims for a "base level" of smart metering and replacement of all 7 million household meters by autumn 2012. When starting from scratch, smart grids are cheaper than conventional systems, and they are helping to electrify regions of sub-Saharan Africa for the first time. The IEA has projected that more than $16 trillion will be spent in pursuit of smart grids worldwide between 2003 and 2030. If consumers are provided direct access to associated benefits, Kurt Yeager projects that a smart grid will open "the door to entrepreneurial innovation which will transform electricity efficiency, reliability and individual consumer service quality beyond even our imagination today."

Source: See endnote 16.

they feed back into it. Guaranteed access to both the electric grid and the market—through either feed-in tariffs (local generators are paid a set price for all the renewable electricity they produce) or net metering (they are paid, generally at the retail rate, for any excess power sent into the grid)—is critical to the expansion of distributed energy and renewables in general, enabling renewable energy not only to add energy supply but also to replace existing fossil fuel sources over time. While most locally generated power will not make use of transmission lines, it will use local distribution systems. An issue still to be resolved is the level of payment for use of the distribution system for the relatively small amount of electricity from local distributed sources.[17]

Smarter Central Power with Large-scale Renewables

Central electric generating stations will continue to be part of the electricity supply system in order to take advantage of an energy resource or to meet large industrial or urban loads. According to the IPCC, installed generating capacity worldwide is now about 2 million MW; it is estimated that demand growth and the need to replace existing plants will require an additional 6 million MW by 2030—at a cost of $5.2 trillion.[18]

Today electricity generation accounts for 41 percent of global primary energy use—meaning total energy use, from coal mine to appliances or other "end uses"—and 44 percent of CO_2 emissions. Thermal power plants typically convert only one third of the energy in fuels to electricity; at least 5–10 percent of that electricity is then lost in transmission, distribution, and voltage adjustments. End-use devices such as computers and appliances are also especially inefficient, and it can take 320 units of energy at a power station to produce one unit of energy in the form of light from an incandescent bulb. Technology available today and just over the horizon can revolutionize such systems, dramatically reducing inefficiencies and associated carbon emissions.[19]

Every country has large-scale domestic sources of renewable energy. Africa has the most in the world. An area covering less than 4 percent of the Sahara Desert, for example, could produce an amount of solar electricity equal to current global electricity demand. The Middle East, India, China, Australia, and the United States also have enormous solar resources. China's wind resources alone could generate far more electricity than that country currently uses, and in the United States wind energy in just a few states could meet total national electricity demand. Vast resources of geothermal, biomass, and ocean energy are also found throughout the world.[20]

Renewables currently provide nearly one fifth of the world's electricity. Although most of this comes from large hydropower, the share from other renewable sources is rising, thanks to growth rates that rival those of the computer and mobile phone industries. In 2007, wind power represented 40 percent of new capacity installations in Europe and 35 percent in the United States. Cumulative installations of solar photovoltaics have grown more than fivefold over the past five years, albeit from a very small base. As economies of scale improve and conventional fuel costs rise, renewables are rapidly becoming cost-competitive. Electricity from wind is cheaper than that from natural gas in many markets and might compete with coal in China by 2015, if not sooner. Solar thermal power now competes with gas peaking plants in California and is close to being economically feasible in China and India. And experts project that PVs will be cost-competitive without subsidies in much of the world within a decade.[21]

Despite such advances, skepticism abounds: some observers claim that renewables lack power density or are too far from demand centers, that they cannot provide baseload power—power available 24 hours a day all year long, that they require 100 percent backup, that they cannot meet more than a small share of global power needs, or that it will be many decades before they play a significant role. Certainly, renewables face significant challenges in achieving large penetrations of the world's electricity system over the time frame required, but all of these hurdles are surmountable.

Large-scale renewable resources far from population centers require new transmission lines. Although this poses a challenge, it is nothing new. The United States, Egypt, Brazil, Canada, China, and Russia transmit power from dams to cities hundreds of miles away, and new lines were required to bring electricity from large nuclear facilities and from coal plants near mines. New transmission infrastructure will be required for all forms of generation as capacity expands, and vulnerable, ailing grids will need to be replaced.[22]

There will be opposition to new power lines in many places, and minimizing the environmental and social impacts will certainly require careful analysis. But innovative technologies, such as high-voltage direct current lines, offer the potential to transmit electricity reliably over enormous distances with lower line losses. Extensive direct current grids have already been erected to balance wind power in Germany and Denmark with hydropower from Norway, for example. Now several European nations are considering a massive grid to North Africa's tremendous solar resource of the Sahara.[23]

In the United States, studies have found that adding new transmission to transport wind energy from the Great Plains to population centers would yield large net economic savings for customers as the benefit of competitive and stable long-term wind power prices outweighs the costs of new transmission. And some of the best renewable resources do not need to travel far. For example, solar in the U.S. Southwest and wind and ocean resources along coastlines offer many large cities the potential to get clean energy while reducing grid congestion and line losses and improving system reliability.[24]

Renewable resources—including biomass, geothermal, ocean thermal, and hydropower—can in fact provide large-scale baseload power, and many already do. New concentrating solar thermal plants in Spain and under construction in the United States can store heat for up to seven hours in molten salts, enabling them to dispatch power on demand and maximize production when it is of greatest value, during hot summer afternoons and early evenings. In coming decades, ocean energy technologies under development will offer power that is baseload or highly predictable. Economical options are being pursued to store wind and PV energy and allow generation of high-value electricity even when the wind is not blowing and the sun is not shining.[25]

For more than a century, pumped hydro and large hydroelectric reservoirs have provided storage for conventional power, enhancing grid stability and balancing demand and supply. They now do the same for renewables. Facilities that store compressed air in underground caverns have operated for years in Alabama in the United States and in Huntorf, Germany, and are under development elsewhere. Low-cost power is used to compress air that can later increase the output of natural gas turbines during peak demand periods. Studies have found that compressed air storage would allow wind power to provide baseload power and that any cost-effective storage option could boost wind's share of the electricity system to more than 80 percent. A diversity of rapidly advancing battery technologies offer great storage potential, particularly in electric vehicles, which could revolutionize how the world produces and uses power and enable all renewables to displace oil. And the value of storage goes beyond renewables, enabling better management of peak demand, providing a more stable grid and better power quality, and in some cases reducing the need for new transmission lines.[26]

Better storage has often been called renewable energy's "Holy Grail." But even without storage, electric utilities are recognizing that individual variable renewable sources like the sun or wind—despite variations in availability from moment to hour to season—can provide as much as 20 percent of a system's electricity, and in some cases more, without serious technical problems. So far no additional backup capacity has been required because existing systems are designed to handle variations in demand and outages of power plants or transmission lines on a routine basis.[27]

The addition of new renewables changes the degree—but not the kind—of variability that utilities face in matching supply with demand. The capacity to absorb large amounts of renewable power is determined primarily by regulatory and market barriers rather than technical constraints. Variability and uncertainty may increase operating costs, but generally by modest amounts, and the overall cost and risk reductions associated with free fuel of most renewables can be significant. It is also worth noting that there are costs to integrating conventional power plants into existing systems, but a lack of studies makes cost comparisons impossible.[28]

Denmark generated 21 percent of its electricity with the wind in 2007, and occasionally wind power meets more than 100 percent of peak demand in parts of western Denmark. Four German states produced more than 30 percent of their electricity with wind power in 2007. In California, renewables make up more than 30 percent of the portfolios of some large utilities. Utilities have balanced supply and demand through the interconnection of grid systems over large regions with a diversity of loads and resources, use of hydropower as temporary storage, dispersal of renewable power plants over large geographic areas, and solar and wind forecasting an hour or a day ahead. These "hybrid grid" tools help utilities to regulate supply, but there is also more they can do to control the demand side—to reduce demand or shift it away from time-insensitive uses.[29]

In the United States, adding new transmission to transport wind energy from the Great Plains to population centers would yield large net economic savings for customers.

Through demand-side management programs, utilities help customers undertake conservation, efficiency, or load shifting to reduce demand or shift it to off-peak periods, when electricity can be generated and transmitted more efficiently, in order to avoid the need for new power plants. Thanks to a Californian program that decoupled transmission utility revenue from sales in 1982, per capita electricity use of the average Californian has remained nearly constant for 25 years and is significantly less than that of the average American. Efficiency improvements enable renewables to more rapidly play a greater role.[30]

But unleashing the full potential of efficiency and renewables will require a modern, more reliable, intelligent grid at the distribution level, where lower voltage power lines transport electricity to (and increasingly from) homes, offices, and other facilities. Smart grid technologies, now being installed in Africa, Asia, Europe, New Zealand, and the United States, will be able to smoothly integrate all types of central plants with renewables, distributed generation, electric vehicles, and electrical storage facilities while enhancing grid reliability. By controlling the flow of electricity electronically, in real time, smart technologies maximize the capacity of existing grid infrastructure and minimize the need for backup and storage, enabling grids to absorb unlimited renewable capacity.[31]

Implementation of smart, hybrid grid systems will require rethinking how the electric industry works in much of the world, but addressing climate change requires such a transformation in any case. Some utilities are already making this transition. The Danish power company DONG is making conventional power plants more flexible so they can be turned down, or even off, when the wind is blowing. "In the old times," explains Chief Executive Anders Eldrup, "wind power was just something we layered on top of our regular production. In the future, wind will provide a big chunk of our baseload production."[32]

A report by the German Aerospace Center (DLR) projects that by 2030 renewables could generate at least 40 percent of national electricity in 13 of the 20 largest economies. (See Figure 4–2.) Hydro, wind, and biomass power will likely achieve the greatest market shares in the near to medium term, with geothermal and particularly solar power playing greater roles in the longer term. By 2050, renewables could contribute at least 50 percent of national electric power in each of the world's large economies, and up to 90 percent in some countries. Some projections

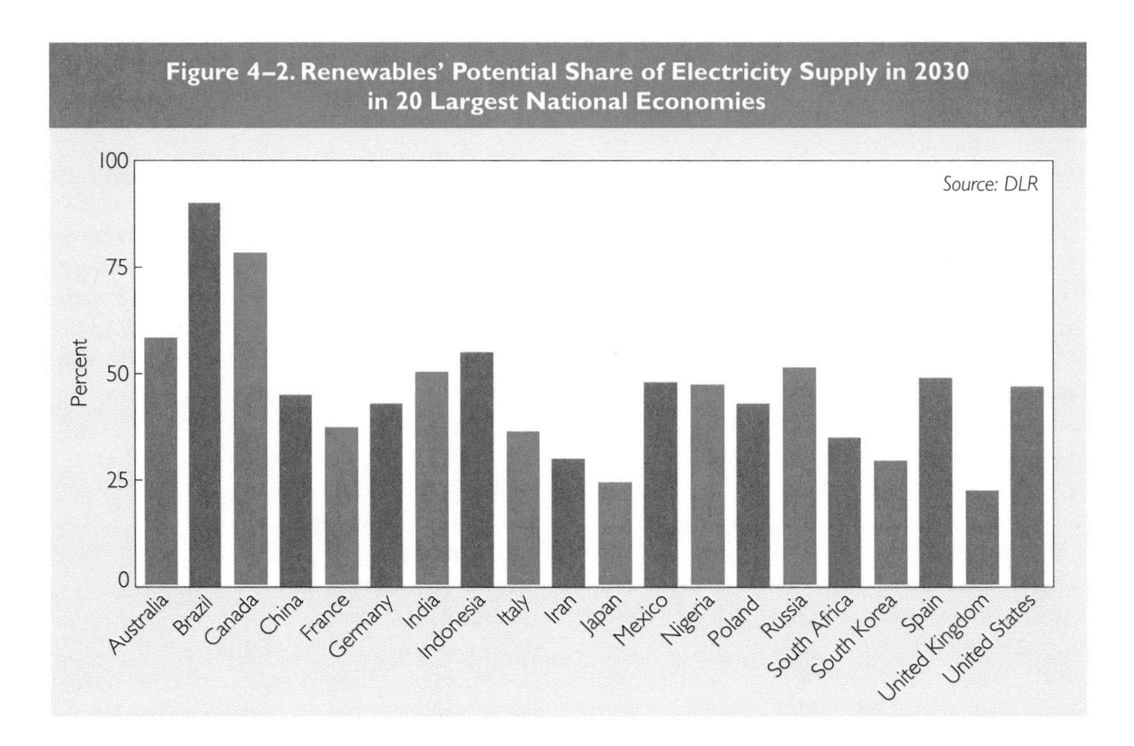

Figure 4–2. Renewables' Potential Share of Electricity Supply in 2030 in 20 Largest National Economies

go even further—a U.S. study projects that by 2050 solar power alone could provide 69 percent of the nation's electricity, plus enough power to fuel 344 million plug-in hybrid vehicles.[33]

By tapping the potential of all renewables, the world can move away from fossil fuels in the next few decades, not only reducing the threat of climate change but also creating a more secure and far less polluting electricity system. The Combined Power Plant, a project linking 36 wind, solar, biomass, and hydropower installations throughout Germany, has already demonstrated that a combination of renewable sources and more-effective control can balance out short-term fluctuations and provide reliable electricity with 100 percent renewable sources. In a recent interview about the future of the electric industry, S. David Freeman, a 50-year veteran of the U.S. electric industry, said "it still will take 25, 30 years to phase out the existing coal-fired plants and have an all-renewable world." But, he concludes, "I'm a utility executive that ran major utilities, and I can tell you there is no reason why the electric-power industry can't be all renewable."[34]

Heating and Cooling with Renewables

Renewable heating and cooling are too often the neglected twins when it comes to climate change and energy policies. They account for 40–50 percent of global energy demand. A large share comes from fossil fuels and is provided inefficiently by electricity or direct combustion.[35]

As early as the Bronze Age, wood was used to turn sand into glass, extract metals from stone, and fuel furnaces to make bronze and pottery. There is evidence that high-temperature geothermal water was used to heat buildings in ancient Pompeii, while Greeks and Romans captured the sun's warmth to do the same job. Today renewable energy and improved efficiency options exist to meet a wide range of heating and cooling needs, from residential and district heating and cooling systems to industrial-scale refrigeration and high-temperature heat. (See Table 4–1.)[36]

Among new renewables, solar heating ranks second only to wind power for meeting world energy demands. China leads the world in the production and use of solar thermal systems, with an estimated 1 in 10 households tapping the sun to heat water; Cyprus, Israel, and Austria top the list for per person use. Solar water heating is mainstream in Israel thanks to a 1980s law requiring its use in new homes. Hybrid solar hot water/photovoltaic systems are now available to capture a large amount of the heat absorbed by PVs, thereby cooling them and increasing their efficiency while simultaneously heating domestic water. One of the first systems sits atop the roof of a central building in Beijing's Olympic Village.[37]

The majority of solar thermal systems in use are for domestic water and space heating, yet solar heating systems—including systems similar to solar heaters for residential buildings and concentrating solar collectors—offer enormous potential for meeting industrial heat demand, particularly at low and medium temperatures (up to 250 degrees Celsius). By late 2007 about 90 solar thermal plants provided process heat for a broad range of industries, from chemical production to desalination and the food and textile industries. Existing plants worldwide represent a tiny fraction of the industrial heat potential available in Europe alone.[38]

Across Europe, the United States, and elsewhere, people are turning to efficient pellet stoves and in some cases using liquid biofuels in boilers to meet heating needs. Between 1980 and 2005, taxes on energy and CO_2 in Sweden drove a major shift from

Table 4–1. Alternatives to Fossil Fuels for Heating and Cooling

Technology	Description	Where Available or Possible
Absorption cooling	Uses a heat source (such as the sun or waste heat from combined heat and power (CHP)) to cool air through an evaporative process; small to large-scale	Anywhere
Bioheat	Heat derived from the combustion of biomass, such as wood or pellets; residential to large-scale	Anywhere close to sustainable wood or other biomass resources
Combined heat and power (cogeneration)	Use of a power plant to produce both heat and electricity; residential to large-scale	Anywhere
Concentrating solar thermal	Uses optical concentrators to focus the sun to provide higher-temperature heat and steam for industrial processes (and thermal electricity production)	Needs clear skies as in Spain, North Africa, parts of China and India, or U.S. Southwest
District heating (or cooling)	Distribution of heating (cooling) from a central generating site, through a piped network, to meet local residential and commercial needs	Possible anywhere for use in urban and campus settings with multiple buildings
Geothermal high-temperature heat	Geothermal steam or hot water used for district water and space heating, warming greenhouses, aquaculture, spas and swimming pools, industrial purposes (and thermal electricity production)	Regions of active or geologically young volcanoes, including Iceland, western North and South America, Philippines, Japan, East Africa
Ground-source heat pump	Pump that makes use of ground-stored solar heat or well water to provide space and water heating/cooling; residential to large-scale	Anywhere
Passive solar heating	Collects solar heat through appropriate building orientation and window placement	Anywhere heating is needed
Passive cooling	Avoids excess heat absorption by designing buildings to reduce passive solar gain, such as avoiding glass and using passive ventilation	Hot, particularly dry, regions
Seawater or lake cooling	Harnesses constant coolness of deep water to provide space cooling (and cold water) to buildings through a piped network	Requires proximity to cold water resource (along deep rivers, lakes, or coastlines)
Solar thermal heat system	Uses the sun's heat to provide space and water heating for buildings and low-temperature heat and hot water for industrial processes	Anywhere

fossil fuels to biomass for district heating, reducing associated emissions to less than one third their 1980 level. Austria and Denmark also rely heavily on biomass to heat homes, farms, and district systems. Poland is replacing coal with biomass for power and heating needs. Biomass can directly replace fossil fuels, and modern wood burners can convert biomass to heat at efficiency rates of up to 90 percent.[39]

Geothermal energy is used for everything from space heating and cooling to warming greenhouses and melting snow on roads and bridges. In France, Iceland, New Zealand, the Philippines, Turkey, the United States, and other countries with high-temperature resources, geothermal heat is used for electricity generation, district heat, and industrial processes like pulp and paper production. Ground-source heat pumps, which can be used virtually anywhere, use the stored solar energy of Earth or well water as a heat sink in summer and heat source in winter. The United States has the world's largest heat pump market, with up to 60,000 systems installed annually.[40]

Because buildings generally require heat as well as electricity, combined heat and power units can be designed to supply both. CHP plants generate electricity and capture remaining heat energy for use in industries, cities, or individual buildings. They convert about 75–80 percent of fuel into useful energy, with efficiencies exceeding 90 percent for the most advanced plants. As a result, even traditional fossil fuel CHP systems can reduce carbon emissions by at least 45 percent. These systems can also make use of absorption chillers for space cooling to lower electricity demand even further. Residential-scale CHP units have been widely available in Japan and Europe for years and were recently introduced in the United States.[41]

Seawater and lake source district cooling systems have been developed for a range of climates, from Kona in Hawaii to Stockholm in Sweden, and can save more than 85 percent of the energy required for conventional air conditioning. The cold waters of Lake Ontario provide district cooling to Toronto in Canada; the system has the capacity to cool more than 3.2 million square meters of building space, avoiding 79,000 tons of CO_2 annually. Many of the world's big cities are near large water bodies, which they could tap for cooling. And as paradoxical as it might seem, solar energy can also provide cooling via the oldest form of air conditioning technology—absorption cooling—with the same devices that provide heat in the winter. While such systems are still relatively costly, several are already in operation, including a solar-driven cooling system in Phitsanulok, Thailand.[42]

Economical heat storage over a wide range of temperatures and time periods can significantly increase the potential of renewable systems. Some storage options are already available and cost-effective, particularly in combination with large-scale district systems. For example, surplus solar heat in summer can be transferred to underground storage for space and water heating in winter.[43]

According to the IEA, "solar water heating, biomass for industrial and domestic heating, deep geothermal heat and shallow geothermal heat pumps are amongst the lowest cost options for reducing both CO_2 emissions and fossil fuel dependency. In many circumstances these technologies offer net savings as compared to conventional heating systems in terms of life-cycle costs." And yet these renewable sources and technologies currently meet only 2–3 percent of total demand.[44]

Attitudes have begun to change as fuel prices rise and countries recognize the enormous potential of renewables. To date, the most successful countries have enacted combinations of policies to address the different barriers facing renewable heating and cooling technologies. These include lack of public awareness, the need to train a work force and educate city planners and architects about integrating renewables, high upfront costs, the "tenant-owner" dilemma (where building owners and inhabitants are different people, and the person who pays does not benefit), and the need for scale.[45]

Cloudy Germany has one of the largest solar thermal heat markets in the world thanks to public awareness of the technology and long-term government investment subsidies. The German state of Baden Württemberg now requires that all building plans for new homes include renewable systems to meet at least 20 percent of space and water heating needs, and as of 1 January 2009 the German federal government requires new buildings to meet at least 15 percent of their heating requirements with renewable sources. Since 2006 Spain has mandated solar systems for all new or renovated buildings, and Hawaii will require solar water heaters on all new homes starting in 2010.[46]

DLR in Germany projects that 12 of the 20 largest economies could meet at least 40 percent of their heating needs with renewables by 2030, representing a significant increase from current shares for most countries. (See Figure 4–3.) By 2050, renewables' share in the majority of these countries could exceed 60 percent, according to DLR and REN21 estimates, with renewables supplying at least 70 percent of heating in some countries.[47]

Waste Not, Want Not

In the natural world, waste from one process provides nutrients for another. Nothing is wasted. The human world, however, functions quite differently. For example, most of the world's power plants convert heat to mechanical energy to electricity; in the process, about two thirds of the primary energy fed into

Figure 4–3. Renewables' Potential Share of Heat Supply in 2030 in 20 Largest National Economies

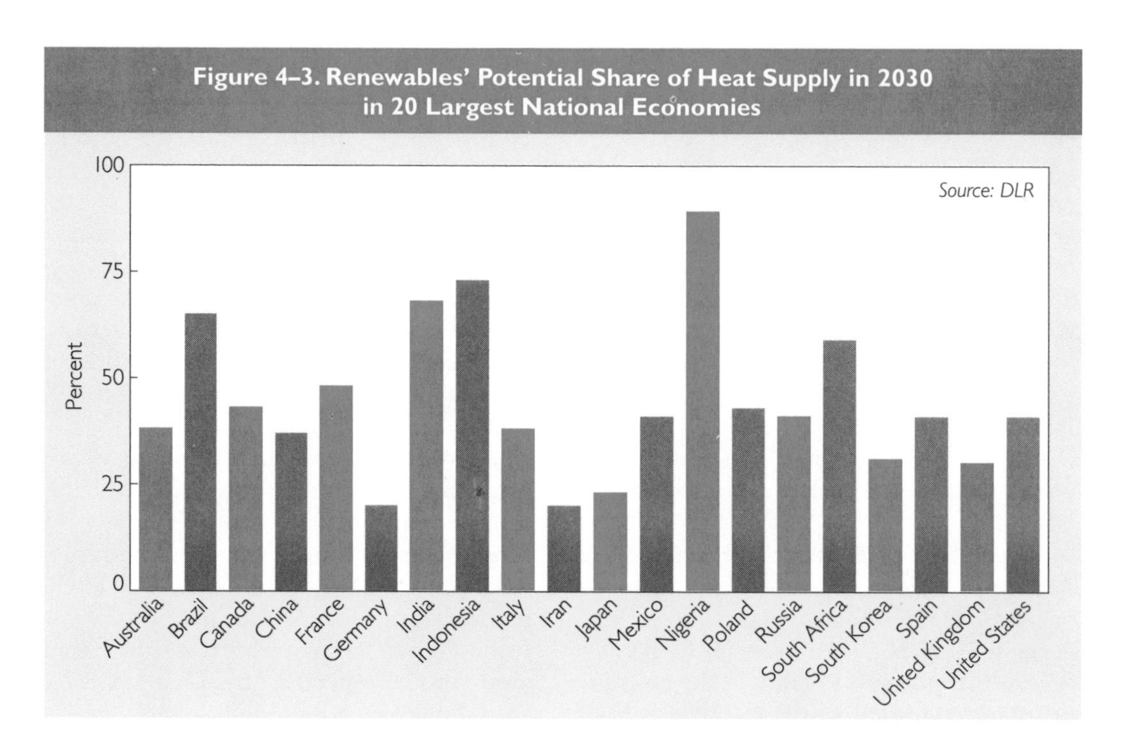

these plants is released into the environment as heat. In Europe, losses from power generation are so great that if they were captured and rerouted they could meet the region's heat demand through district heating. Heat is just one form of waste that could be captured to dramatically increase useful energy without burning more fossil fuels.[48]

A 2005 study by the U.S. Lawrence Berkeley National Laboratory examined 19 different technologies at various scales that can recover energy from waste heat, manure, food industry waste, landfill gas, wastewater, steam and gas pipeline pressure differentials, fuel pipeline leakages and flaring, and numerous other sources. In the United States alone they offer the technical potential to profitably generate almost 100,000 MW of electrical capacity—enough to provide about 19 percent of the nation's electricity in 2002—in addition to useful heat or steam.[49]

Around the world, some of these "wastes" are already being tapped. For example, combined heat and power is used widely in much of northern Europe, with Denmark, Finland, and Russia leading in the shares of national power production. Finland meets about half of its heating needs with district systems, mainly CHP plants. District Energy in the U.S. city of Saint Paul, Minnesota, provides electricity, heating, and cooling to its customers; 70 percent of its fuel is local wood waste.[50]

The world's petrochemical, glass, metal, and other heavy industries offer enormous potential for using waste heat through CHP and through capturing and reusing "cascading" heat for lower temperature uses. Mittal Steel, on the southern shore of Lake Michigan in the United States, captures high-temperature heat released from 250 ovens used to produce coke for its blast furnace; this heat energy, which was formerly vented, today produces 93 MW of electricity plus useful steam. As a result, Mittal saves $23 million and avoids 5 million tons of CO_2 emissions annually.[51]

In China, energy-intensive industries account for almost half of energy use. Nearly 30 percent of large steel furnaces and most cement manufacturers in this country do not capture and reuse waste heat, so the savings potential is enormous. Thus China has been called the "Saudi Arabia of waste heat." A large Baosteel furnace uses waste heat to generate 192,000 kilowatt-hours of electricity a day, enough to meet the needs of more than 43,000 average Chinese. In eastern China, CHP plants are gradually replacing individual kilns and boilers to heat industrial parks and residential facilities clustered with factories.[52]

Anaerobic digesters decompose organic matter in the absence of oxygen to produce biogas for cooking or transport fuel or to generate electricity, as well as create high-quality compost for fertilizer. Biodigesters, fed primarily with animal manure, are widespread throughout India, Nepal, China, and Viet Nam and provide cheap fuel while reducing pollution and diseases caused by untreated waste. On a larger scale, dozens of municipalities in Sweden convert human sewage to biogas for transport fuel; biogas is also available as vehicle fuel in Austria, France, Germany, and Switzerland.[53]

Fats and waste oils can be converted into renewable diesel and jet fuel, which can be transported through existing pipelines. Anything that contains carbon, oxygen, and hydrogen—including construction debris, waste paper, plastic, wood, and lawn trimmings—can be turned into some form of motor fuel today, which has the added benefit of extending the life of landfills. The challenge for many of these technologies is obtaining the capital to scale up and commercialize.[54]

While still in the early stages of development, algae can convert as much as 80 percent of the CO_2 released from coal and natural-gas-fired power plants into biomass. Algae can be used in power plants as fuel or converted into bioethanol, biodiesel, or biogas and provide high-protein feed for livestock and aquaculture. It can grow in polluted or salt water, on nonarable land, or at wastewater treatment facilities. It requires far less water than most biofuel crops, produces several times the biofuels per hectare, and can be productive even in desert regions. Harvesting and processing algae is an energy-intensive process, but it offers the potential to "burn carbon twice"—providing additional energy for each unit of CO_2 emitted and an alternative to long-term physical storage of carbon dioxide.[55]

Using energy more efficiently can reduce emissions even further. For example, lighting accounts for 19 percent of world electricity consumption, yet technologies available today, including compact fluorescent lamps and light-emitting diodes, could halve electricity use for lighting. Realistically, it is feasible to eliminate at least one third of global electricity consumption for lighting simply by changing lightbulbs—saving money and avoiding about 450 million tons of CO_2 in the process. By reducing waste in the production and use of energy, more energy services can be provided with lower carbon dioxide emissions.[56]

Scaling Up Renewables

Some analysts conclude that only very large facilities such as nuclear power, large-scale hydro, or large coal plants with carbon capture and storage can meet the world's rapidly growing energy needs. Renewable energy, it is argued, is too small-scale and too dispersed to make more than a modest contribution. But experiences with renewables in Germany and elsewhere prove otherwise, as described in the next section.

Furthermore, large projects cannot produce any power until construction is completed, which can take a very long time. Consider, for example, that a 1,000-megawatt power facility takes approximately 10 years to complete. If all goes well and it operates at full power in year 11, it will produce almost 8.8 million megawatt-hours of electricity that year. Now consider starting at the same time construction of a modular unit that can produce one tenth as much electricity per year as the single large unit but that begins producing power at the end of year one. This process is repeated each year until 10 modular units have been built and come online in each of 10 years. If each one operates at full capacity, by the end of the eleventh year the modular units will have produced nearly five times as much power as the large unit produces in its first year of operation, and after that the two facilities will produce the same amount annually.

There are demonstrated advantages to modularity when it comes to scaling up production, even with fossil fuels. Most of the thermal electric power capacity introduced over the past 10 years in North America has been natural-gas-fired turbines for several reasons: they have become exceedingly efficient, their unit cost is low because of economies of scale, and they can be produced quickly in modules of 50–100 MW and installed within a year. Rapid installation means a low cost of borrowing and a better match and immediate production of power upon installation. Incentives in the deregulation process have also encouraged installation of these units.[57]

The magnitude of the evolution required is vast, but it is achievable. In 2007 wind power was the largest single source of new capacity in Europe and second only to natural

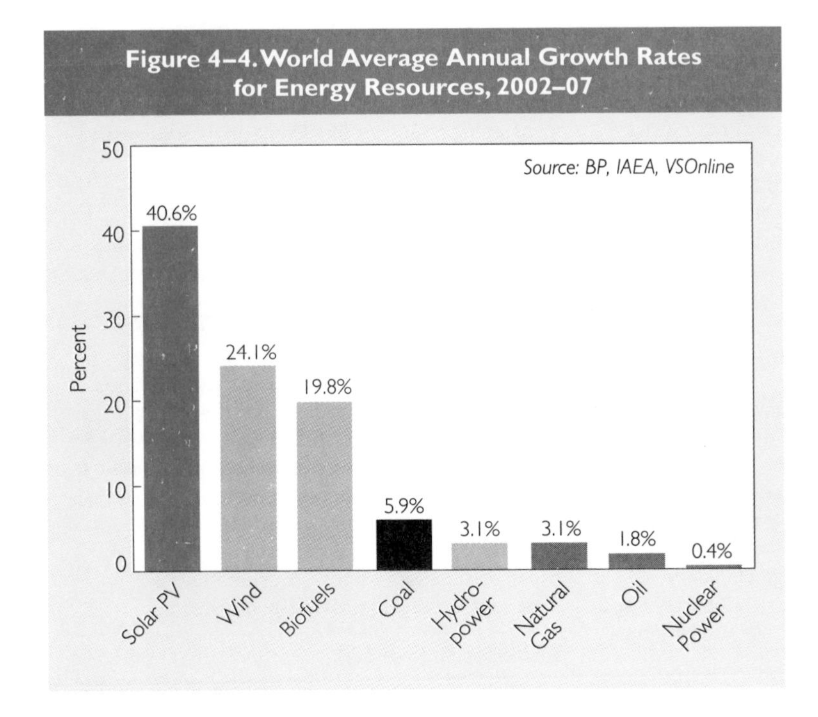

gas in the United States. Globally, even new solar photovoltaic capacity exceeded that of newly installed nuclear power capacity that year. And renewable technologies continue to advance—for example, a new PV technology introduced in 2007 bypasses silicon as a base material, could lower costs by 75 percent, and allows for increased rates of production.[58]

More countries are also joining the transition, promising to push growth rates even higher for the manufacture of and demand for renewables. Indian wind turbine manufacturers are acquiring European and North American suppliers and markets and are now among the top global producers and installers of wind turbines. China was barely in the wind business in 2004 but ranked third after the United States and Spain for new installations in 2007. Similarly, in 2003, China manufactured 9 MW of PV cells—1 percent of the global total. But by 2007, by some estimates, Chinese companies passed Japan and Europe to lead the world in solar PV production. China could account for two thirds of global production by 2010.[59]

Current growth rates indicate that wind, solar, and biomass plants can be manufactured at rates that are comparable to large-scale conventional power projects. (See Figure 4–4.) In 2002–07, photovoltaics grew at an annual average rate exceeding 40 percent and wind's average growth rate topped 24 percent. Annual PV and wind growth rates have actually accelerated in recent years. If current growth rates continue, tapping the wind will generate more electricity than nuclear power in 2020.[60]

Such massive undertakings have succeeded in the past. The U.S. public works projects of the Great Depression, the vast numbers of airplanes and warships built for two world wars, and the enormous number of automobiles manufactured annually provide testimony to possible rates of scaling up. It is a matter of setting priorities and having the political will to establish effective and long-term policies that support a new energy economy. The resources and capabilities exist. By one estimate, if two thirds of U.S. truck production were redirected to the production of wind turbines, about 100,000 MW of wind capacity—the cumulative total installed globally by early 2008—could be manufactured annually in the United States alone.[61]

Of course, energy will be required to move

Box 4–2. Replacing Old Power Plants

Almost all energy-using and energy-producing technologies have natural cycles of capital stock turnover, ranging from 3–4 years for computers to 10–20 years for vehicles and 50–150 years for buildings. Power plants, in contrast, get life extensions. Components degrade at different rates, and if there is no required maximum lifetime they are replaced as needed. Some components could be 40 years old and others two months old, providing a disincentive for utilities to ever retire their plants.

In the United States, power plants constructed before a specific year are exempt from air quality standards unless they are "substantially upgraded." Utilities therefore hold on to old inefficient power plants, retrofitting the minimum required to keep plants operating for as long as possible. As a result, the average efficiency of U.S. coal-burning power plants is only 33 percent and the median age is over 40 years. A similar situation faces heavy coal-producing and coal-consuming countries like China, India, Indonesia, Australia, and Russia.

Economies that successfully reduce their electricity use will have a comparatively easy time retiring older, less efficient, high-emissions plants as demand falls. Those with stable demand must decide which plants to shut down as they wear out or become obsolete and which technologies will replace them. Rapidly expanding economies with a lot of new capacity might not replace plants for some time, but they must make decisions about future additions. To avoid getting locked further into carbon-intensive fuels, low-carbon plants—distributed or central renewable power capacity—should be installed as new capacity is needed. To accelerate the closing of older plants everywhere, governments must set phaseout dates and provide incentives.

Source: See endnote 62.

away from fossil fuels. This is one reason to use the natural capital stock turnover time to replace older plants when they reach the end of their useful lives. (See Box 4–2.) When existing infrastructure is replaced, as much of the energy embodied in the concrete and steel as possible should be recovered through material recycling.[62]

Building massive numbers of new wind, solar, geothermal, and biomass plants and other renewable systems will also require large amounts of energy. But the energy payback periods for renewables are declining as efficiencies increase. They are already relatively short—three to eight months (depending on wind speed) for a wind turbine and one to five years for today's solar PV panels (depending on cell type and location), which have a lifetime of close to 30 years. And once most renewable technologies are built, no further energy is required to extract and transport fuels for them to operate.[63]

Finally, the dramatic improvement in energy efficiency and the matching of supply to demand that is both required and possible today means that the replacement of existing power generators with smaller units capable of delivering comparable energy services with less energy should accelerate in coming years.

Kicking the Habit

Shifting to a sustainable energy system based on efficiency and renewable energy requires replacing an entire complex system. Can such a transformation be accomplished in time to avoid the worst consequences of climate change? Several communities and countries provide hope that it can. Some of the most rapid transitions have taken place at the local level, as seen in Güssing. Many cities are devising innovative means to finance renewables and expand markets. And several countries are demonstrating that transformation can happen quickly even on a national scale.

In the early 1990s Germany had virtually no renewable energy industry and seemed unlikely ever to be in the forefront of these technologies. Yet within a decade this nation had become a world leader, despite the fact that its renewable resources are a fraction of those available in many other countries. In 2000, just over 6.3 percent of Germany's electricity came from renewable sources. Only six years later, this industrial power—the world's leading exporter—generated more than 14 percent of its electricity with renewables, well ahead of official targets for 2010. The fast pace of growth and associated benefits—from new jobs and industries to an improved environment—led the government to set more ambitious targets in 2007. Germany now aims for renewables to generate 30 percent of the country's electricity by 2020 and 45 percent by 2030, meaning renewables will become the largest power source within the next decade.[64]

Germany's experience provides proof that, with a clear sense of direction and effective policies, rapid change is possible. And Germany is not alone. Denmark's economy has grown 75 percent since 1980, while the share of energy from renewables increased from 3 percent to 17 percent by mid-2008. The Danes aim to get 20 percent of their total energy from renewable sources by 2011 and 30 percent by 2030. Costa Rica, Iceland, New Zealand, Norway, and Sweden aim to be carbon-neutral in a matter of decades, relying heavily on efficiency and renewable energy.[65]

What might a low-carbon or even a carbon-free energy future look like? And how might countries with far larger populations than New Zealand or Iceland, or that use much more energy than Germany, make this transition? The United States, for example, has a thousand times as many inhabitants as Iceland does and uses more than one fifth of the world's energy.[66]

Imagine that it is 2030 and that all new buildings in the United States are zero-carbon, the current goal of the American Institute of Architects. A large share of existing buildings have been retrofitted with better insulation, windows, and doors. And all buildings use the most efficient lighting and appliances available. As the Passivhaus Institute projects and other highly efficient buildings already demonstrate, the remaining modest energy supplies for many buildings can be produced on site with renewables or highly efficient systems. Further, industries can dramatically reduce their energy use by eliminating waste and cascading heat from one step to the next, generating more useful energy with the same amount of fuel. By 2030, the resulting energy and economic savings are enormous, and thousands of new local jobs have been created.[67]

Most buildings and factories are still connected through the electric grid, but a modern, smarter, more reliable grid allows utilities to balance the two-way flow of electricity supply and demand in real time. The smart grid in combination with distributed power production and storage—including electric vehicles that charge when the sun shines on PV-covered homes or parking lots, or at night when the wind is blowing—allows even variable renewables to generate a large share of U.S. electricity.

According to a 2008 report by the U.S. Department of Energy, by 2030 the wind could provide 20 percent of U.S. electricity (assuming that U.S. electricity demand increases 39 percent by then). As a result, the nation's CO_2 and other emissions would be significantly lower than they would otherwise be. Tens of thousands of new jobs would be created and rural economies would flourish as wind farms provided new sources of income for landowners and tax revenue for local communities. If fossil fuel prices remain stable over this period (an unlikely assumption), a 20-percent wind portfolio would cost less than an additional 0.06¢ per kilowatt-hour by 2030, or about 50¢ a month for the average household.[68]

One of the most important steps governments can take to address climate change is to eliminate subsidies for conventional fuels and technologies.

Rigorous studies have not yet been carried out for other renewable technologies, but the potential for increasing energy generation with a hybrid electric power generation system that includes wind, solar, biomass, geothermal, small- and large-scale hydropower, and eventually ocean energy is enormous. A 2007 study concluded that efficiency in concert with renewable energy could reduce U.S. carbon emissions 33–44 percent below current levels by 2030. Efficiency improvements could achieve 57 percent of the needed reductions; renewables could provide the rest while generating about half of U.S. electricity. And the study did not consider electricity storage or highly efficient transmission lines for transporting electricity long distances, nor did it include ocean energy or renewable heating.[69]

The United States is rich in renewable resources. But many other nations are as well, and each country on Earth has a diversity of renewable energy sources to draw on. Some of the fastest-growing economies have some of the best resources—for example, China, India, and Brazil have vast solar, wind, biomass, and other renewable resources.[70]

For the world to avoid catastrophic climate change and an insecure economic future, the transition already under way must be accelerated. Success stories must be scaled up, and strategies must be shared across national boundaries. It is important to realize that countries are at different points in their development trajectory and must tailor their approaches to their particular resources and customize technologies to meet specific needs. At the same time, there are several key regulatory and policy changes that, if implemented broadly worldwide, could put humanity on course to steer clear of the worst impacts of climate change.

Putting a price on carbon that increases over time is a critical first step. To encourage an effective transition, most of the revenue generated in the near term can be redirected to help individuals and businesses adjust to higher prices while adopting and advancing the needed technologies. In the 1990s, Denmark began taxing industry for the carbon it emitted and subsidizing environmental innovation with the tax revenues. At the same time, the government made significant investments in renewable energy. The tax gave industry a reason to stop using carbon-intensive fuel, and advances in renewables provided a viable alternative. By 2005, per capita CO_2 emissions in Denmark were almost 15 percent below 1990 levels. But the global price per ton of carbon will have to rise considerably before needed changes and investments come about worldwide, and institutional and regulatory barriers must be overcome with policies that drive the required revolution.[71]

Policies that begin to wring out energy waste and increase efficiency will be critical for reducing demand growth. A combination of financial incentives, such as low-cost loans and tax benefits to purchase renewable and energy-saving technologies, plus continuously tightening efficiency standards for lighting and appliances, is needed. Regulatory barriers to the introduction of distributed energy and CHP generation must be removed, and codes to improve building performance and to use more renewable space conditioning and daylighting must be introduced. Establishing an energy rating system for all buildings at the time of sale, as some European countries have done, would encourage continuous upgrading of existing structures. Training architects, construction tradespeople, and inspectors is essential for designing and constructing more-efficient buildings. Efficiency improvements will reduce energy use and provide life-cycle economic savings as well.[72]

Regulatory systems must foster innovation and motivate vested interests to speed the transition rather than fighting to maintain the status quo. Governments must make it more profitable for electric utilities to invest in renewable energy and efficiency than it is to build new fossil fuel plants or even continue operating old ones. Within the next decade, many power plants in industrial countries will reach the end of their technical lifetimes, and it will be critical to ensure than they are replaced with renewable options. Some countries are starting to phase out subsidies for coal or even its use—for example, the province of Ontario in Canada plans to stop burning coal by 2014—but others have big plans to build new coal plants. It will be critical to minimize their numbers and to enact policies that encourage industrial and rapidly developing countries to blaze new development paths. And governments must work with utilities to upgrade the electric grid so it can use a multiplicity of technologies, both distributed and centralized, and take advantage of active demand management through information technology. Otherwise it will not be possible to take full advantage of renewable energy sources or many energy efficiency measures.[73]

As Germany's experience demonstrates, policies that create markets for renewable technologies can drive dramatic and rapid change. Under the German feed-in tariff, priority grid access combined with a guaranteed market and long-term minimum payments for renewable power have reduced investment risks, making it profitable to invest in renewable technologies and easier to obtain financing. The policy has created nearly 300,000 jobs, strong and broad public support for renewable energy, robust new industries, and significant reductions in CO_2 emissions—all for the cost of a loaf of bread a month for the average German household. In 2007, emissions trading reduced the country's emissions by an estimated 9 million tons; the feed law avoided approximately 79 million tons of CO_2 emissions and is considered Germany's primary climate protection policy. Several studies have determined that feed-in laws are the most effective and economically efficient policy option for advancing renewable electricity generation. Following Germany's lead, more than 40 other countries, states, and provinces have adopted variations of this law.[74]

Although feed laws and other policies that encourage private investment in research and development (R&D) can play a critical role in technology advancement, public R&D funding is also important. According to the International Energy Agency, R&D funding for low-emission technologies including energy efficiency and renewables declined 50 percent between 1980 and 2004. And these technologies continue to receive a relatively small share of R&D funds. Between 2002 and 2007 in the United States, for example, R&D expenditures on energy technologies totaled $11.5 billion, but only 12 percent was directed to all renewable technologies. The vast majority went to nuclear power and fossil fuels.[75]

One of the most important steps governments can take to improve energy markets and address climate change is to eliminate subsidies for conventional fuels and technologies. According to the U.N. Environment Programme, global energy subsidies now approach $400 billion annually, with the vast majority going to fossil fuels. Eliminating fossil fuel subsidies could reduce global CO_2 emissions at least 6 percent between 2000 and 2010 while giving a small boost to the global economy. Recent analysis shows that 96 percent of the annual rise in energy use is occurring in developing countries that subsidize the price of energy at well below world market prices.[76]

Just as the transition to renewables and more efficient energy use has transformed Güssing and other towns, reforming the global energy economy will lead to major changes in national economies and societies. Renewable energy and efficiency improvements provide energy with little to no pollutants, ensuring that air and water will be cleaner, ecosystems stronger, and future generations healthier. They create jobs—today, by conservative estimates, about 2.3 million people worldwide work directly in renewable technology fields or indirectly in supplier industries. And some of the best renewable resources are in some of the poorest regions of the world. In June 2004, at a major conference on renewable energy in Bonn, Germany, government delegates from several African nations claimed that their countries could not develop without renewable energy. Renewable resources are readily available, reliable, and secure, and no battles will ever be waged over access to the wind or sun. As fossil fuel prices continue to rise, renewable energy prices will fall while technologies continue to advance and economies of scale increase.[77]

The dramatic and rapid changes needed to create this new energy economy appear daunting, but remember that the world underwent an energy revolution of comparable scale a century ago. Soon after Thomas Edison improved the electric lamp, skeptics criticized it with comments like this from the President of Stevens Institute: "Everyone acquainted with the subject will recognize it as a conspicuous failure." In 1907, only 8 percent of U.S. homes had electricity. Henry Ford had produced about 3,000 vehicles in his four-year-old factory, and the mass-produced Model T wasn't introduced until 1908. Few of those who supplied town gas for lighting or who met the needs of the extensive market for horse-drawn carriages felt threatened by impending change. Who could have imagined that by the mid-twentieth century virtually every American home—and millions of others around the world—would have electricity and lighting, that the automobile would redefine American lifestyles, and that the economy would be fundamentally transformed as a result?[78]

Fast forward to 2009. Non-hydro renewables generate less than 4 percent of the world's electricity and only a small percentage of its heating and cooling. We are only beginning to construct zero-carbon buildings, and plug-in hybrid vehicles and high-performance electric cars are just making their debut. Yet who can imagine how the mid-twenty-first century global economy will be transformed by more-efficient use of energy and cost-effective renewable energy sources, and how much they will limit the release of greenhouse gases into the atmosphere? We have a once-in-a-century opportunity to make a transformation from an unsustainable economy fueled by poorly distributed fossil fuels to an enduring and secure economy that runs on renewable energy that lasts forever.[79]

CHAPTER 5

Building Resilience

David Dodman, Jessica Ayers, and Saleemul Huq

Climate change is going to present society with a variety of new challenges. Individuals, households, and communities around the world—but particularly in low- and middle-income nations—will all be affected in the coming years and decades. Changes in mean temperature are going to affect food production and water availability, changes in mean sea level will increase coastal inundation, and more-frequent and more-intense extreme events will result in more damage and loss of life from floods and storms. On top of this, rising temperatures can increase the burden of malnutrition, diarrheal illnesses, cardiorespiratory diseases, and infections. These challenges are felt particularly strongly in some of the poorest regions of the world. As Mama Fatuma, a butcher and long-term resident of Njoro Division in Kenya, puts it: "These days we do not know what is happening. Either there is too much rain or none at all. This is not useful to us. When there is too much rain, the floods that result cause us harm. When there is not enough rain, the dry conditions do us harm."[1]

What can be done to reduce the vulnerability of individuals, communities, and countries to the threats of climate change? Farmers in Njoro Division have come up with a wide array of adaptive strategies. They are switching from wheat and potatoes to quick-maturing crops such as beans and maize, and they are planting any time it rains because there is no longer a clear growing season. Their actions have been supported by community groups that have built rain-harvesting tanks and set up savings clubs, while local government agencies have bolstered local resilience by recommending new species and new cultivation techniques to cope with changes in the climate.

A large proportion of the world's popula-

David Dodman is a researcher in the Human Settlements and Climate Change Groups at the International Institute for Environment and Development (IIED) in London, **Jessica Ayers** is a PhD candidate at the London School of Economics, and **Saleemul Huq** directs IIED's Climate Change Group.

tion is vulnerable to changes like those facing farmers in Kenya, but the risks are not distributed evenly. This reflects profound global inequalities: the countries that have profited from high levels of greenhouse gas (GHG) emissions are the ones that will be least affected by climate change, while countries that have made only minimal contributions to the problem will be among the most affected. The uneven distribution of climate change risk mirrors the existing uneven distribution of natural disaster risk—in 2007, Asia was the region hardest hit and most affected by natural disasters, accounting for 37 percent of reported disasters and 90 percent of all the reported victims. Human-induced climate change is likely to have the heaviest impact on small island developing states, the poorest countries in the world, and African nations. Taken together, these countries form a group of 100 nations, collectively housing more than a billion people but with carbon dioxide emissions (excluding South Africa's) accounting for only 3.2 percent of the global total.[2]

Adaptation and resilience not only can reduce the risks from climate change, they can also improve living conditions and meet broader development objectives around the world. This requires a recognition of the many ways in which experiences of vulnerability and resilience differ: between countries; between children, women, and men; and across class and caste within the same society. It requires accepting the importance of the interactions between human and natural systems and realizing that resilient social systems and resilient natural systems—although not the same thing—do exhibit a high degree of co-dependence. It also requires institutional transformations at a variety of scales: local organizations, local governments, national governments, and international organizations all need to become climate-resilient themselves and to work toward creating the frameworks within which individuals, households, communities, and nations can deal with the challenges of climate change.

Vulnerability, Adaptation, and Resilience

Vulnerability, adaptation, and resilience are all deceptively simple concepts with widely varying meanings. Vulnerability is the basic condition that makes adaptation and resilience necessary. In reference to climate change, it is a measure of the degree to which a human or natural system is unable to cope with adverse effects, including changing variability and extremes. It can be seen as an outcome of the seriousness of the stress and the ability of a system to respond to it. Adaptation is a related concept that refers specifically to the adjustments made in natural or human systems in response to actual or expected threats.[3]

Both vulnerability and adaptation are unevenly distributed, and in many cases it is the most vulnerable individuals and communities who are least able to adapt. As the Intergovernmental Panel on Climate Change (IPCC) concludes, "the technical, financial and institutional capacity, and the actual planning and implementation of effective adaptation, is currently quite limited in many regions." When properly implemented, however, adaptation strategies can help limit the loss of life and livelihoods from changes in mean temperatures and the more-frequent and intense extreme climatic events associated with climate change.[4]

The types and scales of adaptation activities are extremely varied (see Table 5–1), and particular strategies will depend on the nature and context of climatic vulnerability. For example, Cavite City in the Philippines is on a peninsula surrounded by three bodies of water, with about half its population on the

Table 5–1. Examples of Planned Adaptation for Different Sectors

Sector	Adaptation Strategies
Water	Expanded rainwater harvesting Water storage and conservation techniques Desalination Increased irrigation efficiency
Agriculture	Adjustment of planting dates and crop variety Crop relocation Improved land management (such as erosion control and soil protection through tree planting)
Infrastructure and settlement	Relocation Improved seawalls and storm surge barriers Creation of wetlands as buffer against sea level rise and flooding
Human health	Improved climate-sensitive disease surveillance and control Improved water supply and sanitation services
Tourism	Diversification of tourism attractions and revenues
Transport	Realignment and relocation of transportation routes Improved standards and planning for infrastructure to cope with warming and damage
Energy	Strengthening of infrastructure Improved energy efficiency Increased use of renewable resources

coast. Cavite experiences on average two tropical cyclones every year, and the city is also affected by drought and sea level rise, all of which are predicted to be greatly exacerbated by climate change. Currently 10 percent of the population is vulnerable to sea level rise, but a one-meter increase would put around two thirds of the population at risk. People have responded through various adaptation strategies, including building houses on stilts, strengthening or reinforcing the physical structure of houses, moving to safer places during extreme weather events, placing sandbags along the shorelines, and taking up alternative income-generating activities locally or in other areas.[5]

Whereas vulnerability is a particular state and adaptation is a set of activities in response to it, resilience is a less distinct concept. In engineering it means the ability of a material to return to its original state after being subjected to a force; in ecology it often means the time taken for a system to return to a state of equilibrium. Both of these meanings have been applied to human systems, in an analysis that focuses on the ability of individuals, households, and nations to return to "normal" after disrupting events. The legacy of these definitions can be seen in the Fourth Assessment Report of the IPCC, which defines resilience as "the ability of a social or ecological system to absorb disturbances while retaining the same basic structure and ways of functioning, the capacity for self-organization, and the capacity to adapt to stress and change."[6]

But is resilience of this type really desirable? Is it acceptable to return to the "same basic structure" in which some 1 billion people live on less than $1 per day, in which there are 350–500 million cases of malaria each year, and in which around half of the people living in African and Asian cities lack adequate water and sanitation facilities? With this in mind, perhaps it is more appropriate to consider resilience as a process, a way of functioning, that enables not only coping with added shocks and stresses but also addressing the myriad challenges that constrain lives and livelihoods and facilitating more general improvements to the quality of human lives. Resilience as a process therefore needs to take into account the economic, social, psychological, physical, and environmental factors that are necessary for humans to survive and thrive.[7]

Many aspects of resilience are closely associated with a holistic approach to development. Individuals who have access to adequate food, clean water, health care, and education will inevitably be better prepared to deal with a variety of shocks and stresses—including those arising from climate change. Communities and cities that are served by appropriate infrastructure—particularly water, sanitation, and drainage—will also be more resilient to these shocks. Indeed, one of the most significant reasons poor people in developing countries are more at risk from climate change is because they are inadequately served by the day-to-day services that are taken for granted in more-affluent locations.

Resistance and resilience are shaped by a person's access to rights, resources, and assets.

Resilience of this kind requires a variety of components, all of which must be present to different extents, but some of which are more pertinent to human systems or to ecological systems. An overriding component is appropriate human-natural relations: human systems that do not exceed the capacities of the natural systems in which they are located, and natural systems that are not unduly threatened by human activities. Both human and natural systems require a capacity for self-organization in order to deal with threats, and diversity is a key element for both types of system too. As shocks and stresses create changes from the norm, a capacity for learning (to deal with novel threats) and innovation (coming up with new solutions) are also important.

At a practical level, people rely on a variety of assets and entitlements to support themselves in difficult times, including human, natural, financial, social, and physical capital. Resistance and resilience are shaped by a person's access to rights, resources, and assets. Mobilizing these assets, however, requires local institutions and national and international systems of governance to provide the framework within which they can function. All these disparate elements—the ecological, the individual, and the institutional—need to come together in a mutually reinforcing manner to help individuals, households, and communities cope with change, including climate change.[8]

A recent adaptation program in five villages in Kabilas VDC in the Chitwan district of Nepal run by the nongovernmental organization (NGO) Practical Action demonstrates well the necessary links between assets, communities, and institutions in building resilience. The main activities focused on improving access to resources and assets, which involved small livestock distribution, vegetable farm demonstrations, kitchen gardening and organic farming, seed and fruit sapling distribution, and sloping agricultural land technologies. A subsequent evaluation suggests that the project had a positive influence on food production and income generation: vegetable production increased in the project area threefold, and farmers are able to sell surplus vegetables. Some 10 percent of farmers now grow vegetables on a small-scale commercial basis, compared with none prior to the project.[9]

The project was linked to local government structures through representatives of each project community being members of a broader Climate Change Impacts and Disaster Management group, which was registered with the District Administration Office. This group was responsible for coordination and implementation of the project in different villages, encouraging local institutional support for the scaling up of project benefits for replication in other communities. The success of the project was ultimately limited, however, by a lack of access to national financial markets. And there was a lack of support for marketing the surplus agricultural products generated by the project. As a result, people have complained that they have been unable to receive reasonable enough prices for their produce to affect their incomes significantly.[10]

Linking Ecological and Social Resilience

There are many linkages between social and ecological resilience. Human livelihoods and settlements rely on the resources and services provided by natural systems, whether located nearby or far away—a concept often referred to as the "ecological footprint." In the face of environmental changes, the resilience of communities that rely heavily on particular ecosystems will in part be determined by the capacity of those ecosystems to buffer against, recover from, and adapt to these changes and continue to provide the ecosystem services essential for human livelihoods and societal development. In turn, ecological resilience is influenced by the sustainability of the human behavior that has any impact on the environment: an ecosystem is more resilient when resources are used sustainably and its capacity is not exceeded.[11]

Climate is one of the most important factors influencing habitats and ecosystems and the abundance, distribution, and behavior of species. Climate change therefore carries severe implications for the sustainability of the world's ecosystems, and these impacts are already beginning to show. The ability of many ecosystems to adapt naturally is likely to be exceeded during this century by an unprecedented combination of change in climate and other global changes (including land use changes, pollution, and overexploitation of resources). If global mean surface temperatures increase by more than 2–3 degrees Celsius, an estimated 20–30 percent of plant and animal species will be at increasingly high risks of extinction; substantial changes in the structure and function of terrestrial ecosystems are also expected.[12]

The resilience of ecosystems to climate change depends in large part on the stresses, human and otherwise, that are already being faced. Natural systems will be better able to adapt if other stresses are minimized. For example, chronic overfishing, blast fishing techniques, and the pollution of water around coral reefs in South Asia have made them more vulnerable to cyclones and warmer sea temperatures. In this sense, social resilience to climate change may sometimes be at odds with ecological resilience: human adaptive strategies for socioeconomic development may increase pressure on marine and terrestrial ecosystems through changes in land management practices, shifts in cultivation and livestock production, and changes in irrigation patterns. In addition, more resilient and developed communities may have a greater capacity to exploit natural resources to support their adaptive strategies.[13]

Attempting to address ecological resilience from this standpoint might encourage a return to protectionist approaches to conservation in an attempt to minimize the impacts of people on nature. But that could actually decrease social resilience where people are forced away from certain ecosystem services, which may in turn put greater pressure on alternative natural resources that had previously been managed sustainably. Further, such an approach to building ecological resilience overlooks the close relationship between environmental and social vulnerability to climate change and fails to acknowledge the notion of resilience as a process rather than a return to a "stable" state. First, climate change is bringing new pressures to ecosystems, so attempting to "restore" ecosystems in the context of a changing climate is inappropriate. For example, the notion of "invasive species" may become redundant as many species expand and retreat in reaction to changing climate patterns. Second, approaches to building ecological resilience must also focus on the socioeconomic development of the communities dependent on the ecosystem.

The close association between people and the environment is perhaps most apparent in poor rural societies in low-income countries that rely directly on ecological systems for environmental goods and services. Socioeconomic development can reduce dependence on single ecosystems by paving the way for diversification of livelihood activities, while reliance on a narrow range of resources can lead to social and economic stresses on livelihood systems, thereby constraining development. This is not a fixed relationship, however. Many people living in rural areas use diverse approaches to meet their basic needs, although these often rely strongly on land-based resources. At the same time, mono-crop agriculture often allows large landowners to become wealthy on the basis of a reliance on a single activity. This highlights an important point: the strength of association between ecological and social resilience is closely tied to the development context and is mediated by a variety of other factors, including wealth, ownership of land and the means of production, and social networks. In situations where human activity has such direct implications for ecosystem resilience, and where the type and level of these human activities are determined by the resilience of the ecosystem and the wider institutional context, integrated approaches to building resilience are essential.

This social resilience-ecological resilience-development nexus is evident in the coastal fishing communities in the Straits of Malacca in peninsular Malaysia. Following the Nagasaki Spirit oil spill in 1992, a study in Kuala Teriang, Malaysia, found that only 4 percent of individuals from non-fishing households reported any disruption in their activities or other impacts, including the ability to obtain fish for meals. In contrast, losses due to the oil spill were reported by 90 percent of fishers' households. The concentration of impacts in these households demonstrates the fishers' particularly high vulnerability, in part because their livelihood was tied to coastal resources. The resilience of the communities to the oil spill therefore depended both directly on ecosystem resilience (the ability of the coastal ecosystem to buffer against and recover from the impacts of the oil spill) and on the alternative livelihood options open to the fishers as the ecosystem regenerated. In turn, the resilience of the ecosystem was related to the extent to which social systems continued to exploit it in times of stress.[14]

So although building ecological resilience will certainly contribute to social resilience in resource-dependent communities, it is not enough. The institutional factors that tie a society to dependence on a narrow range of ecosystems must also be addressed. Adaptation strategies that focus on either social or ecosystem resilience must take into account the tight interactions between them—and then aim to address both. This requires a holistic approach that addresses the institutional barriers to sustainable development and livelihood diversification, as well as sound and participatory natural resource management strategies.

When considering how to build resilience in the face of climate change, therefore, it is necessary to consider not only the direct impacts of a changing climate on the environment but also the implications this has for social resilience, the feedbacks on ecological vulnerability it may entail, and the wider institutional mechanisms that can enable this cycle to be broken. Building ecological resilience is essential, although not sufficient, for achieving social resilience. But achieving social resilience through sustainable development is essential for reducing pressures on ecosystems so they can adapt in the face of climate change.

Building Rural Livelihoods That Are More Resilient

As climate change has an impact on ecosystems, the livelihoods and well-being of those reliant on the functioning of those systems is clearly threatened. This vulnerability is particularly worrying because 75 percent of the 1.2 billion people who survive on less than $1 per day live and work in rural areas of developing countries. They lack the institutional and financial capacity to cope with the impacts of climate change, and they already suffer problems associated with subsistence production, such as isolated location, small farm size, informal land tenure, low levels of technology, and narrow employment options, in addition to unpredictable and uneven exposure to world markets.[15]

Rural households engaged in subsistence and smallholder agriculture in developing countries have been identified as one of the groups most sensitive to the impacts of climate change because of their high dependence on a climate-sensitive sector within ecologically fragile zones. It is impossible to forecast the impacts on rural households accurately, as livelihood systems are small, complex, diverse, and context-specific and involve a variety of crop and livestock species. Nevertheless, the recent IPCC report on impacts, adaptation, and vulnerability identified several likely impacts of climate change on rural smallholdings in developing countries:

- increased likelihood of crop failure;
- increased diseases and mortality of livestock and forced sales of livestock at disadvantageous prices;
- the sale of other assets, indebtedness, emigration, and dependence on food relief; and
- eventual worsening of human development indicators.[16]

In Bangladesh, for example, agriculture employs more than half of the labor force. Temperature and rainfall changes associated with climate change have already begun to affect crop production in many parts of the country, and the area of arable land is decreasing. On average during the period 1962–88 Bangladesh lost about half a million tons of rice annually as a result of floods, the equivalent of nearly 30 percent of the country's average annual food grain imports. Future climate change trends are set to worsen agricultural conditions; a study by the International Rice Research Institute showed that a 1 degree Celsius increase in night temperatures during the growing season would reduce global rice yields by 10 percent. Such temperature increases are well within the predicted ranges for global warming.[17]

Although local farmers in Bangladesh are generally unaware of the implications of climate change on their livelihoods, they are noticing changes in seasons and rainfall patterns. They have observed that planting seasons have shifted and are shorter and earlier than before; in addition, unusual extremes of temperature—including cold snaps and heat waves—are damaging crops. Temperatures in Bangladesh increased about 1 degree Celsius in May and half a degree in November between 1985 and 1998, and further temperature increases are expected. Extremes are also increasing, and winter temperatures as low as 5 degrees Celsius were recorded in January

2007—reportedly the lowest in 38 years.[18]

The effort to build resilience to climate change in rural livelihoods is one of the best illustrations of the close relationship between ecological resilience, social resilience, and the development and institutional context. There are a range of adaptation options, many of which are already under way in developing and industrial countries, with their effectiveness varying from only marginally reducing negative impacts to bringing about positive benefits for the communities involved. Clearly, the latter is desirable in terms of being resilient.[19]

Many autonomous adaptations to climate change are already occurring in rural areas, where livelihood systems experience a number of interlocking stressors other than climate change and where the most appropriate strategies will incorporate local knowledge and take a livelihoods-first approach. These include altering the timing or location of cropping activities and adopting water storage and conservation strategies in times of water stress. Poor smallholders in areas of ecological fragility tend to have extensive knowledge of options for coping with adverse environmental conditions and shocks. Further, local farmers often have intricate systems of gathering and interpreting weather patterns and adapting their seasonal farming practices accordingly.[20]

In northeast Tanzania, for example, local farmers use very specific indicators to predict the beginning of the rains: increases in temperature; lightning; changing behavioral patterns of birds, insects, and mammals; and three different types of plant changes (flowering, new leaves, and grass wilting). Indicators for the end of rains rely on meteorological factors such as steady rainfall and wind strength, but also fauna and flora signals such as bee swarms and the ripening of seeds. In the same region, the intensity and quantity of rainfall are predicted through local assessments of the distribution of rains, fogs, and sunshine periods.[21]

Strengthening and adapting existing local and indigenous coping strategies is key to enabling and empowering communities to enhance their own resilience. Community-based adaptation strategies use participatory tools to identify, assist, and implement community-based development activities that contribute to resilience building in rural areas in regions where adaptive capacity depends as much on livelihoods indicators as it does on climatic changes.

In Humbane village in Gwanda, Zimbabwe, for instance, traditional methods of rainwater harvesting are being communicated and aided by the NGO Practical Action in order to help rural communities adapt to increasing drought conditions. Rainwater is being captured as it falls and retained in the soil or in tanks below ground so that it can be used later as a source of clean water. By constructing ridges of soil along the contours of fields, rainfall is held back from running off the hard-baked soils too quickly, so that even when rain levels are low, families can harvest enough food.[22]

Tias Sibanda, one of the first farmers on his ward to build contours to conserve rainwater, has experienced significant improvements in the harvest on his farm. Before the program, he used to plant maize on 4.5 hectares but often had nothing to harvest because of droughts. Following the project, he had two crops of maize, yielding 1.5 tons and then 0.75 tons. This meant that Sibanda did not have to buy food that year and even had enough left over to last until the next season, saving about $400—or the equivalent of about 12 goats—on food.[23]

However, resilience building is only effective insofar as it is supported by wider institutional and fiscal support mechanisms.

Efforts to build rural resilience need to address the social processes that have caused particularly vulnerable groups to be vulnerable in the first place and then consider how climate-related stresses can exacerbate risks to rural livelihoods. Higher-level structures are needed to mediate increasing competition for resources, protect poor households against marginalization by powerful actors, and coordinate responses to increasing climate variability. This would reduce the risk that household decisionmaking, which is typically focused on economic opportunities rather than climate-related risk, results in "maladaptive" practices.[24]

All of this requires adaptive institutions—rural institutions that encourage and facilitate resilience and that are themselves resilient in the face of climate change. Institutions and their organizational forms shape the adaptation practices of the rural poor. In order to craft external interventions that can build rural resilience, it is important to understand how local institutions can respond.[25]

The activities of the Direction Nationale de la Météorologie (DNM) in Mali show how government institutions can help rural communities manage climate risks. For the past 25 years, the DNM has been providing climate-related information directly to farmers, helping them to measure climate variables themselves so they can incorporate this information into their decisionmaking. The project has evolved into an extensive collaboration between government agencies and research institutions, media, extension services, and farmers, forming a strong institutional base on which the challenges of climate change can be faced by helping smallholders make the most efficient decisions.[26]

Testimonies from farmers indicate that following the DNM initiative they felt less exposed to the uncertainties of a changing climate and more confident about investing in improved seeds, fertilizers, and pesticides, all of which boost production. Results from the cropping season also show that crop yields and farmers' incomes were higher in DNM program fields than in those where it was not used. However, it is difficult to prove that agrometeorological information is the main reason for the increased yields. Further, measuring the success of any adaptation program is inherently problematic. The impacts are mainly in the form of negative outcomes that have not happened: the houses that have not been destroyed, the people who have not suffered from diseases, and the children who are not malnourished.[27]

Building Urban Areas That Are More Resilient

While cities are often blamed inaccurately for producing the bulk of climate-changing activities, there is little doubt that they are centers of climate vulnerability. Hundreds of millions of urban dwellers in low- and middle-income nations are at risk from the direct and indirect impacts of climate change. As the number of people living in cities and towns has grown—more than half of the world's population now lives in urban areas—so too has the concentration of residents in vulnerable settings. U.N. estimates suggest that at least 900 million urban dwellers in low- and middle-income nations "live in poverty," a situation exacerbated by the greater need in urban areas to pay for housing, water, access to toilets, health care, education for children, and traveling to and from work. Yet this concentration of people and economic activities also provides the potential for effective adaptation, improved resilience, and the chance to meet broader development needs.[28]

Building urban resilience is important, first, because of the scale of the population at risk: a large and growing proportion of those most at risk from climate change live in urban areas. Second, it is important because of the potential economic costs of not having effective adaptation strategies: successful national economies depend on well-functioning and resilient urban centers. Third, it is important because of the vulnerability of these large urban populations to a variety of hazards that will result from climate change, including extreme weather events, floods, and water shortages.

The city of Manizales in Colombia has taken steps to build resilience, particularly by not letting rapidly growing low-income populations settle on dangerous sites.

Urban residents are vulnerable to a wide range of climate change impacts. Changes in temperature may worsen air quality and increase energy demands for heating or cooling, changes in precipitation will increase the risk of flooding and landslides, and sea level rise will lead to coastal flooding and the salinization of water sources. More-frequent and more-intense extreme events—such as tropical cyclones, drought, and heat waves—will also affect human health and well-being. In addition, there will be changes in the types of threats cities are exposed to as people move away from stressed rural habitats and as biological changes mean disease carriers can survive in a wider area.

All these threats come together in cities like Dhaka, the capital of Bangladesh. Its population has grown more than twentyfold in the last 50 years, and it now has more than 10 million inhabitants. Severe floods, most recently in 2007, have had major economic impacts: damaging houses and infrastructure and reducing manufacturing productivity through power outages, increased congestion, and poor health among the work force. Large sections of the city are only a few meters above sea level, and the combination of sea level rise and increasingly frequent and intense storms is likely to greatly increase these risks.[29]

The challenges are often even more acute in smaller urban centers. Elsewhere in Bangladesh, Khulna is a coastal city with a population of 1.2 million. Large parts of the city are frequently waterlogged after heavy rainfall, and there are problems with the salinization of surface water. Despite being neglected by policymakers and researchers, small- and intermediate-sized cities house an increasing proportion of the world's urban population, and the changes that are made—or not made—in these places will have substantial implications for resilience to climate change in future years.[30]

What can be done to build resilience in these settings? Resilience will require improving urban infrastructure, creating more effective and pro-poor structures of governance, and building the capacity of individuals and communities to address these new challenges and move beyond them. In some ways, building infrastructural resilience is the most straightforward aspect of this. After all, cities around the world exist in hostile natural surroundings: much of lower Manhattan is land that has been reclaimed from the sea, and London is protected from major flood events by the Thames Barrier. Even in wealthy countries, however, these measures may not be sufficient to deal with the most extreme events, as was horribly evident in the aftermath of Hurricane Katrina in New Orleans.

In many of the most vulnerable cities, the financial resources to provide this sort of protection are not available. Thus it is necessary to place a high priority on ensuring that the systems that facilitate resilience—at the urban, community, and household scale—are adapted to take into account the threats of climate change. Indeed, effective adaptation is all about the quality of local knowledge and about local capacity and the willingness to act, although this has to take place in the context of transparent and effective systems of local and national governance.

City and municipal governments have a key role to play in facilitating urban resilience. (See Box 5–1.) They participate directly, through the provision and maintenance of infrastructure, but even more important they participate indirectly through encouraging and supporting particular activities by individuals and private enterprises. Municipal authorities often have responsibility for land use planning, which needs to ensure that low-income groups can find affordable land for housing that is not on a site vulnerable to climate change. They also often have responsibility for enforcing building codes, and they can ensure that buildings and infrastructure take account of climate change risks without imposing unaffordable costs on low-income urban residents. Urban resilience can also be facilitated through the adoption of pro-poor strategies that enable individuals to develop sustainable and resilient livelihoods. Indeed, having a solid economic base is one of the main ways to help households cope with the shocks and stresses that will become more frequent as a result of climate change.[31]

All these urban actions to build resilience rely on the active engagement of other local stakeholders and a supportive national government. Higher levels of government have key roles in building urban resilience as they provide the legislative, financial, and institutional basis within which urban authorities, the private sector, civil society, and other stakeholders act to adapt to climate change. A supportive legal system can bolster locally developed responses and provide appropriate guidelines for stakeholders to build resilience at the most appropriate scale. Unfortunately, many bilateral aid agencies and multilateral development banks do not recognize the importance of local authorities in this process and fail to provide adequate support to increase local competence and willingness to act. Redressing this situation would provide a substantial boost to building urban resilience.[32]

Recent activities in Durban, one of South Africa's largest cities, illustrate the practical ways that forward-thinking urban institutions can help cities become more resilient to climate change. The Environmental Management Department in eThekwini Municipality (an expansion of what was Durban Municipality) initiated a Climate Protection Programme in 2004. This has included building an understanding of global and regional climate change science and then translating that into the implications of climate change for Durban. The city developed a Headline Climate Change Adaptation Strategy to highlight how key sectors should begin responding to unavoidable climate change. Most important, the municipality has incorporated climate change into long-term city planning to address the vulnerability of key sectors such as health, water and sanitation, coastal infrastructure, disaster management, and biodiversity. Adaptation strategies of this type yield few obvious short-term benefits but will generate greater rewards as the effects of climate change are increasingly felt.[33]

The city of Manizales in Colombia has also taken steps to build resilience, particularly by not letting rapidly growing low-income populations settle on dangerous sites. Its population was growing rapidly, with high levels of spontaneous settlement in areas at risk from floods and landslides. Local authorities, universities, NGOs, and communities worked together to develop programs aimed not only at reducing risk but also at improv-

Box 5–1. Protecting Watersheds to Build Urban Resilience

In a warmer world, water supply challenges will require new ways of thinking about resilience that go beyond the engineering of pipes and ditches to new nonstructural land management approaches that work with nature to protect the quality and quantity of the resource. Among the pioneers of such new thinking was New York City, which in the early 1990s rejected a proposal to build more water filtration plants in favor of buying and protecting forested land well beyond city lines in the upstream watershed of the Hudson River.

Having a forested watershed may not guarantee more water flow; trees, after all, transpire vast amounts of water into the atmosphere. But the soils that healthy forests develop tend to ensure water filtration, as well as to filter out sediment and impurities from water that flows into rivers. The spongelike forest soil, the product of high carbon content in combination with healthy microbial communities as leaves and other tree litter decay, also holds water and thus moderates the extremes of stream flow—which could temper the water-flow extremes that follow the melting of mountain glaciers.

Some communities are building new institutions rather than water infrastructure in hopes of reducing their vulnerability to hydrological extremes. One such community is Quito, the capital of Ecuador. Set in a bowl-shaped basin in the northern Andes, Quito receives most of its water from mountain grasslands that have long been considered virtual water factories because of their capacity to turn the melting snow and the cold, humid air above glaciers into stream flows that make their way into the metropolis.

As human activities strained the water supply from these grasslands at the end of the last century, however, Quito began investing in an innovative public-private partnership to protect and manage the grassland-covered watersheds above the city. The Quito Watershed Protection Fund (known as FONAG, from its name in Spanish) is funded through a 1.25-percent tax on municipal water in the metropolitan area, supplemented by payments by electrical utilities and donations from private water users. A diversity of outside donors, both domestic and international, have contributed as well. The money raised finances the conservation and protection of the grasslands, wetlands, and upstream forests and natural areas.

FONAG finances predominantly long-term activities—community park rangers for protected areas, reforestation, environmental education, outreach and training, and hydrology. Some short-term interventions or projects, such as sustainable production activities and handcrafts, are also supported in order to ensure innovative approaches and promote continuous learning and improvement. With assets of $5.5 million, FONAG had a budget in 2008 of $2.9 million.

One of the Fund's challenges is that hydrological science and information have yet to catch up to the need to work with nature in supplying metropolitan water. River systems are a complex product of physiology, hydrology, biology, and human demands. Data are only beginning to come in that can demonstrate the cost-effectiveness of FONAG's efforts. Trained human resources in the new field are scarce, and institutional capacity is in its infancy. There is no certainty that taking preventive measures today will ameliorate the impacts of future climate change.

Despite these challenges, other cities in Ecuador and in neighboring Colombia and Peru are beginning to replicate the FONAG model of public-private partnerships that aim to conserve clean and abundant water supplies. The days when water resources were assumed to be both renewable and "always there" are fading, especially as dwindling glaciers remind those who depend on high-mountain water that yesterday's climate is no guarantee of tomorrow's.

—*Marta Echavarria, Ecodecisiòn*

Source: See endnote 31.

ing the living standards of the poor. Between 1990 and 1992, some 2,320 dwelling units were built for people in the lowest-income groups, reducing the number of households in high-risk zones by 63 percent and allowing 360 hectares of land to be reforested as eco-parks with strong environmental education components.[34]

Unfortunately, actions of this type are not widespread and are particularly rare in small- and medium-sized cities. Where cities and urban institutions have addressed climate change issues, this has usually been from the perspective of mitigation, which involves limiting GHG emissions, particularly carbon dioxide and methane, to reduce further climate change. Adaptation is far more complicated to measure, resolve, and bring about. On a more positive note, however, many of the strategies to build urban resilience to climate change also represent good practice for urban development more generally. More-responsive local governments, improved infrastructure, and better systems for disaster preparedness are all key for improving the quality of life for urban residents more generally, as well as for building resilience.[35]

Financing Resilience

One major constraint on building more-resilient local and national institutions has been the limited funds available. The costs of adapting to climate change are huge: while accurate calculations are difficult, the World Bank estimates the amount at between $10 billion and $40 billion annually for "climate-proofing" investments in developing countries. This estimate has been criticized, however, for failing to take into account several additional factors, such as climate-proofing existing supplies of natural and physical capital where no new investment was planned, financing new investments specifically to deal with climate change, and the adaptation costs faced by households and communities. Taking these into account, Oxfam recently put the figure at over $50 billion annually.[36]

Currently, funding for adaptation falls woefully short of these figures. There are two main avenues for financing resilience in developing countries: formal climate change financing mechanisms under the United Nations Framework Convention on Climate Change (UNFCCC) and various national mechanisms for official development assistance (ODA).

Funding for adaptation under the UNFCCC is currently disseminated through four funding streams: the Least Developed Countries Fund (LDCF), established to help developing countries prepare and implement National Adaptation Programmes of Action; the Special Climate Change Fund (SCCF), intended to support a number of climate change activities such as mitigation and technology transfer, but with adaptation as a top priority; the Global Environment Facility (GEF) Trust Fund's Strategic Priority for Adaptation (SPA), which pilots operational approaches to adaptation; and the Adaptation Fund (AF), which under the Kyoto Protocol is intended to help developing countries carry out "concrete" adaptation activities.[37]

The LDCF, SCCF, and Trust Fund are relatively small funds based on voluntary pledges and contributions from donors. All three are managed by the GEF, the primary financial mechanism used by the UNFCCC. The LDCF and SCCF only amount to around $114 million in received allocations. The GEF Trust Fund SPA contains $50 million to support adaptation pilot projects. The Adaptation Fund of the Kyoto Protocol is financed by a 2-percent levy on transactions in the Clean Development Mechanism (CDM), the market-based initiative under the protocol that allows countries with greenhouse gas reduction targets to generate emission "credits" from projects that offset emissions in developing countries and produce sustainable development benefits. The Adaptation Fund has the potential to generate by far the largest amount of funds for adaptation; the revenue generated from the CDM levy alone is projected to be between $160 million and $950 million. There is also talk of applying the levy to international air travel, which could generate $4–10 billion annually.[38]

The funds managed by the GEF have been heavily criticized for failing to meet the needs of vulnerable developing countries. These nations have expressed concern over difficulties in getting funds for adaptation through the GEF due to burdensome criteria for reporting, additionality, and co-financing, which most vulnerable developing countries simply do not have the capacity to meet. In addition, while international adaptation efforts have delivered some information, resources, and capacity building support, they have yet to facilitate significant on-the-ground implementation, technology development or access, or the establishment of robust national institutions to carry the adaptation agenda forward.[39]

The Adaptation Fund is the most promising financing vehicle, not only because it has the potential to generate the greatest amount of money but also because of its unique governance structure, decided upon at the Conference of the Parties to the UNFCCC in Bali in 2007. The fund is not managed by the GEF. It has its own independent board with representation from the five U.N. regions as well as special seats for least developed countries and small island developing states. The GEF provides secretariat services on an interim basis. Further, countries can make submissions for funding directly to the Adaptation Fund as opposed to going through designated implementing agencies (as is the case with the GEF funds), and governments can also designate their own implementing agencies (such as NGOs) to make funding submissions. It is hoped that this governance structure will minimize problems of accessibility and increase the effectiveness of climate change financing for adaptation. In addition, the fund is designed to fund concrete adaptation action on the ground, which has the potential to contribute significantly to building resilience. However, funding through the Adaptation Fund is not yet operational, and even when it is it will still fall short of meeting the full finance costs of adaptation.

A second option for funding resilience building is through existing ODA financing. Given the close relationship between development objectives and building climate change resilience, funding adaptation this way might appear to make sense: sustainable development reduces vulnerability to climate change, while resilience-building activities often contribute to the broader goals of sustainable development. Reaching the Millennium Development Goals, for example—reducing poverty, improving living conditions in urban and rural settlements, providing general education and health services, and providing access to financing, markets, and technologies—will improve the livelihoods of the most vulnerable and in turn improve their resilience regarding climate change. A recent analysis of ODA activities reported by members of the Organisation for Economic Co-operation and Development found that more than 60 percent of development assistance could be relevant to building adaptive capacity and facilitating adaptation. To date, bilateral programs have committed more than $110 million to more than 50 adaptation projects in 29 countries.[40]

Although ODA contributions can complement UNFCCC actions on adaptation, they cannot be seen as a means of "plugging the gap" in adaptation financing. Fundamentally, the responsibility for helping the most vulnerable countries cope with the impacts of climate change ought to be in addition to existing aid commitments. Industrial-country financing for adaptation should be based on the "polluter pays principle," which attributes the costs of pollution abatement to polluters without subsidy, pointing toward responsibility-based rather than burden-based criteria. As highlighted by Oxfam, ActionAid, and many other NGOs advocating for greater UNFCCC funding for adaptation, these monies should not be donated to poor countries as "aid" but are owed as compensation from high-emissions countries to those that are most vulnerable to the impacts of climate change. So although there is clearly a role for development institutions in enhancing adaptive capacity, responsibility for adaptation does not lie with these institutions, particularly when it may compete with other development objectives in partner countries.[41]

At the same time, however, there is certainly room for complementarities. The UNFCCC explicitly provides support for adaptation to climate change rather than climatic variability. In climate negotiations, the distinction is important, informing political questions regarding costs and burden sharing. Yet building resilience through development can address a much broader range of factors contributing to vulnerability on the ground than targeted climate change interventions alone. Further, development assistance can invest in capacity building in partner countries in order to facilitate UNFCCC-financed activities. For example, donors are well positioned to strengthen national capacity, while development practitioners and disaster risk reduction practitioners have a wealth of experience in dealing with reducing vulnerability to climate hazards and extremes at local, subnational, and national scales.[42]

ODA can therefore be used to add value to the formal UNFCCC mechanisms through development that contributes to resilience. But regardless of development investments in resilience building, funding through the UNFCCC must be scaled up significantly, and existing funding needs to be more accessible, in order to come close to meeting the huge challenge of building climate change resilience in vulnerable developing countries.

The responsibility for helping the most vulnerable countries cope with the impacts of climate change ought to be in addition to existing aid commitments.

At the country level, there is a need to think carefully about the delivery mechanisms for this funding. Experience with adaptation funding under the UNFCCC has shown that national institutional responsibilities for adaptation are unclear and sometimes competing. With the proliferation of bilateral, multidonor, and convention funds, it is vital to avoid duplication of efforts and to ensure consistency in approach.[43]

One solution that is currently being piloted in Bangladesh is the development of a country-owned multidonor trust fund to receive all funding for adaptation from different national and multilateral climate change funds. This fund was launched in London in September 2008 and will consist of contributions from the Bangladeshi, British, Danish, and Dutch governments as well as from the World Bank. It is hoped that this framework will significantly reduce transaction costs for global and bilateral funds and could pave the way for large flows of money in the future, while ensuring proper institutional structures, governance, management, and targeting of funds at the national level. The scale and scope of the fund are currently under discussion, but it is hoped that the fund would be accessible to government agencies, NGOs, and the private companies that design and implement climate change mitigation and adaptation projects.[44]

It is also important that the institutions and agencies that channel funds to vulnerable people on the ground are given careful consideration. The role of government institutions is clearly vital, and mainstreaming adaptation through existing national, sectoral, and local development plans is one way to build the resilience of government institutions and services, and, by extension, of the people who use them. However, this is not always appropriate—for example, under weak governments or in situations where local government capacity and accountability are flawed. In such circumstances, building resilience through NGOs and private institutions may be more appropriate.

Early NGO efforts to enhance resilience of vulnerable communities have proved promising, but they have also been limited in scale and scope where they have not been closely linked to and supported by governmental plans and institutions. Another avenue that is receiving increasing attention is the private sector, particularly in relation to technology transfer and insurance schemes. One example is index-based insurance, still in the formative stages, which is intended to facilitate adaptation by farmers by discouraging risk-averse behavior. Crops are insured against weather patterns rather than crop losses, which also reduces perverse incentives to underproduce.

The social enterprise BASIX has been piloting an index-based insurance program in India, which initially has grown from 230 customers to around 12,000 in 2006–07. However, an early review of this program shows that it has not fulfilled expectations for encouraging adaptive strategies of poor and vulnerable farmers. One reason is the high cost of the product (5–12 percent of the value insured), which reduces the coverage that clients can afford. The vast majority of clients insure only their inputs, not the projected value of their crops. Coverage is therefore not sufficient to encourage risk-taking. Further, reinsurers are increasing prices on these policies because of growing climate change concerns, putting the product even further out of the reach of many potential customers.[45]

The private sector may produce some options for financing resilience building, but these have so far been shown to be limited, particularly for enabling adaptation by the most vulnerable. Therefore attention is currently best focused on improving the capacity of existing local institutions that have knowledge and a history of working with the most vulnerable, in order to lessen gaps between local and national processes and to ensure that financial resources reach those who can use them best. This requires the involvement of local groups and civil society organizations with the knowledge and capacity to act, as well as a willingness among governments to work with lower-income groups.[46]

Linking Mitigation and Adaptation

Although this chapter has focused on responding to the impacts of climate change, mitigation is another response strategy that is both necessary and urgent to ensure long-term resilience. As Tom Wilbanks and colleagues note, "if mitigation can be successful in keeping impacts at a lower level, adaptation can be successful in coping with more of the resulting impacts". Until recently, mitigation and adaptation have been considered separately in climate change science and policy. Mitigation has been treated as an issue for industrial countries, which carry the greatest responsibility for climate change, while adaptation is seen as a priority for developing countries, where the capacity to mitigate is low and vulnerability is high.[47]

But in order to maximize global resilience against climate change, any post-Kyoto arrangements will have to engage developing countries in the mitigation agenda. And mitigation strategies that offset carbon in developing countries through carbon trading mechanisms such as the CDM and the voluntary carbon market have the potential to bring sustainable development benefits, which can contribute to climate change adaptation and resilience. Taking this further, attention has recently started to focus on exploring synergies between mitigation and adaptation, to see if they can be achieved together, contributing to both short-term local and long-term global resilience.[48]

Linking mitigation and adaptation at the national and sectoral level is problematic, because the needed actions and policies involve different sectors. Mitigation actions tend to focus on transport, industry, and energy, while adaptation decisionmakers usually focus on the most immediately vulnerable sectors such as agriculture, land use, forestry, and coastal zone management. There is some potential for overlap at the sectoral level, however: for example, adaptation policies on agriculture, land use, and forestry have implications for carbon dioxide sequestration and avoided methane emissions.[49]

Achieving synergies between mitigation and adaptation strategies is most fruitful at the project level, where the activities are linked in very specific ways. In Dhaka, for instance, a CDM mitigation project uses organic waste to produce compost. This reduces methane emissions by diverting organic waste from landfills (where anaerobic processes occur that generate higher levels of methane) to a composting plant (where aerobic processes occur).[50]

This mitigation project has clear potential for contributing to climate change resilience in rural areas. The impacts of climate change will include agroecosystem stresses in drought-prone areas in Bangladesh. Thus, enhancing soil organic matter content through organic manure to increase the moisture retention and fertility of soil both reduces the vulnerability to drought and increases the carbon sequestration rates of crops. Linking mitigation and adaptation in this way contributes to both long-term and short-term ecological and social resilience. The composting projects contribute to global resilience by reducing GHG emissions directly through preventing the generation of methane and indirectly through contributing to the carbon sequestration capabilities of crops. They also build local resilience through soil improvement in drought-prone areas, as poverty is exacerbated when climate change reduces the flows of ecosystem services.[51]

An integrated approach could therefore go some way toward bridging the gap between the development and adaptation priorities of developing countries and the need for mitigation at the global level. In addition, this will increase the relevance of mitigation for the most vulnerable developing countries, moving beyond the perception of mitigation as only an issue for industrial nations and helping to engage even the poorest developing countries in global mitigation efforts. Building resilience requires climate change response measures that bring together integrated climate change, development, and resource management concepts to build adaptive capacity. Linking mitigation at the project level is one way of achieving such an integrated approach, to build local and global resilience now and for the future.[52]

Bouncing Forward to Greater Resilience

Low- and middle-income countries—and particularly their poorest inhabitants—are at the frontline of climate change. The political difficulties in drawing attention to their plight and the practical difficulties of measuring increases in resilience present serious challenges.

As discussed at the beginning of this chapter, building resilience to climate change is not only about ensuring that individuals, communities, and nations can maintain their current situation. It is also about improving living standards in a way that does not worsen the problems of climate change. When billions of people around the world lack adequate water supplies, building resilience involves extending the provision of water and sanitation. When millions of children die of preventable diseases each year, building resilience means improving child mortality figures rather than merely preventing them from increasing. Rather than thinking about resilience as "bouncing back" from shocks and stresses, it is perhaps more useful to think of it as "bouncing forward" to a state where shocks and stresses can be dealt with more efficiently and successfully and with less damage to individual lives and livelihoods.

Bouncing forward will require new commitments from influential actors at a variety of scales, including NGOs, local and national governments, and international bodies. The good news is that building resilience to climate change will also help address many of the environmental, health, and developmental challenges facing the world today. And as the farmers in Njoro Division in Kenya, the climate scientists in Mali's Direction Nationale de La Météorologie, and the planners in Durban have all shown, practical actions to support human ingenuity can yield impressive results.

CHAPTER 6

Sealing the Deal to Save the Climate

Robert Engelman

The atmosphere is kind. It takes the carbon dioxide (CO_2) and other heat-trapping greenhouse gases that humans create and disperses them equally all over the world. But that is also its cruelty. The accumulation of these waste gases over the decades, disproportionately from industrial countries but increasingly from some rapidly developing ones, is overwhelming the planet's energy balance and heating up its surface. This accumulation must end, but how that will happen is hard to imagine. The mechanisms needed must engage all humanity in ways that are manifestly fair to all.

Saving the global climate and protecting ecosystems in a warming world must become a national interest for each of nearly 200 sovereign states. Negotiating a successful treaty that achieves this will be a diplomatic feat unlike any in history, given the stark inequalities in per capita emissions levels and income—and all the harder given that solving the climate problem will likely require some real sacrifice.

This is nothing like war, in which military might defeats the enemy and dictates the peace. Rather it is an emergency with long-term risks comparable to world war but requiring the surrender of no one and the cooperation of all. An economically and demographically diverse world of 6.7 billion people reaches for more energy, food, mobility, and creature comforts even as it enters the early stages of human-driven warming. And that world grows by 78 million people each year.[1]

In a tragedy of the commons as big as all outdoors, each country benefits directly from actions within its borders that release greenhouse gases, but the emissions themselves dissipate into thin air and spread their impacts globally. The atmosphere recognizes no borders and considers no molecule an illegal immigrant. And there is an added twist of inequity to this commons: the people least responsible for loading the air with heat-trapping gases tend also to be the ones most vulnerable to the impacts of the warming now beginning. (See Box 6–1.)[2]

Defying the natural imbalance of national

Box 6–1. Equity and the Response to a Changing Climate

Many scientists expect that poor countries with little responsibility for today's climate instability will be hit hard by climate change. This asymmetry of circumstance prompts a pressing question: Can climate treaties be built on strong principles of fairness?

In truth, equity already plays a role, albeit a limited one, in climate agreements. The Kyoto Protocol, for example, is based on the principle of "common but differentiated responsibilities," which recognizes different obligations for parties in different economic and emissions positions. And the Kyoto negotiating positions of many countries—from France and Iran to Brazil and Estonia—incorporated specific equity dimensions.

But fairness concerns are likely to assume a higher profile in future climate negotiations as the demands of climate stabilization become more burdensome. Two nagging questions in particular have equity at their core: How should rights to emit greenhouse gases be allocated? And who should bear the costs of emissions reductions and adaptation to climate change?

A broad range of answers is given to these questions—each grounded in one or more climate equity principles. On emissions rights, for example, two very different principles are often cited by proponents of allocation schemes:

- The Egalitarian Principle states that every person worldwide should have the same emission allowance. This principle gives populous countries the greatest number of emissions rights. India, for example, with 3.8 times as many people as the United States, would be entitled to 3.8 times the emissions allowance available to the United States.
- The Sovereignty Principle argues that all nations should reduce their emissions by the same percentage amount. Large emitters would make large absolute reductions of greenhouse gases, while low-volume emitters would make smaller absolute reductions. Thus under an agreement to reduce emissions of carbon dioxide by, say, 10 percent, the United States would cut output by some 579 million tons of CO_2, while India would reduce its emissions by 141 million tons.

Two other principles are often invoked to determine the economic burden of curbing

and global interests, many countries—especially those of the European Union (EU) and, impressively, China—have been acting in recent years to slow the growth of their emissions. Representatives of most of the world's governments have been meeting regularly since the late 1980s to craft ways in which all countries can agree to stop changing the planet's climate. Most nations—although not the historically largest emitter, the United States—ratified an international climate agreement termed the Kyoto Protocol, which went into force in 2005. The agreement requires industrial-country signatories to control emissions of carbon dioxide and five other key greenhouse gases to levels somewhat below (or in a few cases somewhat above) those recorded in 1990.[3]

What Will It Cost?

The requirements to control greenhouse gases have economically benefited some developing countries that signed the Kyoto Protocol but are not obligated by it to cut their own emissions. And they probably have meant the avoidance of some emissions that would have occurred. By official count, trading in 2006 and 2007 in emerging worldwide carbon markets—a novel mechanism that has arisen from international climate agreements—will prevent an estimated 1.5 billion tons of CO_2-equivalent emissions. This is less than 2 percent of global emissions in those two years—not enough to noticeably slow the warming in progress, but possibly a start.[4]

In working toward the Kyoto goals, about $19.5 billion moved from industrial to devel-

Box 6–1. continued

climate change for different nations:

- The Polluter Pays Principle asserts that climate-related economic burdens should be borne by nations according to their contribution of greenhouse gases over the years. Since 1950 the United States has emitted about 10 times as much CO_2 as India; using this historical baseline suggests that the U.S. bill for dealing with climate costs should be about 10 times greater than India's. (The difference would be greater still if the baseline were set at 1750, roughly the start of the Industrial Revolution.)
- The Ability to Pay Principle argues that the burden should be borne by nations according to their level of wealth. If gross domestic product figures are used to determine how much each country pays, the U.S. responsibility would be some 12 times greater than that of India.

A 2006 survey of climate negotiators from a broad range of nations revealed that the vast majority believe equity considerations should figure in climate negotiations. The survey found a relatively high degree of support for the Polluter Pays and the Ability to Pay Principles, and a relatively low degree of support for the Sovereignty Principle, consistent with a general sense in the international community that wealthy historical emitters should pay more and poor countries should pay less.

In the end, agreement on emissions allocations may require a mixture of different principles. Some analysts, for example, see egalitarianism as a desirable long-term equity goal, with other principles used to transition to an egalitarian outcome.

These four equity principles address only the distributional dimension of climate equity concerns. Other principles are used to assess the equity of outcomes (how fair is the result of climate negotiations?) and of process (how fair is the procedure by which deals are negotiated?). The result is a thicket of principles, often conflicting, that will compete for policymakers' attention as climate negotiations unfold in the years ahead.

—Gary Gardner

Source: See endnote 2.

oping countries during those two years. (This figure, while impressive, is less than a fifth as much as the money transferred annually from industrial countries in development assistance—$107 billion in 2005—and is dwarfed by the remittances that immigrants send to their home countries, which totaled $300 billion in 2006). These payments have come through the Kyoto Protocol's Clean Development Mechanism (CDM), designed to reward industrial nations for emissions reductions they effectively purchase from developing countries by sponsoring energy development projects that are less emissions-intensive than would have been constructed otherwise.[5]

A worldwide network of carbon markets worth $64 billion in 2007 has developed, with $50 billion of that moving through the European Union's Emissions Trading Scheme. Both these numbers are more than double their values in the previous year. Officially they imply the retirement, avoidance, or other offsetting of 3.0 billion tons of CO_2-equivalent emissions. As with other high-finance instruments, however, emissions credits are often held and resold multiple times, so the emissions avoidance that underlies many credits might not become real for years.[6]

One little-noted source of greenhouse emissions reductions is an international environmental agreement not directly related to climate change: the Montreal Protocol, which went into force in 1989. Countries agreed to phase out the production of gases that eat away the atmospheric ozone shielding the world from hazardous levels of the sun's ultraviolet radiation. Since these gases powerfully add to the warming of Earth's surface, phasing them out offers a double benefit. Some of the gases now moving into production to replace those that deplete ozone also trap heat, however. So the final impact of the Montreal Protocol on climate depends to a large extent on future production levels of these newer greenhouse gases.

All this said, greenhouse gas emissions have been rising significantly—and, in recent years, at an accelerating pace—despite ongoing diplomatic efforts and the growth of a market designed to reduce CO_2 emissions. The leading economy in this greenhouse emissions boom is now China, the world's most populous and economically dynamic country. The government there has given priority to the development of renewable energy and has committed to reducing the carbon-dioxide intensity of its economy. Yet coal-reliant China has singlehandedly accounted for two thirds of the world's growth in carbon dioxide emissions from electric power generation since 2000. This is probably the best example of one of the problems that most hinders a global climate solution. The United States and other industrial countries account for an estimated 76 percent of all greenhouse gas emissions from 1850 to 2002. But developing countries—with their more rapidly growing population and economies—will drive the bulk of the buildup expected in the future.[7]

Vast tracts of new forests and a conversion of most of the world's farms to practices that allow soil to capture and store atmospheric carbon could remove some of the buildup of carbon dioxide. (See Chapter 3.) As climate change raises the risk of forest fires and droughts, however, it will be hard to be certain that carbon stays securely locked away in farms and forests. Such approaches nonetheless offer one exit strategy for the CO_2 already in the atmosphere. But they need to be paid for by the wealthier countries that are responsible for most past emissions. And to prevent as many future emissions as possible, the world's wealthier countries will need to finance much or even most of the reductions needed in poorer countries—whether these reductions come from avoiding deforestation and land degradation or constructing wind turbines rather than coal-fired power plants—as well as those achieved within their own borders.

The $19.5 billion provided in 2006 and 2007 by a few industrial countries for emissions reductions in a few developing countries helps blaze a path toward the reductions the world needs. But the path must very soon become—to use an inappropriate metaphor—a multilane highway. And this highway awaits construction, even as industrial countries themselves need to invest massively to boost energy efficiency at home, shift from fossil fuels, and develop climate-friendly ways to produce food, goods, and services.[8]

Among the most respected estimators of the total global costs of this transition is Nicholas Stern of the London School of Economics and Political Science, who pegs the needed spending at 2 percent of gross global product for decades to come. That works out to more than $1 trillion a year—a daunting figure, but smaller than the $1.5 trillion that oil consumers send annually to oil producers, and much less than the $4.1 trillion the world spends on health. These comparisons help put in perspective the public relations challenge of financing a truly significant reduction of climate change risk. Yes, improving energy efficiency and shifting from fossil fuels helps countries deal with high energy prices, avoid pollution, and build energy independence. But based on current experience, these motivations fall far short of what will be needed to really "save the climate." Will most people come to see reducing that risk as comparable in importance to their own need for good health?[9]

Without more insistent public outcries about the risks climate change poses, that will not happen soon—maybe not until impacts are much more severe and the process is all the harder to stop. In that future, the world may face the real and incalculable long-term costs of past inaction and think wistfully about lost opportunities to invest in emissions prevention. Still, the upfront costs of effective prevention today seem huge, with uncertain benefits. And the size of the needed financing is only one among many obstacles to arriving at a workable world climate pact.

Who Will Emit?

Given the challenges, it is not surprising that the current negotiating process on climate change is forbidding in its complexity and far from any certainty of success. Even on financial issues that many governments take more seriously than climate change, negotiations sometimes founder. In July 2007 a round of world trade talks that had continued for seven years suddenly collapsed in unbridgeable disagreement, with no prospect they would start up again anytime soon.[10]

But the round of intergovernmental climate talks now in progress under the auspices of the United Nations is the only game on the planet likely to lead to cuts in global emissions on the scale needed. It deserves public attention and political support despite the seemingly impenetrable raft of proposed mechanisms and the tortuous frustrations of working toward an agreement. Given the past resistance of the U.S. government to any international action or commitments on emissions reductions, the new president taking office in January 2009 has an important opportunity. He can demonstrate the leadership the world needs to work out an effective agreement to save not just the global climate but perhaps human civilization itself, in negotiations that will culminate in Copenhagen in late November 2009.

No one knows how much the world can warm above preindustrial levels before the changes become truly catastrophic. But some scientific assessments and their acceptance by the European Union, the U.N. Development Programme, and others suggests that the risk of climate catastrophe approaches an intolerable level if the world's average temperature fails to stay within 2 degrees Celsius (3.6 degrees Fahrenheit) of the preindustrial global average. This is about 1.2 degrees Celsius above the current average temperature. Significant climate risks may lurk even in more-modest temperature increases, especially if they are sustained over time. (See Chapter 2.) Most of that possible safety valve of 1.2 degrees may literally already be baked into the world's existing system, however, continuing to drive more storms, droughts, and sea level rise even if emissions ended immediately. The window of avoiding potential climate catastrophe is thus closing quickly.[11]

Humanity needs eventually to shrink net greenhouse gas emissions to zero, with flows out of the atmosphere balancing flows in. And since the biosphere cannot infinitely absorb these gases out of the atmosphere, in order to avoid continued human-induced climate change the world presumably must someday have negligible emissions of greenhouse gases. Yet all combustion releases heat-trapping CO_2 into the air. All molecules of more than two atoms—from water vapor to methane to the polyatomic industrial gases used in refrigerators and air conditioners—trap Earth's solar heat before it escapes into space and send it back down to the surface. Most people would consider a zero-emission society impossible if it were not essential to a reasonable hope that civilization will continue.

Suppose the world collectively decided to allow 500 billion more tons of CO_2-equivalent emissions before reaching that zero-emissions point. If the world then fairly allocated those precious remaining tons, who would get what? Who would do the allocating, who would enforce it, and how?

Since the vast majority of greenhouse gas emissions now come from the countries and regions that are demographic and economic giants—the United States, the EU, Russia, and Japan among industrial countries and China, India, and Brazil among developing ones—the early participation of these countries in a global atmospheric stabilization program is essential. Over the long term, however, there is no alternative to engaging all countries in a global climate alliance. Absolving smaller or less economically significant ones from the task would risk the evolution of a two-tiered world that would inevitably draw greenhouse-gas-intensive development and possibly even people to the excluded countries. That could not work for long. And besides, all countries and all people have a right and a need to participate in deciding how to resolve this crisis.

Lessons Learned, Time Lost

The upward trends in greenhouse gas emissions over the last two decades trace tracks of lost time. More than two decades have passed since prominent climate scientists first began calling news media and public attention to the growing urgency of the problem. While the signature of human-induced warming is now clearer than it was then, the basic science and the riskiness of stuffing ever more heat-trapping gases into the atmosphere has never been in doubt among the world's leading scientists.

In the late 1980s, the world experienced a test run for the climate talks to come, as nations negotiated and then ratified the Vienna Convention for the Protection of the Ozone Layer and then its subsidiary, the Montreal Protocol on Substances that Deplete the Ozone Layer. With the backing of President Ronald Reagan and most of the world's major producers of the regulated gases, the protocol provided a system by which industrial countries phased out the ozone-depleting gases quickly.[12]

Though rarely recalled today, the Montreal Protocol offers lessons for the climate negotiations of 2009. The U.S. government and chemical manufacturers strongly supported the phaseout of ozone-depleting gases. The agreement allowed developing countries a later timetable and established a global fund to funnel them needed financing from industrial countries. The fund to date has spent $2.3 billion. The agreement defined the dividing line between the two groups by per capita production and consumption. Although the climate problem is far larger and more complex than ozone depletion, each of the elements that help this treaty succeed could contribute to an effective climate agreement.[13]

By 1994, most of the world's nations, including the United States, had ratified and put into force the United Nations Framework Convention on Climate Change, first agreed to at the United Nations Conference on Environment and Development in 1992. That treaty expressed two key principles that have guided global climate negotiations ever since. First, humanity should "achieve...stabilization of greenhouse gas concentrations in the atmosphere at a level that would prevent dangerous anthropogenic [human-induced] interference with the climate system." Second, countries should respond "in accordance with their common but differentiated responsibilities and respective capabilities and their social and economic conditions." In short, stop climate change before it is too late, and expect the longest-running and worst climate offenders—the wealthier and more industrialized countries—to step up to the head of the line to fix the problem.[14]

Three years later, most of the world's nations agreed in Kyoto, Japan, to the protocol to the climate change convention. (In diplomacy, protocols are supplements or amendments to existing conventions; either may be called a treaty.) The Kyoto Protocol aimed to drive down the greenhouse gas emissions of industrial countries as a first step in what was planned to be a two-phase process comparable to that of the Montreal Protocol.[15]

In negotiating the agreement, industrial countries volunteered emission targets in 2012 based on a percentage of what each country's emissions had been in 1990. These targets—averaging a 5-percent emissions reduction among participating countries—were originally intended to be achieved by actual emissions cuts in those countries. To ease fears that such cuts might be too onerous and expensive, however, more flexible mechanisms were allowed—and these quickly became the favored approaches to compliance.[16]

Under the terms of the Protocol, participating industrial countries can trade unneeded emission allotments among themselves or work together jointly on projects that promise to cut emissions in any other participating industrial country. (These cuts, called Joint Implementation, are done within the European Union and in formerly communist countries like Russia and the Ukraine, where aging and energy-inefficient capital equipment can be improved at a relatively low cost.) Or they can invest in projects that achieve the needed reductions in developing countries through the Clean Development Mechanism, which then can sell those reductions as carbon credits to the investing country.

The CDM is the only inducement for emissions reductions in developing countries. For understandable reasons, purchasers of the emissions credits it offers have been drawn mostly to large-scale projects in countries capable of offering such opportunities. Practically speaking, this means a heavy tilt toward China, India, and a handful of other Asian powers, with little activity in Latin America or sub-Saharan Africa. On top of that, critics have noted that the CDM has produced windfall profits for some investors while failing so far to take much of a bite out of global greenhouse gas emissions. These problems are now well recognized, however, and any new climate agreement is likely to reform this mechanism so that it covers many more emissions-saving activities and reaches many more countries. Or perhaps negotiators will craft new approaches altogether to encourage emissions reductions in developing countries that industrial ones will pay for.[17]

Though rarely recalled today, the Montreal Protocol offers lessons for the climate negotiations of 2009.

Like all treaties, the protocol is binding, but penalties for unachieved emissions reductions were deferred into an unknown future. Those who fail to comply must face proportionally greater emissions-reduction obligations following the first "commitment period" from 2008 to 2012. But those obligations and any later commitment period, of course, remain to be negotiated. Some countries, especially in Europe, with its mature economies and generally stable populations, are on track to meet their commitments. Others are experiencing emissions growth that will make the objective much harder. Environmentalists are suing the government of Canada, for example, in an effort to get it to take its Kyoto promises more seriously.[18]

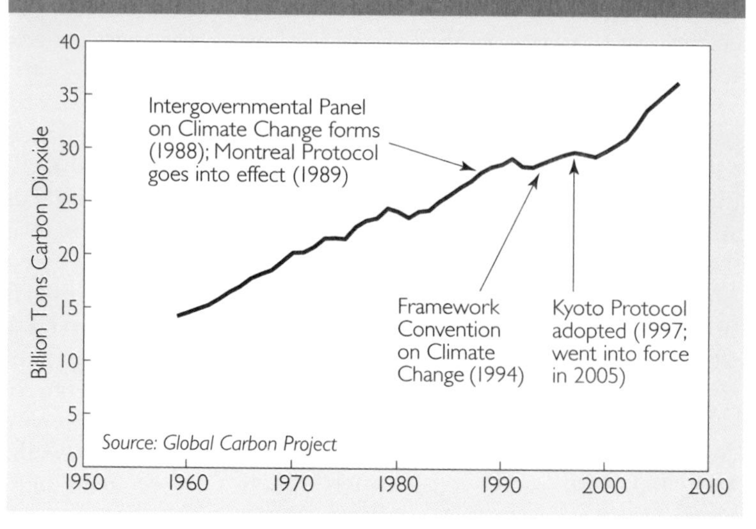

The idea that industrial countries would move first on climate change was firmly rooted in principles accepted in the Montreal Protocol and the Framework Convention on Climate Change. But the Kyoto Protocol's perceived "free ride" for developing countries—some of them now becoming major emitters—provided a rationale for the United States to reject the protocol after initially signing it. The country's substantial emissions were thus left unfettered. U.S. ratification would have been far from easy anyway. Even before U.S. delegates in Kyoto signed the new document, back in Washington the Senate voted 95–0 to oppose its ratification on the grounds it would hurt the U.S. economy and leave developing countries, without comparable commitments, at an unfair economic advantage.[19]

At the time, U.S. emissions were tops in the world. China, rapidly industrializing and with four times the U.S. population of 305 million, has since overtaken the United States in CO_2 emissions from fossil fuel combustion and cement production. But it will be many years before any nation approaches the United States in cumulative greenhouse gas emissions. The country's unwillingness to commit to emissions reductions despite this fact is undoubtedly the greatest single obstacle to international action on the problem. Yet with a new president in office already having declared his willingness to limit emissions, 2009 is the most promising year for real action since ratification of the climate convention in 1994.[20]

Although some emissions have undoubtedly been avoided, none of the scientific and diplomatic efforts on climate has had an obvious impact on the overall global increase in carbon dioxide emissions. (See Figure 6–1.) Although less well documented, the story is similar for other gases and for carbon dioxide from deforestation and land degradation.[21]

There is, however, a real victory for which both the Montreal and Kyoto Protocols deserve thanks. Atmospheric concentrations of greenhouse gases would have grown even faster had neither treaty gone into effect. New international institutions and financial instruments designed to reduce global emissions are riding gingerly forward on training wheels. Chief among the Kyoto Protocol's accomplishments is the remarkable emergence of carbon markets described earlier, which has as its valued commodity, in effect, bad things—carbon dioxide emissions—that are not happening. Global emissions levels have nonetheless so far responded more to the vagaries of the global economy than to diplomacy. The world needs much more effective mechanisms for reversing course in greenhouse gas emissions as rapidly and dramatically as possible, beginning now.[22]

State of Play

Seemingly undaunted by these challenges, today's climate negotiators are building on the mixed outcomes of the Kyoto Protocol to craft a strategy for moving forward. Despite the absence of the United States, parties to the protocol continue to strengthen its provisions and have committed to improving and expanding the carbon trading, Clean Development Mechanism, and other emissions-reducing tools to which it gave birth. The CDM and its governing board, for example, already are moving to shift toward ongoing programs of emissions reductions in a diversity of developing countries.

The European Union, for its part, is modifying its ambitious Emissions Trading Scheme. This is a cap and trade approach in which total industrial emissions (amounting to about 45 percent of CO_2 emissions in the EU) are restricted, the limited emissions are allocated among companies, and unused allocations can be traded for whatever the market will bear. The system has drawn outrage from environmentalists and consumers because of its free allocations of emissions among electric utilities and industries, which provided windfall profits to companies as the value of carbon credits rose. Architects of the system have promised to shift to one of auctioning allocations, with any revenue to be used for climate or other public benefits.[23]

Japan, Canada, and New Zealand also participate in Kyoto Protocol–based carbon trading. And states, provinces, and cities in the United States and other industrial nations are experimenting with their own Kyoto-style emissions-reducing mechanisms and commitments. The province of British Columbia and the city of Boulder in Colorado are taxing carbon, returning the revenue to residents through reductions in other taxes. A carbon-trading exchange in Chicago deals with voluntary but binding commitments from a range of companies, communities, and organizations. In September 2008, six northeastern states in the United States sponsored a regional auction of carbon dioxide emission rights for the power generation sector. New South Wales in Australia since 2003 has been requiring utilities to offset any emissions that occur beyond regulated limits. Although these subnational efforts exclude transportation and other greenhouse emissions and the cost of emissions are generally low (little more than $3 per ton of carbon in the U.S. example), it is worth noting that jurisdictions are working to reduce their emissions with no certainty of global or national mechanisms to reward such early efforts.[24]

In December 2007, climate negotiators agreed at a major conference in Bali, Indonesia, on a plan and timetable for working toward a protocol to succeed Kyoto when its first commitment period ends in 2012. One resolution of the Bali Action Plan was to continue the focus of global climate negotiations on four main areas:

- mitigation, a term covering efforts to reduce emissions below what they would otherwise be, especially through energy efficiency and a transition to low-carbon energy production, as well as avoiding deforestation in developing countries;
- adaptation to the climate change that is already on the way, bringing rising sea levels and more-severe weather patterns;

- technology transfer from industrial to developing countries to facilitate and help pay for these efforts in countries that otherwise may not be able to afford them, or in some cases transfers between developing countries; and
- financing for poorer countries provided by wealthier ones and potentially a pool of all nations, for the three activities agreed upon.

Some analysts of the state of play add to this list "vision," an overarching statement about what the negotiations are designed to achieve and how they will do so.[25]

The Bali Action Plan for the first time expressed the objective that all parties—indeed, all human beings—will reduce emissions.

The conference also clarified that major departures from the overall architecture of the climate change convention and the Kyoto Protocol were unlikely. Thus the major division of responsibilities to act between industrial and developing countries would remain.

Yet the Bali Action Plan also for the first time expressed the objective that all parties—indeed, all human beings—will reduce emissions. Given how much variation exists in emissions and development within each group, some proposals aired at Bali envision breaking the two groups into subcategories—at least in terms of the commitments they would be asked to make. These could distinguish former communist states in Eastern Europe from wealthier industrial countries, for instance, or rapidly industrializing or oil-producing developing nations from those of sub-Saharan Africa. Such subgroups might have their own differentiated responsibilities and timetables.

One way or another, the Bali conference reiterated, poorer and less industrialized countries are not likely to move as soon as wealthier and more industrialized ones must in committing to emissions cuts or taking responsibility for the financing needed when they do make such cuts. Conferees at a follow-up workshop in Bangkok supported continuation of the market-based carbon trading mechanisms of Kyoto, such as the CDM, while pledging to refine them to improve their reach and effectiveness. Those decisions signaled to the world's business leaders, noted Yvo de Boer, executive secretary of the U.N. secretariat administering the climate negotiations, that "long-term certainty [will] guide their investments over the coming years."[26]

Knowing a new U.S. president would take office in January 2009, the negotiators fast-tracked the issues of financing and technology transfer for a climate conference in Poznan, Poland, in December 2008 and saved most details of the more crucial and difficult issues of mitigation and adaptation for Copenhagen in late 2009. When the U.S. delegation in Bali blocked consensus on the imperative for emissions caps in industrial countries, the negotiators regrouped and instead established a working group to address critical issue areas prior to Copenhagen. (Such working groups often do the time-consuming brainstorming and bargaining needed to pave the way for negotiating conferences.) The Bali Action Plan made clear to governments and to global capital markets that the basic Kyoto approach of setting binding national emissions targets would move forward, but with more stringent targets and longer timelines, that the international carbon market would be expanded, and that the controversial Clean Development Mechanism would be reviewed and modified.

Some concepts moved forward in evolutionary leaps at Bali. More engagement in the carbon market from developing countries seemed likely after a reiterated commitment to financing climate change adaptation activities through a 2-percent levy on CDM transactions. Discussion moved forward on the idea of emissions cuts negotiated within important industrial sectors—electric utilities, steel and aluminum production, aviation, shipping, or even land transportation. Helped by governments, companies in these sectors would pledge an overall emissions cap for their industry and then work together across national borders to invest in and secure the needed reductions where they could be achieved most cheaply—most often, probably, in less wealthy countries, where the industrial infrastructure is less modern and efficient. By May 2008, China indicated its interest in this approach—a breakthrough from the developing country with by far the largest industrial sectors.[27]

The sector concept, while controversial because it could undermine more comprehensive emissions reduction strategies, is appealing on several fronts. Almost all global greenhouse gas emissions can be categorized by sector (although some fit into more than one sector). About a fifth of all emissions can be attributed to the production processes of specific industries, such as chemicals, cement, and iron and steel. A cap and trade approach within such sectors could thus produce significant emissions savings while funneling private investment into the industrial capital stock of developing countries.[28]

In many sectors, a small handful of countries are responsible for the majority of emissions, reducing the number of actors and simplifying the mechanism's structure. And sectoral agreements and mechanisms can provide important guidance for the more comprehensive and ambitious cap and trade approaches likely to form the basis of long-term efforts to reduce greenhouse gas emissions. Although the details of such agreements remain to be worked out, there is enough support for the idea that a sectoral approach seems a plausible candidate as an element in a future protocol.

The development that provided the most excitement at Bali was a new willingness by developing countries to consider reductions in the destruction of forests and land degradation if these could be financed by industrial countries. Again, the details remain to be worked out. The most contentious question is whether to allow such reductions to compete with reductions in fossil fuel emissions in international carbon markets. But the potential for synergistic benefits is obvious. An estimated 23 percent of all global carbon dioxide emissions come from deforestation and other changes in land use, a proportion just a bit larger than the CO_2 emissions of the United States or China (which account for about 20 percent of the world total each). Reducing the emissions associated with these activities would directly contribute to the preservation of forest-based biodiversity, reductions in soil erosion, and reductions in landslides and flooding in mountain communities. (See Chapter 3.) The need for reductions in fossil fuel and comparable industrial emissions would nonetheless remain.[29]

New Directions

In the Bali discussions and in the months that followed, central themes emerged or gained momentum. Outside of the United States, most countries appeared to support a timetable under which industrial countries focus in the years after 2012 on "hard" emissions caps, which have been made easier to attain through carbon trading mechanisms such as the strengthened Clean Development Mechanism. The 2007 report of the Intergovernmental Panel on Climate Change suggested that in order to have a reasonable chance of permanently restraining global warming to no more than 2.4 degrees Celsius (4.3 degrees Fahrenheit) above preindustrial levels. As noted earlier, some scientists believe this is too high a threshold, but even by this standard the world must reduce its CO_2 equivalent emissions by 50–85 percent of 2000 levels by the middle of this century.[30]

To make that possible, industrial countries would need to slash their own emissions by 25–40 percent by 2020. The European Union has committed to 20 percent cuts from a 1990 emissions base by that year, while saying it would aim for a 30-percent cut if joined in comparable efforts by the United States and other industrial powers. (The lack of consensus that the EU commitment reflects about what year to use as a basis for future reductions is just one of the complicating factors in acting globally on climate change.) Such commitments are crucial, because it is these rather than international treaties per se that will lead to real emissions reductions through the legislation that countries enact—with the European Union's emissions trading scheme the best model of this dynamic.[31]

The U.S. Congress, despite the Senate's refusal to ratify the Kyoto Protocol, briefly considered legislation in 2008 that would have capped a significant proportion of U.S. carbon dioxide emissions while rewarding developing countries for reducing greenhouse gas emissions from deforestation and land degradation. Many U.S. climate activists found the proposed legislation flawed, but its consideration was a sign that the United States will someday enact emissions-reducing laws, especially as a new administration puts its stamp on U.S. policy. (Both major presidential candidates in 2008 supported a U.S. commitment to emissions reductions, with cap and trade mechanisms the preferred approach.)[32]

At whatever point industrial countries make binding commitments, rapidly developing countries such as China, India, and Brazil will find themselves under pressure to declare their own pledges—though perhaps with a few years' allowance before taking specific actions. "Commitment" is a difficult word for most developing countries to use, given their proportionally smaller responsibility for filling the atmosphere with heat-trapping gases. By taking on "no lose" objectives, at least to slow the growth of greenhouse gas emissions, developing countries can engage in the global process. They might pledge to reduce the "carbon intensity" of each unit of economic activity, as China has. Such efforts can defuse accusations from wealthier countries that the poorer ones are increasing their emissions rapidly but face no obligations whatsoever.

The ideal mechanisms for developing countries would offer strong incentives, with financing provided mostly by wealthy countries, eventually perhaps backed by modest prodding "sticks" such as trade restrictions or finance "carrots." And, as described further later, one concept worth exploring is for developing countries to contribute climate-related financing in proportion to their well-off populations, above certain generous thresholds.

Some analysts speak hopefully, borrowing a phrase from the U.S.-led occupation of Iraq, of a "coalition of the willing," implying a voluntary approach to emissions reductions even by industrial countries. Developing countries and environmental organizations, however, tend to see the voluntary approach in wealthy countries as too little, too late. Long-time major emitters that decline to push down their emissions as rapidly as possible will need to "lose" something, beyond the respect of other countries and unspecified future penalties along the lines described in the Kyoto Protocol, given how critical these emissions reductions are. But what those "sticks" would be remains to be debated.[33]

On the positive side, an obvious synchronicity between emissions cuts and new sources of financing arises in the concept of cap and trade—if countries auction the allocation of emissions rights. Those auctions, supplemented possibly by revenue from a parallel carbon tax, could raise substantial revenue, which could then be directed toward both domestic and foreign efforts to reduce emissions further and to adapt to ongoing climate change.

Meanwhile, critical questions await discussion at the 2009 Copenhagen meeting and the working conferences leading up to it. How is climate change adaptation defined, for example? How is the concept separate from overall economic development, which certainly would help countries better adapt to all environmental change, including climate change? How can developing countries be assured that funding provided specifically for their climate change adaptation efforts is not simply subtracted from existing development assistance? And what specific investments and activities will truly enable countries to improve their resilience to the possibly devastating impacts of human-induced global warming?

The questions are equally challenging on the issue of technology transfer. Most technology transfer is a business matter. Willing sellers of a new technology find willing buyers who can afford it. But technology that facilitates reduction of greenhouse gas emissions is a different matter altogether. Clearly, affordability should not be an obstacle if the technology serves the global good of emissions reductions.

Just as clearly, inventors and others need incentives to innovate. Someone will need to fund technology transfers, and a balance will need to be struck on such critical issues as patent law and intellectual property rights to secure the widest dissemination of useful technologies at the lowest possible costs. Progress made on such questions in distributing anti-retroviral drugs to treat HIV/AIDS in developing countries offers hopeful signs, and innovative climate-related technology deployment mechanisms are now the subject of negotiations ahead of the Copenhagen conference.

Near the end of 2008, formal country and regional proposals began to emerge that were aimed at both the reduction of greenhouse gas emissions and the financing of adaptation to inevitable climate change. (See Box 6–2.) Academics and nongovernmental organizations (NGOs) were putting forward ideas as well, and an even greater number and variety will emerge in the months leading up to Copenhagen.[34]

Recognition of the importance of adaptation financing is growing rapidly, even among climate activists who once saw attention to this issue as a distraction from the needed preventive measures to stop climate change. The reason for this shift is sobering: there is no avoiding significant and damaging impacts from the greenhouse gases already in the atmosphere, and the poorest and least responsible will fare the worst. They will need much help. The just solution to this dilemma is that historic emitters must not just help but must compensate those who suffer through little or no fault of their own. Turning this obvious principle into actual financing instruments and real money, however, is another matter.

The Real Deal

To step onto an emissions path likely to offer some safety, humanity needs to cap and then start shrinking global emissions within just over a decade, however much the world grows demographically and economically. Every country will need to do its part. But

Box 6–2. Government Proposals for Climate Change Mitigation, Adaptation, and Technology Transfer

Government proposals for financing climate change programs that could be included in a new protocol to the Framework Convention on Climate Change began emerging after the Bali Conference of the Parties in late 2007.

China and the Group of 77 (G-77, a U.N coalition of developing countries, now with 130 members) propose a financial mechanism that would link private and public funding sources to the spending needs of governments, in order to reduce potential fragmentation in financing related to climate change needs. A governing board with equal representation from developing and industrial nations would determine how much funding would be allocated for programs on adaptation, mitigation, and technology transfer.

Funding would be additional to current official development assistance (which generally consists of direct grants and comparable support to promote economic development in developing countries). The majority of funds would come from industrial nations and would be offered as grants rather than loans. The level of funding would be set at 0.5–1 percent of the gross national product of industrial countries as a group.

In addition, China and the G-77 propose a separate technology transfer financing mechanism called the Multilateral Climate Technology Fund. This would finance activities in developing countries related to clean energy technology research, development, diffusion, and transfer. The fund would operate under the Conference of the Parties to the climate change treaty.

Mexico proposes a Comprehensive World Climate Change Fund, which would include mitigation, adaptation, and technology transfer activities. All countries—industrial and developing—would contribute to this fund. Withdrawals would be limited to countries that contribute and would be determined by a formula based on current GHG emissions, population, and gross domestic product.

In its initial phase, the Comprehensive World Climate Change Fund would aim to mobilize and spend no less than $10 billion a year. Mechanisms that could mobilize financial resources include auctioning permits in domestic cap and trade systems in industrial countries and taxing air travel. Mexico proposes that part of the fund be set aside for the benefit of the poorest countries, as they will be most affected by climate change. Governance of the fund would be transparent and inclusive: all countries would have an equal voice in the governing structure.

Switzerland proposes a funding scheme for climate adaptation based on a global carbon tax of $2 per ton of carbon dioxide emitted, in accordance with "common but differentiated responsibilities," a phrase that harks back to the climate change convention. Countries emitting

what is each country's part? That is what negotiators must decide—and keep deciding as both the global climate and the world's nations evolve. And negotiators must weigh the relative importance of past, present, and future emissions in assigning responsibility for the problem. They must also decide how to weigh the economic capacity of each country when asking for commitments to act.

Well-verified data on emissions is critically important—where they come from, how they influence atmospheric concentrations of greenhouse gases, what sinks remove greenhouse gases from the air, and how securely they do so. The emerging currency for the negotiations is carbon dioxide equivalence, but as yet there are no databases that carefully track emissions from all nations using this measure. It will take effort to produce an authoritative database. But until that is accomplished, how can the world's countries be assured that their collective emissions reduction efforts are succeeding?

The Kyoto Protocol addresses the six most important greenhouse gases and gas categories—carbon dioxide, the number one

Box 6–2. continued

less than 1.5 tons of carbon dioxide per person per year would be exempt from the tax. Estimated overall revenues from the funding scheme would be $48.5 billion annually. Of that, $18.4 billion would be for a Multilateral Adaptation Fund. Revenues collected in each country from a global carbon tax would be paid into the fund based on its level of economic development. High-income countries would pay 60 percent of their revenues to the fund. Medium-income countries would pay 30 percent, and low-income countries would pay 15 percent.

India proposes a New Global Fund for Adaptation. Industrial nations would contribute 0.3–1 percent of their gross domestic product and the monies would be especially used for adaptation activities in developing countries. The fund would be financed by both private and public sources.

South Africa, representing a coalition of African governments called the Africa Group, proposes scaling up adaptation funding by more than 100 times what is now available. Financial resources would be beyond existing funds under the United Nations Convention. The Africa Group proposes that a work program on adaptation be based on an assessment of its costs for developing countries, and the group would facilitate the implementation of adaptation strategies and programs through financing and capacity building.

In terms of adaptation financing, the *European Union* would focus on expanding the global carbon market, leveraging private investment flows, and making financing predictable and timed to the needs of developing countries. In addition, the EU strategy would consider auctioning emissions allowances, introducing taxes on aviation and shipping, and instituting a global tax on CO_2 emissions.

Norway proposes that adaptation needs under the climate convention be met through auctioning a share of "assigned amount units"—portions of allowed emissions—of all industrial countries. Companies in countries obliged to cap national emissions could buy these certificates to help them reach their emissions targets. Revenues from a system of auctioning emission allowances in the shipping sector would fund adaptation activities in developing countries.

Under a proposal by *Brazil*, industrial countries would finance a new Clean Development Fund that would aim to finance the costs of climate adaptation for developing countries. Brazil proposes that adaptation funding be increased considerably and focus on building the capacity of developing countries to translate climate adaptation information into actions, designating national and regional centers of vulnerability, and mapping climate vulnerability in light of national economic and social indicators.

—*Ambika Chawla*

Source: See endnote 34.

offender, released during numerous human activities; methane, released by agriculture and from landfills and leaky natural gas pipes; nitrous oxide, released in agriculture production; sulfur hexafluoride, used in electricity production; hydrofluorocarbons, which replaced chlorofluorocarbons in cooling and refrigeration; and perfluorocarbons, used in medical applications. Many other industrial gases that trap atmospheric heat remain outside of any negotiated framework and are not currently even monitored. Some have quite high global warming potentials molecule per molecule, but all are now so thinly distributed in the atmosphere that they collectively make relatively insignificant contributions to global warming in comparison to the main regulated gases. This could change, however, as production of any of these gases grows.

A new protocol to specify what will follow the Kyoto first commitment period could engage all countries in a globally transparent effort to monitor emissions of as many significant greenhouse gases as possible. Financed primarily by industrial countries, the effort could capture the imagination of young people concerned about the global climate they will inherit, and it could stimulate education and scientific advancement all over the world.

As these efforts proceed, the world will need to evolve beyond the antiquated and overly simplistic division of all countries into the categories "industrial" and "developing" that has characterized climate negotiations since the drafting of the Framework Convention in the early 1990s. A relic of the post-colonial landscape that took shape after World War II, this bifurcation fails to capture the wide diversity of responsibility (past and current emissions, including on a per capita basis) and capability (national and per capita income and wealth) of the world's nearly 200 nations. In particular, it fails to distinguish rapidly industrializing countries such as China and India from those more slowly developing countries that are still far from contributing substantially to Earth's greenhouse gas buildup.

Dealing with global climate change in a world of nations will require industrial and rapidly industrializing countries to cap their greenhouse gas emissions within the next decade—and then to steadily reduce the totals toward zero. Even poorer countries would eventually need to follow. But how many national leaders will agree to an emissions allotment that allows their citizens a lower average level of emissions than those of other countries—especially if those countries earlier contributed much more to the atmosphere's total greenhouse gas load?

Many observers who peer far enough into the future of global climate regulation have acknowledged that ultimately either climate emissions will need to be roughly equal on a per capita basis or countries that emit more than the global per capita average will need to compensate those that emit less. Nicholas Stern has acknowledged that annual "global average per capita emissions...will—as a matter of basic arithmetic—need to be around two tons by 2050," based on a world population of 9 billion by then and using carbon dioxide equivalence as his measurement unit. "This figure is so low that there is little scope for any large group to depart significantly above or below it."[35]

The leaders of India and Germany called attention in the summer of 2007 to the importance of per capita emissions parity—or at least fairness. Both suggested that a new climate pact allow emissions from developing countries to rise until they converged with those of industrial countries (which would presumably be decreasing rapidly), at which point both groups of countries could reduce their per capita emissions in tandem. "What kind of measure do we use to create a just world?" German Chancellor Angela Merkel asked.[36]

Moreover, given the historically greater responsibility of industrial countries for most of the buildup of greenhouse gases in the atmosphere, could even a true convergence of future per capita emission levels constitute "full payment" to the less wealthy countries for a changed climate? In 1997, Brazil proposed a plan by which country responsibilities to address climate change were made proportional to their historical contribution to the problem. The idea made no headway on the international stage. In 2005, researchers at the World Resources Institute revisited the suggestion and concluded that assigning historic responsibility depends significantly on the starting date of the history selected. Global data would not support a definitive comparison for periods earlier than 1990, the researchers added, as that is when systematic national emissions monitoring began.[37]

Most analysts who follow the process would argue that a climate agreement based on either per capita emissions allocations or historical cumulative emissions is unlikely to emerge from the Copenhagen conference. Not only would industrial countries understandably fear the implications for them, but even some developing countries have reason to worry they would join the ranks of the "high emitters" if the per capita emission dividing line were set low enough to force radical emissions reductions. The urgency of rapidly slashing emissions will need to become much more obvious to many more people before such approaches can be taken seriously.

Over the long term, as Nicholas Stern recognized, there is no real alternative to convergence on roughly equal per capita emissions at very low levels. Zero net emissions globally at some point in the future will, of course, mean zero net emissions per person. So it becomes all the more critical to keep thinking about how this convergence could eventually come to be—and, if possible, to help the process along.

Equity and the End of Emissions

We have choices to make. Bringing greenhouse gas emissions down to a fraction of current levels will take an ongoing worldwide effort that engages all nations and touches all lives. We can fail to slash emissions, or fail even to try. We can try risky geoengineering schemes or simply hope to brave the heat and storms to come. Or we can adopt a positive attitude about preventing future emissions and adapting collectively to past ones, and we can get to work.

We live in exciting times and can rise to the occasion. We have handed ourselves a problem we can solve only by learning new ways to live and to cooperate for a common goal. It could be a good thing. But by any measure the 10 months leading up to the Copenhagen negotiations on the next climate agreement offer one last opening—any other 10 months might come too late—to seal a deal that can save the global climate for the next century and beyond.

One proposal gaining attention in advance of Copenhagen sets out to integrate emissions reductions and climate change adaptation with a "right to sustainable development." Called Greenhouse Development Rights and jointly developed by a U.S. group, EcoEquity, and the Stockholm Environment Institute, the concept is designed to share in fair ways the burden of cutting greenhouse gas emissions while shielding the poor from potentially high costs. It would base climate-related obligations on a national Responsibility and Capacity Indicator. Responsibility would reflect each country's contribution to the climate problem and be defined in terms of cumulative per capita greenhouse gas emissions from a specific date, perhaps 1990. Capacity would reflect each country's ability to help deal with the climate problem without sacrificing necessities and be defined in terms of national income.[38]

The indicator index combines these two pillars of the climate convention with a simple but critical adjustment: income below a "development threshold" of $7,500 per capita does not count in the calculation of capacity, and emissions corresponding to consumption below that income threshold do not count in the calculation of responsibility. This figure, the proposal developers note, is modestly higher than a global poverty line, to reflect a level of welfare that is beyond basic needs, though well short of today's levels of "affluent" consumption.[39]

The Greenhouse Development Rights framework thereby accommodates developing countries' claim that their development and poverty eradication must trump solving the climate problem. But it does so in a nuanced way. It assesses capacity and responsibility at the level of individuals, in a manner that takes explicit account of the unequal distribution of income within countries. It thus confronts a key obstacle to negotiating an agreement that few other proposals even acknowledge: many reasonably wealthy and high-emitting individuals live in poor countries. Their income above $7,500 per person per year would count in assessing each country's capacity to respond to climate change.

This graduated approach to climate-change-related obligations eliminates the need for a simplistic division of the world into industrial and developing countries. While it deviates from a division of the world's countries established in the Framework Convention on Climate Change and fortified in subsequent negotiations, it also takes the negotiations beyond one of the key stumbling blocks. After all, there is no reason why living in a country with an average income at the poverty level should excuse wealthy and high-emitting people from curtailing their emissions and contributing toward climate change adaptation efforts.

The nuanced treatment of countries' real differences, and the focus on a right to development and the principles of capacity and responsibility, may prove the ultimate strength of this and similar future approaches. Requiring developing countries to take on commitments only in proportion to the responsibilities and capacities of their wealthy and high-emitting populations offers the potential for a compromise that a diversity of countries could eventually endorse.

In practical terms, the emissions cuts needed to avoid a warming in the range of 2 degrees Celsius or more would be so radical that under the Greenhouse Development Rights proposal the world's wealthier countries and individuals would have to finance emissions reductions in low-income countries long after the emissions in industrial nations bottomed out near zero. Will the wealthy and fortunate ever take on such obligations to save the world's climate? As the proposal's authors note, if they won't, no one else will.

Taking on such obligations will be more likely if wealthier countries and a climate pact itself can ease and make economically attractive a rapid transition to energy efficiency and renewable sources. There are plenty of attractive options governments and private-sector investors can move forward aggressively and immediately—especially improvements in energy efficiency and electrical power generation through wind, solar energy, and geothermal energy. (See Chapter 4.) People do not really want carbon-based power per se, after all; what they want is power itself, whether at the flip of a light switch or the turn of a key in the ignition of the family car.

One promising mechanism to kick-start this shift, at least in the electricity production sector, is a concept known as feed-in tariffs or renewable energy payments. Already more than 40 nations, states, and provinces have enacted feed-in laws. These generally guarantee anyone who produces electricity with renewable sources priority access to the electricity grid and long-term premium payments for their electricity, thus reducing the insecurity of investment in renewable sources and technologies. Another approach, even simpler, is to root out and close off all government incentives that boost combustion of carbon-based fuels and other greenhouse-gas-intense activities. In the 1990s, a World Bank report estimated that such subsidies cost taxpayers an estimated $210 billion a year and prompted 7 percent of all global CO_2 emissions.[40]

An idea that still waits to be more prominently touted is the concept of "shadow carbon pricing." Ideally, a climate agreement should contribute to a uniform high and rising global price for carbon dioxide that would both discourage release of the gases into the air and raise revenues for adaptation and further emissions reductions. But until the world's nations are ready for such a step, institutions from the World Bank to NGOs should pick a number—any number, almost—to define an imaginary or shadow price for a ton of the gas. Then the shadow carbon cost of any activity, from building a power plant to driving a gas-guzzler to the local convenience store, could be calculated and publicized. The point? Simply to educate the public about how deeply greenhouse gas emissions are embedded in daily living and the global economy and to prepare the way for eventual real costs applied to these emissions.

As human-induced climate change becomes increasingly palpable everywhere, people in all walks of life will grow weary of unfulfilled promises to reduce greenhouse gases at the margins. With enough public pressure, nations may find ways to push each other into action commensurate to the threat. In today's globalized society, few countries can manage without free trade, but trade should be freest among the nations that jointly commit to act forcefully to save the climate. The task is doable; of all the hundreds of scientists presenting diverse opinions on the climate problem, no prominent one has spoken up to say it is already too late to act.

The world needs to prepare to work cooperatively to adapt for serious and disruptive climate change beyond what has already been seen—while still preventing potentially cataclysmic changes. The approach may combine both cap and trade mechanisms, within and among countries and industrial sectors, and domestically focused carbon taxes. The latter may be refunded to people as dividends, thereby softening the regressive nature of the tax and building a constituency for the needed global anti-carbon price tilt. Also needed, even in an era of higher prices on carbon, may be some old-fashioned regulation of energy and industry practices where such governmental nudges can make an important difference at low cost.

Ideally, a climate agreement should contribute to a uniform high and rising global price for carbon dioxide that would discourage release of the gases and raise revenues for adaptation and further emissions reductions.

There is nothing inherently incompatible about applying all three of these diverse approaches—cap and trade, carbon taxes, and regulation—to the task of wringing carbon dioxide and other greenhouse gases out of the growing global economy. Nor is there any reason that industrial countries should not take on the lion's share of the load in helping developing countries reduce both their emissions and their vulnerability to human-induced climate change—with an understanding that wealthy people in developing countries have special responsibilities as well and that eventually economic development will both empower and obligate most of the world to radically reduce greenhouse gas emissions.

Perhaps this will turn into a world of fortified nations dealing individually with a warming climate and rising seas as best as they can while defending themselves against desperate neighbors. But as Hurricane Katrina in 2005 and the heat wave that killed thousands in France two years earlier demonstrate, the wealthiest nations are quite vulnerable to extreme weather events. Ultimately, to reduce climate risk the world will need to work toward a negotiated framework based on the equal right of all people to use the common atmosphere while advancing themselves economically. Even in the near term, the climate negotiating process could inspire—perhaps among a coalition of NGOs—development of a metric similar to shadow carbon pricing that builds an ongoing tally of who uses what "atmospheric space" on a per capita basis for a future allocation process that remains to be imagined.

This approach could be called "no loss—for the present—but no promises about the future." Simply by raising public awareness that everyone will need some day to contribute financing in proportion to excessive emissions today, and by developing an accounting system to illustrate and measure the growing burden of future payments, it may be possible to stimulate new pressure to shift away from carbon-based energies and create new innovations in carbon trading. That is just one unconventional idea to help unravel the post–Kyoto Protocol negotiations puzzle. There will be many more.

It helps that shifting away from fossil fuels will also mean shifting away from their rising costs as demand outstrips shrinking supplies as well as shifting away from the immense human and environmental costs of coal mining (and mining accidents), oil drilling, oil spills, and air pollution and the respiratory problems it causes. It helps, too, that some of the most abundant renewable energy resources—intense sun and high winds—can be found in developing countries.

In addressing the climate change that humans are causing, people may learn lessons to help them face the many other problems that stem from humanity's growing presence and appetite on a resource-constrained planet. While Earth and its envelope of air are fixed, there are no known limitations on the social sphere. In the century to come, people may well have to retreat from rising seas, to recycle most wastewater, to restore and cultivate ravaged soils, and to build cities that can survive brutal storms.

But if we act soon, shrewdly and with a commitment to fairness for all, there may still be time to keep nature and ourselves intact and even thriving despite the changes we will see. We may step safely into a manageably warming world, with a new appreciation of our common humanity and what we can accomplish together.

Climate Change Reference Guide and Glossary

Alice McKeown and Gary Gardner

At the heart of climate change is the greenhouse effect, in which molecules of various gases trap heat in Earth's atmosphere and keep it warm enough to support life. Carbon dioxide and other "greenhouse gases" (GHGs) are an important part of Earth's natural cycles, but human activities are boosting their concentrations in the atmosphere to dangerous levels. The result is rising global temperatures and an unstable climate that threatens humans, economies, and ecosystems.

Sources of Climate Change

Global Emissions of Greenhouse Gases

The primary human-generated greenhouse gases are carbon dioxide, methane, chlorofluorocarbons (fluoride gases), and nitrous oxide. Greenhouse gases are only one source of climate change; aerosols such as black carbon and land use changes such as deforestation also affect warming.[1]

Greenhouse Gas		Generated by
Carbon Dioxide (CO_2)		Fossil fuel combustion, land clearing for agriculture
Methane (CH_4)		Livestock production, extraction of fossil fuels, rice cultivation, biomass burning, landfills, sewage
Nitrous Oxide (N_2O)		Industrial processes, fertilizer use, land clearing
F gases	Hydrofluorocarbons (HFCs)	Leakage from refrigerators, aerosols, air conditioners
	Perfluorocarbons	Aluminum production, semiconductor industry
	Sulfur Hexafluoride (SF_6)	Electrical insulation, magnesium smelting

Share of Global Emissions, in Carbon Dioxide Equivalent, 2004

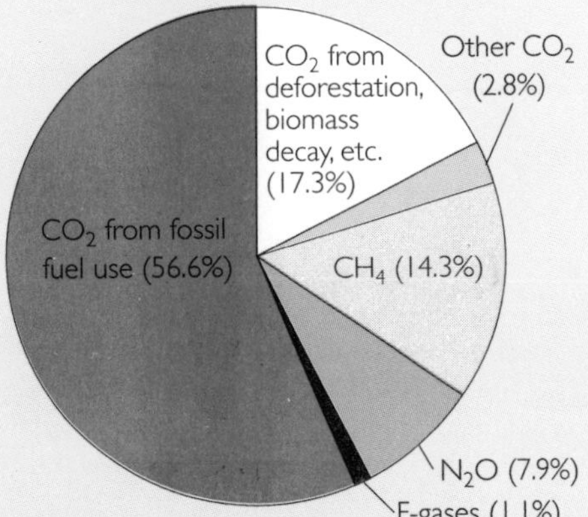

Source: IPCC

Greenhouse Gas Sources, by Sector

Greenhouse gases come from a broad range of human activities, including energy use, changes in land use (such as deforestation), and agriculture.[2]

Source	Sample Emission-generating Activities
Energy Supply	Generation of primary energy supplies, chiefly from fossil fuels; production of fuels for electricity, transportation, and heat; includes extraction and refining
Industry	Production of metals, pulp and paper, cement, and chemical production
Forestry	Deforestation, decomposition of biomass that remains after logging
Agriculture	Crop and livestock production
Transport	Travel by car, plane, train, or ship
Residential and Commercial Buildings	Heating, cooling, and electricity
Waste	Landfills, incineration, wastewater

Emissions by Sector, in Carbon Dioxide Equivalent, 2004

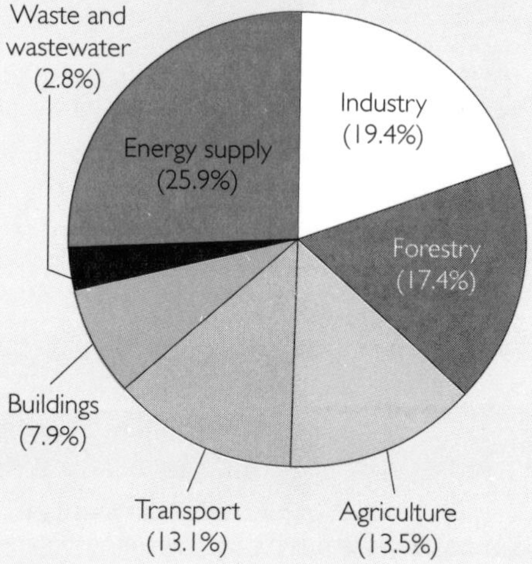

Waste and wastewater (2.8%)
Industry (19.4%)
Energy supply (25.9%)
Forestry (17.4%)
Buildings (7.9%)
Transport (13.1%)
Agriculture (13.5%)

Source: IPCC

Measuring Climate Change

The Carbon Cycle

Carbon flows among land, sea, and the atmosphere. But human activities since the mid-eighteenth century have changed carbon flows in ways that have lasting implications for the climate. This graphic depicts changes to global carbon flows in the 1990s relative to the preindustrial state.[3]

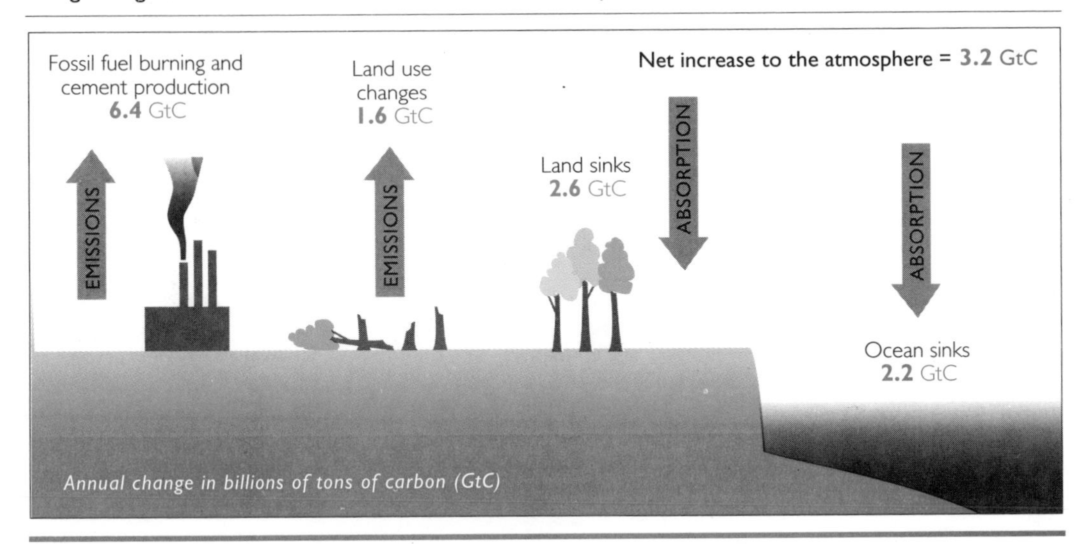

Temperature Conversion

Changes to global temperature caused by climate change are usually measured in degrees Celsius. One degree Celsius is equal to 1.8 degrees Fahrenheit—meaning that a 2-degree Celsius rise is 3.6 degrees Fahrenheit. Actual temperature readings in the different scales are easily compared when placed side by side.

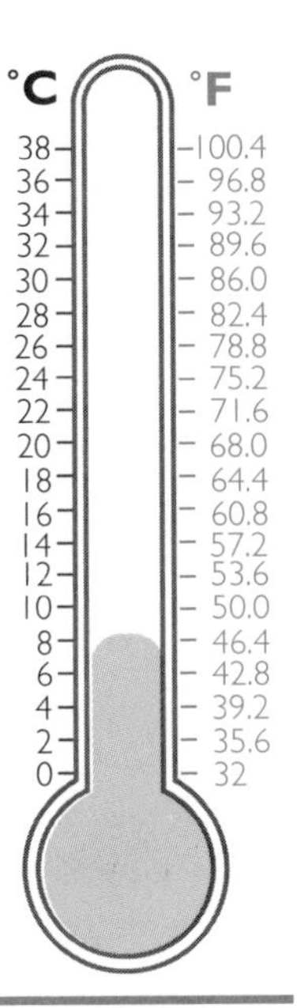

Carbon, Carbon Dioxide, and Carbon Dioxide Equivalents

Carbon, the basis of life on Earth, is at the center of the climate crisis. Carbon is found in solid, liquid, and gaseous form. CO_2 is the most prevalent of human-generated greenhouse gases. CO_2 is so dominant that all other greenhouse gases are evaluated in terms of their equivalency to CO_2.

Indicator	Carbon	Carbon Dioxide	Carbon Dioxide Equivalent
Molecular makeup	One atom of carbon.	One atom of carbon and two atoms of oxygen.	A measurement, not a chemical element, so no molecular formula.
Symbol	C	CO_2	CO_{2eq} or CO_{2e}
Description	Carbon cycles among land, sea, air, and biological systems and is the building block of many but not all greenhouse gases.	A gaseous form of carbon, CO_2 is the breath people exhale, the fizz in soda—and part of the exhaust from burning fossil fuels. Most human carbon emissions are in the form of CO_2.	A unit of measurement that allows the global warming contribution of greenhouse gases to be compared with each other, even if they have a different molecular makeup.
Calculation	One ton of carbon = 3.67 tons of carbon dioxide.	Not typically converted to other units. Measured as emissions and as a concentration in the atmosphere.	Quantity of a greenhouse gas multiplied by its global warming potential.

Global Warming Potential of Selected Greenhouse Gases

Global warming potential (GWP) expresses a gas's heat-trapping power relative to carbon dioxide over a particular time period (this Table uses the common 100-year frame). GWP allows observers to compare the contributions to climate change made by various greenhouse gases that have different warming effects and life spans. A methane molecule, for example, has 25 times the warming potential of a carbon dioxide molecule, and some gases are hundreds or thousands of times more powerful.[4]

Greenhouse Gas	Global Warming Potential
Carbon Dioxide	1
Methane	25
Nitrous Oxide	298
Hydrofluorocarbons	124 – 14,800
Perfluorocarbons	7,390 – 12,200
Sulfur Hexafluoride	22,800

Top 10 CO_2-Emitting Nations, Total and Per Person, 2005

National emissions levels vary greatly. Among the top 10 emitters, the United States generates 12 times more CO_2 than Italy does. The 10 leading emitters generate many more times the emissions of most developing countries, although emissions in those countries are rising rapidly and could soon overtake the annual emissions in industrial countries. The top 10 emitting nations also exhibit a broad range of emissions per person. Wealthy countries tend to emit more carbon dioxide per person than poor countries do.[5]

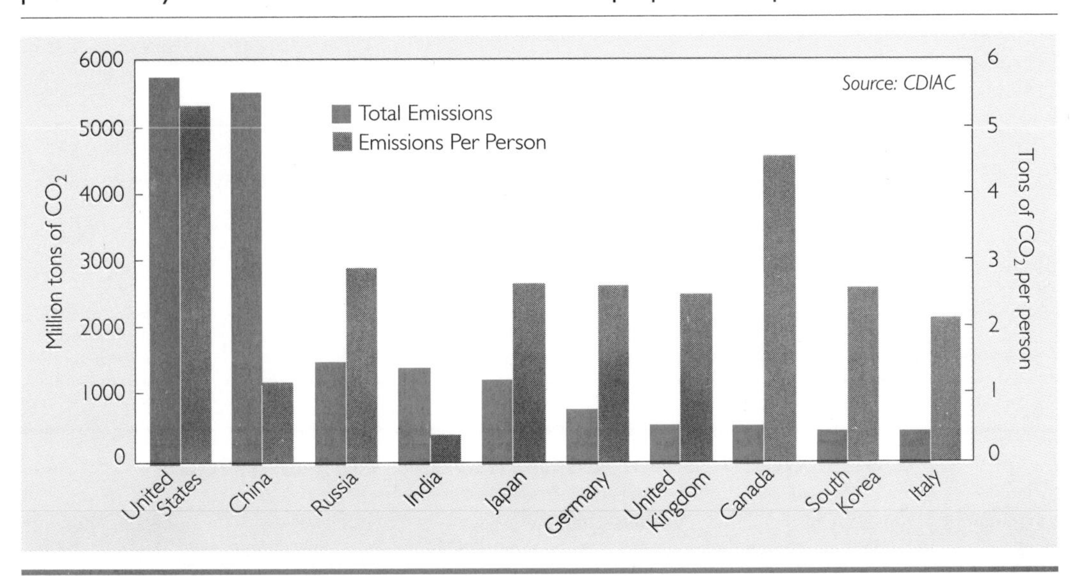

Top 10 CO_2-Emitting Nations' Share of Global CO_2 Emissions, 1950–2005

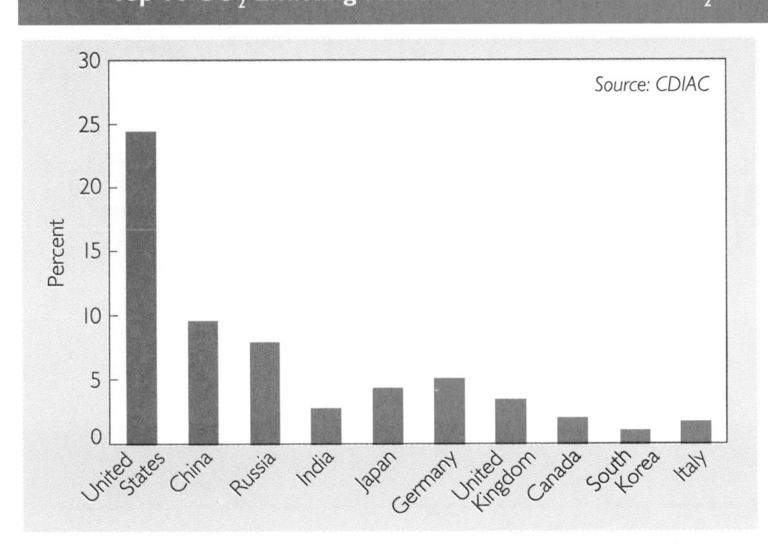

Over time, early industrializing nations typically have emitted more carbon dioxide to the atmosphere than nations that industrialized later.[6]

Concentration of CO_2 in Earth's Atmosphere, 1744–2007

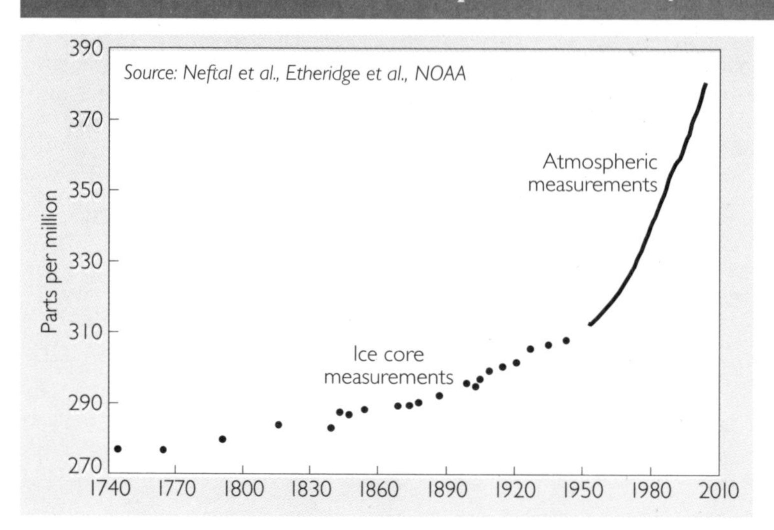

Since the mid-eighteenth century fossil fuel use and cement production have released billions of tons of CO_2 to the atmosphere. Carbon dioxide levels in the atmosphere before the Industrial Revolution were some 280 parts per million (ppm). By 2007, levels had reached 384 ppm—a 37-percent increase.[7]

Consequences of Greenhouse Gas Buildup

Average Global Temperature at Earth's Surface, 1880–2007

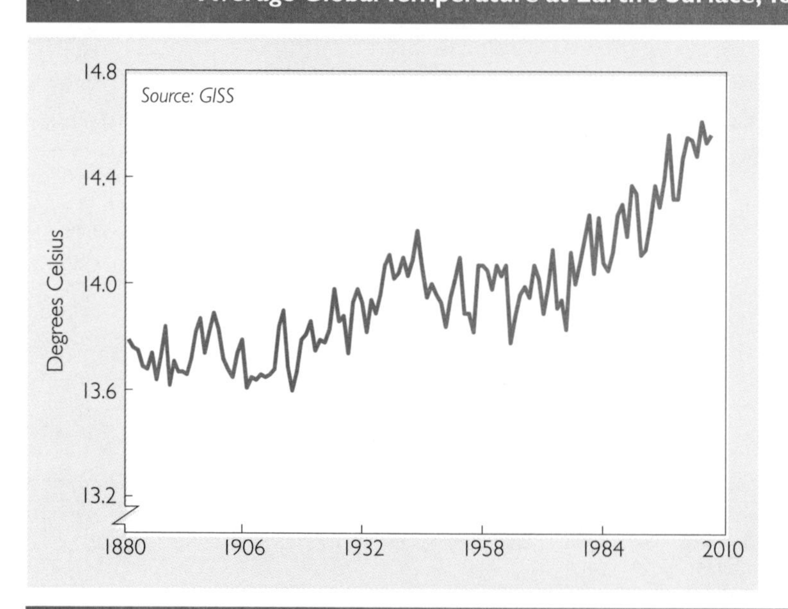

Average global temperature increased by 0.74 degrees Celsius between 1906 and 2005. The Intergovernmental Panel on Climate Change (IPCC) predicts an additional rise of 1.8–4.0 degrees Celsius this century, depending on how much and how soon greenhouse gas emissions are curbed.[8]

The 10 Warmest Years on Record, 1880–2007

Direct temperature readings dating back to the nineteenth century show that the last 10 years had 8 of the 10 warmest years on record.[9]

Ranking	Year
1	2005
2	1998
3	2002
4	2003
5	2007
6	2006
7	2004
8	2001
9	1997
10	1995

Climate Tipping Elements

Scientists believe that several "climate tipping elements" could destabilize the planet's climate by setting off chain reactions—"positive feedbacks"—that accelerate other climate changes. Once a tipping element is triggered by crossing a threshold or tipping point, there is no turning back even if all greenhouse gas emissions were to end. Some tipping elements, such as the loss of Arctic summer sea ice, may be triggered within the next decade if climate change continues at the same rate. Others—the collapse of the Atlantic ocean current, for instance—are thought to be many decades away.[10]

Tipping Element	Expected Consequences
Loss of Arctic summer sea ice	Higher average global temperatures and changes to ecosystems
Melting of Greenland ice sheet	Global sea level rise up to 7 meters and higher average global temperatures
Collapse of West Antarctic ice sheet	Global sea level rise up to 5 meters and higher average global temperatures
Collapse of the Atlantic ocean current	Disruptions to Gulf Stream and changes to weather patterns
Increase in El Niño events	Changes to weather patterns, including increased droughts, especially in Southeast Asia
Dieback of boreal forest	Severe changes to boreal forest ecosystems
Dieback of Amazon forest	Massive extinctions and decreased rainfall
Changes to the Indian summer monsoon	Widespread drought and changes to weather patterns
Changes to the Sahara/ Sahel and the West African monsoon	Changes to weather patterns, including potential greening of the Sahara/Sahel—one of the few positive tipping elements

Expected Impacts of an Unstable Climate

System or Condition	Changes
Fresh Water 	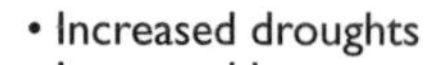 • Increased droughts • Increased heavy precipitation events and flooding • Decreased drinking and fresh-water supplies and availability • Glacier melt decline • Increased salinization of freshwater sources
Ecosystems	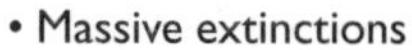 • Massive extinctions 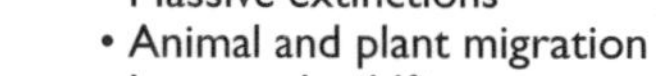 • Animal and plant migration 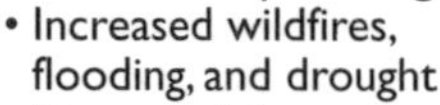 • Increased wildfires, flooding, and drought • Decreased forest coverage, expanding arid lands, and other similar changes • Ocean acidification and coral reef bleaching • Spread of exotic, invasive plants and animals
Food and Agriculture	 • Reduced crop yields • Shifting growing zones • Increasing hunger and malnutrition • Declining fish yields
Health 	• Increased deaths due to floods, heat waves, storms, fires, and drought • Changes in the distribution of certain infectious diseases, including malaria • Increased cardiorespiratory diseases • Increased disease spread from contaminated and polluted drinking water supplies • Increased diarrheal disease • Increased malnutrition
Coasts 	• Increased coastal flooding, especially in low-lying islands and heavily populated delta regions • Increased soil erosion • Increased intensity and strength of tropical storms

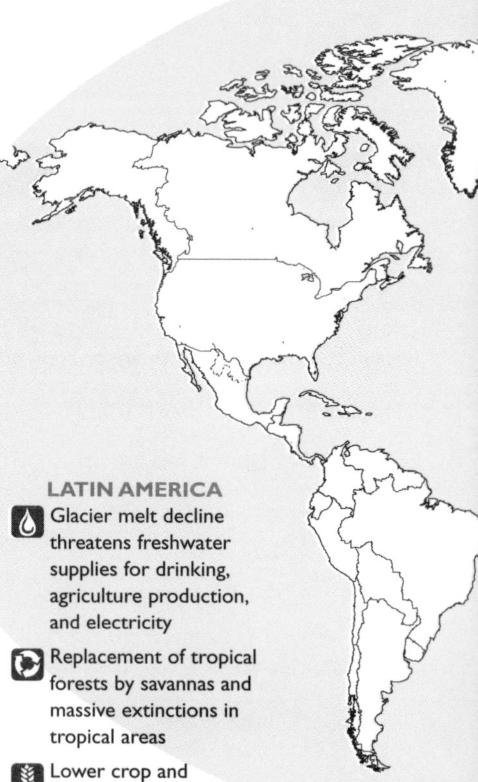

Climate changes are already occurring today and will continue to accelerate as greenhouse gas concentrations rise over time. While climate change is global, the impacts are felt differently from region to region.[11]

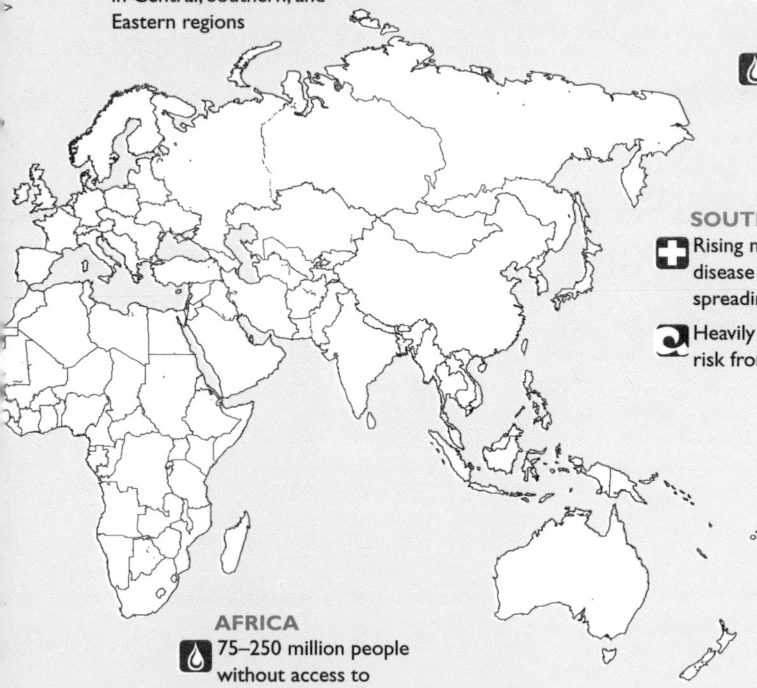

EUROPE

- Coastal flooding, more frequent inland flash floods, and mountain glacier melt
- Widespread extinctions and species loss
- Declining crop production in the South with potential increases in the North
- Growing risk of deaths from heat waves, especially in Central, Southern, and Eastern regions

ASIA

- 1 billion people at risk from decreasing freshwater supplies

SOUTH AND EAST ASIA

- Rising mortality from diarrheal disease and potential massive spreading of cholera
- Heavily populated regions at risk from flooding

AFRICA

- 75–250 million people without access to fresh water by 2020
- Severe reductions in crop yields and fisheries production
- Heavily populated delta regions at risk from flooding

AUSTRALIA AND NEW ZEALAND

- Widespread lack of access to fresh water
- Significant loss of biodiversity, including Great Barrier Reef
- Heavily populated coastal regions at risk from flooding and strong storms

Avoiding Dangerous Effects of Climate Change

Scientists talk about several potential climate stabilization levels that could help minimize the negative effects of climate change. Policymakers rally around these different stabilization points, using them to develop policies to rein in greenhouse gas emissions. But not everyone agrees on the same stabilization points, and recent studies indicate that the levels may need to be lower than once believed.[12]

Potential Stabilization Points	Details
Global temperature increase of 2 degrees Celsius	According to the IPCC, the risks and threats of climate change increase dramatically when global temperature rises more than 2 degrees Celsius (3.6 degrees Fahrenheit). Government leaders and nongovernmental organizations have embraced 2 degrees as the maximum rise allowable if the worst effects of climate change are to be avoided.
Global greenhouse gas reductions of 15–20 percent below baseline levels within the next 10–20 years	Reduction needed to limit global temperature rise to 2–3 degrees Celsius, according to the IPCC. This goal suggests that carbon dioxide concentrations must peak by 2015–20 and then fall. Many policymakers use a variation of this number to set guidelines for action.
Atmospheric CO_2 at 350 ppm	NASA climate scientist James Hansen and his colleagues argue that many global warming tipping points have already been passed. Although current concentrations of CO_2 in the atmosphere exceed 380 parts per million, these scientists believe that atmospheric concentrations need to drop to 350 ppm or lower as soon as possible.
Atmospheric CO_2 at 450–550 ppm	U.K. economist Nicholas Stern advises that the uppermost stabilization levels for atmospheric concentrations of CO_2 should not exceed 450–550 parts per million in order to avoid global economic collapse. Based on climate models, this stabilization point takes into account predictions about technological developments and the time needed for widespread action.

The Diplomatic Road to Copenhagen

Fifteen years after international climate negotiations began at the Rio Earth Summit in 1992, and 10 years after the Kyoto Protocol was completed, the Bali Road Map and Action Plan outlined the steps needed to reach a new, post-Kyoto climate treaty in Copenhagen by the end of 2009. Beyond 2009, international negotiations on climate will likely continue in order to set new emission reduction targets, adapt to scientific advances, and adjust to a changing climate.

Event	Date	Meeting
United Nations Framework Convention on Climate Change adopted	JUNE 1992	Rio de Janeiro Earth Summit
Kyoto Protocol adopted to control greenhouse gas emissions through 2012	DECEMBER 1997	Kyoto Meeting
Kyoto Protocol enters into force	FEBRUARY 2005	
The Bali Road Map and Action Plan outline the steps needed to reach a new international climate treaty by the end of 2009	DECEMBER 2007	Bali Meeting
Groundwork for the new agreement		2008 Meetings in Bangkok, Bonn, Accra, and Poznan
		2009 Meetings
Target date for agreement on a new international climate treaty	DECEMBER 2009	Copenhagen Meeting

Additional Information

Intergovernmental Panel on Climate Change: www.ipcc.ch

United Nations Environment Programme: www.unep.org/themes/climatechange

United Nations Framework Convention on Climate Change: www.unfccc.int

Carbon Dioxide Information Analysis Center: cdiac.ornl.gov/faq.html

Glossary: 38 Key Terms for Understanding Climate Change

Adaptation: Changes in policies and practices designed to deal with climate threats and risks. Adaptation can refer to changes that protect livelihoods, prevent loss of lives, or protect economic assets and the environment. Examples include changing agricultural crops to deal with changing seasons and weather patterns, increasing water conservation to deal with changing rainfall levels, and developing medicines and preventive behaviors to deal with spreading diseases.

Additionality: Emissions reductions that are greater than would have occurred under a business-as-usual scenario. For example, in order for emission credits to be awarded, projects under the Clean Development Mechanism and Joint Implementation must show that any emissions reductions are in addition to what would have occurred without the project. Additionality can also be used to describe other added benefits from the projects, including funding, investment, and technology.

Annex countries: Groups of nations (for example, Annex 1 or Annex B) with different obligations under international climate agreements. Under the U.N. Framework Convention on Climate Change, Annex 1 countries include industrial countries and economies in transition that agreed to reduce their greenhouse gas emissions to 1990 levels collectively. Annex 2 countries are industrial countries that committed to help developing countries by providing them with technology, financial assistance, and other resources. Annex B countries have assigned emission reduction targets under the Kyoto Protocol. The category non-Annex 1 includes countries that are the most vulnerable to climate change. Some countries are included in more than one Annex.

Anthropogenic emissions: Greenhouse gas emissions that are caused by human activities. Also includes emissions of GHG precursors and aerosols.

Atmospheric concentration: A measure used by climate scientists to register the level of greenhouse gases in Earth's atmosphere. Atmospheric concentration is most often measured in parts per million of carbon dioxide and can be tracked over time to understand trends and make projections.

Baseline: A level or year against which subsequent greenhouse gas emission levels and concentrations are measured, especially in the context of emission reductions. For example, the Kyoto Protocol calls for 5-percent reductions in human-caused greenhouse gases below 1990 levels (the baseline) by the 2008–12 period.

Black carbon: Soot and other aerosol particles that come from the incomplete combustion of fossil fuels. Black carbon increases atmospheric warming by lowering the reflectivity of snow, clouds, and other surfaces and by absorbing heat from the sun. Some scientists believe that black carbon plays a large role in climate change and that reducing it may be one of the best opportunities to slow climate change in the short run.

Cap and trade: An approach to limiting greenhouse gas emissions that sets a maximum emissions level (a cap) for a region or nation and that requires participating emitters to obtain permits to pollute. Companies or governmental jurisdictions with extra pollution permits can sell or trade them to parties whose permits are insufficient to cover their full emissions.

Carbon capture and storage (CCS): A process in which carbon dioxide is separated and captured during energy production or industrial processes and subsequently stored (often by pumping it underground) rather than released into the atmosphere. Also known as carbon capture and sequestration.

Carbon dioxide (CO_2): The most widespread greenhouse gas. CO_2 is released to the atmosphere through natural and human activities, including fossil fuel and biomass burning, industrial processes, and changes to land use, among others.

Carbon dioxide equivalent (CO_{2eq}): A unit of measurement used to compare the climate effects of all greenhouse gases to each other. CO_{2eq} is calculated by multiplying the quantity of a greenhouse gas by its global warming potential.

Carbon dioxide intensity and carbon dioxide per capita: Alternatives to total emissions for measuring a nation's greenhouse gas emissions. Carbon intensity measures emissions per unit of gross domestic product. CO_2 per capita measures emissions per person. Both measures can be used to look at emission differences between nations. For example, while China has recently taken the lead in total greenhouse gas emissions, its per capita emissions level is far lower than that in most industrial countries.

Carbon tax: A tax levied on carbon dioxide emissions that aims to reduce the total amount of greenhouse gas emissions by setting a price on pollution. A carbon tax can be used independently or in conjunction with other emissions controls such as a carbon cap. The tax generates revenue that can be used to underwrite further emissions reductions, technology development, cost relief for consumers, or other initiatives.

Clean Development Mechanism (CDM): A mechanism under the Kyoto Protocol that allows industrial countries to meet their emission reduction targets by investing in low- or no-emission projects in developing nations. The CDM also aims to stimulate investment in developing countries.

Conference of the Parties (COP): Regular meetings of governments that have signed an international treaty to discuss its status and possible revision. The fifteenth COP of the UNFCCC will be held in Copenhagen 30 November – 11 December 2009.

Emission Reduction Unit (ERU): One metric ton of carbon dioxide equivalent that is reduced or sequestered. Under the Clean Development Mechanism, industrial countries earn certified emission reduction units (CERs) for projects in developing countries that can be applied toward their national reduction targets. Countries can also earn emission reduction units under the Joint Implementation mechanism.

Emission trading: A market approach to reducing greenhouse gas emissions. Trading allows parties that emit less than their allowed emissions to trade or sell excess pollution credits to other parties that emit more than they are allowed. The European Union Emissions Trading Scheme (EU-ETS) is a mandatory emission trading scheme currently in place; the Chicago Climate Exchange (CCX) is a voluntary trading program.

Forcing: Changes to the climate system that are caused by natural (volcanic eruptions, for example) or human-caused (such as greenhouse gas emissions) factors. Scientifically, radiative forcing measures changes to the natural energy balance of Earth's atmosphere that affect surface temperature. So named because it measures incoming solar radiation against outgoing thermal radiation, radiative forcing is expressed as a rate of energy change in watts per square meter. Human-caused forcing factors like greenhouse gases have a positive radiative forcing and cause surface temperature to heat. Other such factors, including some aerosols, have a negative radiative forcing and cause surface temperature to cool.

Global warming potential (GWP): A measurement of the relative strength and potency of a greenhouse gas as well as its projected life span in the atmosphere. GWP is based on carbon dioxide, the most common greenhouse gas, and allows comparisons among different greenhouse gases.

Greenhouse development rights: Within the context of climate change obligations, the principle that all societies have a fundamental right to reduce poverty, achieve food security, increase literacy and education rates, and pursue other development goals. Societies or countries below a certain income level are excluded from greenhouse gas emission reduction scenarios and are expected to concentrate their resources on raising their standard of living rather than lowering emissions.

Greenhouse gases (GHGs): Atmospheric gases that cause climate change by trapping heat from the sun in Earth's atmosphere—that is, produce the greenhouse effect. The most common greenhouse gases are carbon dioxide, methane, nitrous oxide, ozone, and water vapor.

Intergovernmental Panel on Climate Change (IPCC): The international scientific body established by the World Meteorological Organization and the U.N. Environment Programme in 1988 to provide an objective and neutral source of information on climate change. The IPCC releases periodic assessment reports that are reviewed and approved by experts and governments.

Joint Implementation (JI): An initiative of the Kyoto Protocol that allows industrial countries to earn emission reduction credits by investing in reduction projects in other industrial countries. JI is related to the Clean Development Mechanism, which involves reduction projects in developing countries. Many JI projects are located in Eastern Europe.

Kyoto Protocol: A binding agreement that requires 37 countries and the European Community to reduce their human-caused greenhouse gas emissions 5 percent collectively from 1990 levels in the period 2008–12. It was adopted in 1997 under the U.N. Framework Convention on Climate Change and lays out specific steps countries must take to comply. More than 180 countries have signed the protocol, which entered into force on 16 February 2005.

Land use, land use change, and forestry (LULUCF): Land use is the set of activities that occur on any given parcel of land, such as grazing, forestry, or urban living. Changes to land use such as converting forestland to agriculture can release significant amounts of greenhouse gases. These activities are considered during climate negotiations and when planning emission reductions.

Mean sea level: The average global sea level over time. Mean sea level eliminates variations due to tides, waves, and other disturbances. Sea level is affected by the shape of ocean basins, changes in water quantity, and changes in water density. Climate change is expected to raise sea level by increasing glacier melts and sea temperatures.

Mitigation: Policies and behaviors designed to reduce greenhouse gases and increase carbon sinks.

Models, predictions, and pathways: Tools for analyzing alternative climate futures. Scientists use climate and atmospheric modeling to understand how the climate works and how greenhouse gas concentrations and other triggers lead to climate change. Models help scientists make predictions about climate changes resulting from biological, physical, and chemical variables such as greenhouse gas emissions and land use changes. Emission pathway scenarios are developed to understand what emission limits are needed to meet climate stabilization points, such as avoiding a 2-degree rise in surface temperature.

Parts per million (ppm): A ratio-based measure of the concentration of greenhouse gases in the atmosphere. Carbon dioxide is usually measured in parts per million; in 2007 the atmospheric concentration of carbon dioxide passed 384 ppm, an increase of more than 100 ppm since 1750. Other less widespread greenhouse gases may be measured in parts per billion or parts per trillion.

Peak date: The year that atmospheric concentrations of greenhouse gases must stop growing and begin declining if a given target concentration is to be achieved.

Reducing emissions from deforestation and degradation (REDD): A policy that aims to reduce greenhouse gas emissions from deforestation and forest degradation. In principle, REDD provides financial incentives for countries to maintain and preserve forestlands as carbon sinks rather than cutting them down. In December 2007, climate change negotiators in Bali agreed to consider including REDD as part of a new climate change agreement.

Resilience: The ability of natural or human systems to survive in the face of great change. To be resilient, a system must be able to adapt to changing circumstances and develop new ways to thrive. In ecological terms, resilience has been used to describe the ability of natural systems to return to equilibrium after adapting to changes. In climate change, resilience can also convey the capacity and ability of society to make necessary adaptations to a changing world—and not necessarily structures that will carry forward the status quo. In this perspective, resilience affords an opportunity to make systemic changes during adaptation, such as addressing social inequalities.

Sink: An activity, mechanism, or process that removes greenhouse gases, their precursors, or other small aerosols from the atmosphere. Removals typically occur in forests (which remove carbon dioxide from the atmosphere during photosynthesis), soils, and oceans.

Stabilization: The point at which the climate is stable and not undergoing additional systemic changes. Often discussed as carbon dioxide stabilization and measured as concentration of carbon dioxide in the atmosphere.

Surface temperature (global): An estimate of the average surface air temperature across the globe. When estimating climate change over time, only abnormal changes to the mean surface temperature—not daily, seasonal, or other common variations—are measured. Global surface temperature is most commonly expressed as a combination of land and sea temperature.

Technology transfer: The flow of knowledge, equipment, and resources among stakeholders that helps countries, communities, firms, or other entities adapt to or mitigate climate change.

UNFCCC: United Nations Framework Convention on Climate Change. Adopted on 9 May 1992 and signed at the Rio de Janeiro Earth Summit, the convention established general principles to stabilize greenhouse gas concentrations and prevent dangerous human-caused interference with the climate system. The treaty includes requirements such as preparing national inventories of GHG emissions and a commitment to reduce emissions to 1990 levels. The convention has nearly universal membership, with more than 190 signatory countries.

Vulnerability: The degree to which an ecosystem or society faces survival risks due to adverse climate changes. Vulnerability includes susceptibility as well as the ability to adapt. The level of vulnerability determines whether an ecosystem or society can be resilient in the face of climate change.

Notes

State of the World: A Year in Review

October 2007. National Snow and Ice Data Center, "Arctic Sea Ice Shatters All Previous Record Lows," press release (Boulder, CO: 1 October 2007); Nobel Institute, "The Nobel Peace Prize for 2007," press release (Oslo: 12 October 2007); "October 2007 California Wildfires," at en.wikipedia.org/wiki/California_wildfires_of_October_2007; Beverley Balkau et al., "International Day for the Evaluation of Abdominal Obesity (IDEA): A Study of Waist Circumference, Cardiovascular Disease, and Diabetes Mellitus in 168,000 Primary Care Patients in 63 Countries," *Circulation*, 23 October 2007, pp. 1942–51; "China Birth Defects Soar Due to Pollution—Report," *Reuters*, 29 October 2007.

November 2007. Meryl J. Williams, *Enmeshed: Australia and Southeast Asia's Fisheries* (Sydney, Australia: Lowy Institute for International Policy, 7 November 2007); "Russian Oil Tanker Splits in Half," *BBC News Online*, 11 November 2007; International Federation of Red Cross and Red Crescent Societies, "Bangladesh: Threat of Disease Looms over Isolated Communities," press release (Geneva: 23 November 2007); Bonobo Conservation Initiative, "Massive New Rainforest Reserve Established in the Democratic Republic of Congo," press release (Washington, DC: 20 November 2007); Asian Development Bank, "Developing Asian Countries Must Rethink Water Management to Avoid Crisis," press release (Manila: 29 November 2007).

December 2007. Organisation for Economic Co-operation and Development, "Climate Change Could Triple Population at Risk from Coastal Flooding by 2070, Finds OECD," press release (Paris: 4 December 2007); U.N. Environment Programme (UNEP), "Silver Lining to Climate Change—Green Jobs," press release (Nairobi: 6 December 2007); WWF, "Penguins in Peril as Climate Warms, WWF," press release (Bali, Indonesia: 10 December 2007); U.N. Framework Convention on Climate Change, "UN Breakthrough on Climate Change Reached in Bali," press release (Bali, Indonesia: 15 December 2007); "China Reels from Worst Drought in a Decade," *Reuters*, 21 December 2007.

January 2008. Ruth David, "Tata Shows Off Its $2,500 Car At New Delhi Auto Expo," *Forbes.com*, 10 January 2008; The University of Sheffield, "Record Warm Summers Cause Extreme Ice Melt in Greenland," press release (Sheffield, U.K.: 25 January 2008); Stuart Grudgings, "Amazon Deforestation Surging Again—Scientist," *Reuters*, 18 January 2008; Point Carbon, "Global Carbon Market Grows 80% in 2007," press release (Oslo: 18 January 2008).

February 2008. TRAFFIC International, "South Asia Commits to Regional Co-operation in Controlling Wildlife Trade," press release (Kathmandu: 7 February 2008); Svalbard Global Seed Vault, "Arctic Seed Vault Opens Doors for 100 Million Seeds," press release (Longyearbyen, Norway: 26 February 2008); REN21, "Renewable Energy Accelerates Meteoric Rise. 2007 Global Status Report Shows Perceptions Lag Reality," press release (Paris: 27 February 2008); Timothy Searchinger et al., "Use of U.S. Croplands for Biofuels Increases Greenhouse Gases Through Emissions from Land-Use Change," *Science,* 29 February 2008, pp. 1238–40; Joseph Fargione et al., "Land Clearing and the Biofuel Carbon Debt," *Science*, 29 February 2008, pp. 1235–38.

March 2008. Jad Mouawad, "Oil Prices Pass Record Set in '80s, but Then Recede," *New York Times*, 3 March 2008; UNEP, "Meltdown in the Mountains," press release (Nairobi: 16 March 2008); Natural Resources Defense Council, "American West Heating Nearly Twice as Fast as Rest of World, New Analysis Shows," press release (San Francisco: 27 March 2008); Earth Hour, "See the Difference You Made," www.earthhour.org, viewed 4 September 2008.

April 2008. Environmental Defense Fund, "Analysis Shows Effective Action on Climate Change Will Have Little Impact on U.S. Economy," press release (Washington, DC: 21 April 2008); "San Francisco Hits 70 Percent City Recycling Rate," *Environment News Service*, 23 April 2008.

May 2008. Matthew Weaver, "Cyclone Nargis: One Month On, US Accuses Burma of Criminal Neglect," (London) *Guardian*, 2 June 2008; U.N. Food and Agriculture Organization (FAO), "Intact Mangroves Could Have Reduced Nargis Damage," press release (Rome: 15 May 2008); "Wenchuan Earthquake Has Already Caused 69,196 Fatalities and 18,379 Missing" (in Chinese), *Sina.com*, 6 July 2008; "More Than 4.8 Million Homeless in Sichuan Quake: Official," *Agence France-Presse*, 16 May 2008; "Brazil's Amazon Minister Resigns," *BBC News Online*, 14 May 2008; Stephen Foley, "Oil Veteran Boone Pickens Makes $2bn Gamble on Wind Farm in Texas," (London) *The Independent*, 16 May 2008; Linda Sieg, Johannes Ebeling, and Maï Yasué, "Generating Carbon Finance Through Avoided Deforestation and Its Potential to Create Climatic, Conservation and Human Development Benefits," *Philosophical Transactions of the Royal Society B*, 27 May 2008, pp. 1917–24.

June 2008. Yingling Liu, "Plastic Bag Ban Trumps Market and Consumer Efforts," *China Watch* (Worldwatch Institute), 30 June 2008; World Food Programme, "World Food Crisis Summit: WFP Scales Up Urgent Food Assistance in 62 Countries Worldwide," press release (Rome: 4 June 2008); "Gas Price Record Reaches $4 a Gallon," *CNNMoney.com*, 8 June 2008; Netherlands Environmental Assessment Agency, "China Contributing Two Thirds to Increase in CO_2 Emissions," press release (Bilthoven, The Netherlands: 13 June 2008).

July 2008. FAO, "Land Degradation on the Rise; One Fourth of the World's Population Affected, Says New Study," press release (Rome: 2 July 2008); "Oil Hits New High on Iran Fears," *BBC News Online*, 11 July 2008; Rights and Resources Initiative, "New Studies Predict Record Land Grab as Demand Soars for New Sources of Food, Energy and Wood Fiber," press release (London: 14 July 2008); Convention on International Trade in Endangered Species of Wild Fauna and Flora, "Ivory Sales Get the Go-ahead," press release (Geneva: 16 July 2008); Andrew Revkin, "World Bank Should Improve Environmental Record, Review Says," *International Herald Tribune*, 22 July 2008.

August 2008. Keith Johnson, "Green Games: Beijing's Cleaning the Air—Indoors," *WallStreetJournal.com*, 7 August 2008; Pacific Gas and Electric Company, "PG&E Signs Historic 800 Mw Photovoltaic Solar Agreements with Optisolar and Sunpower," press release (San Francisco: 14 August 2008); Robert J. Diaz and Rutger Rosenberg, "Spreading Dead Zones and Consequences for Marine Ecosystems," *Science*, 15 August 2008, pp. 926–29; UNEP, "Cutting Fossil Fuel Subsidies Can Cut Greenhouse Gas Emissions Says UN Environment Report," press release (Nairobi: 26 August 2008).

September 2008. C. L. Weber et al., "The Contribution of Chinese Exports to Climate Change," *Energy Policy*, September 2008, pp. 3572–77; American Wind Energy Association, "U.S. Wind Energy Installations Surpass 20,000 Megawatts," press release (Washington, DC: 3 September 2008); Regional Greenhouse Gas Initiative, Inc., "RGGI States' First CO_2 Auction Off to a Strong Start," press release (New York: 29 September 2008); "Congress Allows Offshore Oil Drilling Ban to Expire," *Environment News Service*, 30 September 2008.

Chapter 1. The Perfect Storm

1. U.S. National Ice Center, "Northwest Passage Open 21 August 2008," press release (Suitland, MD: 21 August 2008); U.S. National Ice Center, "The Northern Sea Route (Northeast Passage) Appears 'Open'," press release (Suitland, MD: 5 September 2008).

2. Paul Crutzen and Eugene F. Stoermer, "The Anthropocene," *Global Change Newsletter* (International Geosphere-Biosphere Programme), 1 May 2000, pp. 17–18.

3. Intergovernmental Panel on Climate Change, *Climate Change 2007: Synthesis Report* (Geneva: 2007).

4. James Hansen, Testimony Before U.S. Senate Committee on Energy and Natural Resources, Washington, DC, 23 June 1988.

5. United Nations, *United Nations Framework Convention on Climate Change*, at unfccc.int/essential_background/convention/background/items/1349.php; United Nations, *Kyoto Protocol to the United Nations Framework Convention on Climate Change*, 1998, at unfccc.int/resource/docs/convkp/kpeng.pdf.

6. "Turn Down the Heat at The Hague—Kyoto Protocol for Reduction of Greenhouse Gases in Danger of Nullification," *The Ecologist*, November 2000.

7. Office of the Press Secretary, "President Bush Discusses Global Climate Change," The White House, Washington, DC, 11 June 2001; "Revealed: How Oil Giant Influenced Bush," (London) *Guardian*, 8 June 2005.

8. Population data from Population Division, *World Population Prospects: The 2006 Revision* (New York: United Nations, 2008); emissions from Carbon Dioxide Information Analysis Center (CDIAC), at cdiac.ornl.gov, viewed 8 October 2008.

9. Emissions from CDIAC, op. cit. note 8; Population Division, op. cit. note 8.

10. Joanna Lewis, Georgetown University, "China: Energy Use, Emissions Trends, and Forecasts," presentation at U.S.-China Climate Dialogue, sponsored by the Center for American Progress, Heinrich Böll Foundation, and Worldwatch Institute, Washington, DC, 16 September 2008; Jay S. Gregg, Robert J. Andress, and Gregg Marland, "China: Emissions Pattern of the World Leader in CO_2 Emissions from Fossil Fuel Consumption and Cement Production," *Geophysical Research Letters*, 24 April 2008.

11. Global Carbon Project, *Carbon Budget and Trends 2007* (Canberra, Australia: September 2008); Alexei Barrionuevo, "Brazil Rainforest Analysis Sets Off Political Debate," *New York Times*, 25 May 2008.

12. James Hansen, Twentieth Anniversary of the "Hansen Hearing," sponsored by Worldwatch Institute, National Press Club, Washington, DC, 23 June 2008.

13. Martin Parry, Osvaldo Canziani, and Jean Palutikof, "Key IPCC Conclusion on Climate Change Impacts and Adaptations," *WMO Bulletin*, April 2008.

14. J. Hansen et al., "Target Atmospheric CO_2: Where Should Humanity Aim?" *Open Atmospheric Science Journal*, 2008.

15. Ian Traynor and David Gow, "EU Promises 20% Carbon Reduction by 2020," (London) *Guardian*, 21 February 2007; Tim Johnston, "Australia's Prime Minister Defeated After Four Terms," *New York Times*, 24 November 2007.

16. Michael Northrop, David Sassoon, and Ken Colburn, "Governors on the March," *Environmental Finance*, June 2008; U.S. Climate Action Partnership, *A Call For Action: Consensus Principles and Recommendations from the U.S. Climate Action Partnership* (Washington, DC: 2007).

17. Prime Minister's Council on Climate Change, *National Action Plan on Climate Change* (Delhi: Government of India, June 2008); Wang Youling and Zhou Wei, "To Break the Bottleneck on Energy and Environment, China Making

Overhaul of Energy Conservation Law," *Xinhua News Agency*, 28 October 2007; Zhu Jianhong, "Policy Analysis: One-vote Veto on Energy Saving and Emission Reduction," *People's Daily*, 30 November 2007.

18. Aaron Sachs, "Population Growth Steady," in Worldwatch Institute, *Vital Signs 1995* (New York: W. W. Norton & Company, 1995), p. 94.

19. Rattan Lal, cited in Eleanor Milne, "Soil Organic Carbon," *The Encyclopedia of Earth*, at www.eoearth.org/article/Soil_organic_carbon, viewed 25 July 2008.

20. Calculations on per capita carbon dioxide emissions by Worldwatch based on Population Division, op. cit. note 8, and on CDIAC, op. cit. note 8; Nicholas Stern, *Key Elements of a Global Deal on Climate Change* (London: London School of Economics and Political Science, undated).

21. Stern, op. cit. note 20.

22. "John Gardner's Writings," Public Broadcasting Service, at www.pbs.org/johngardner/sections/writings.html.

Chapter 2. A Safe Landing for the Climate

1. Intergovernmental Panel on Climate Change (IPCC), *Climate Change 2007: Synthesis Report. Summary for Policymakers* (Geneva: 2007); C. Rosenzweig et al., "Attributing Physical and Biological Impacts to Anthropogenic Climate Change," *Nature*, 15 May 2008, pp. 353–57; J. Stroeve et al., "Arctic Sea Ice Extent Plummets in 2007," *EOS Transactions*, 8 January 2008; M. R. Raupach et al., "Global and Regional Drivers of Accelerating CO_2 Emissions," *Proceedings of the National Academy of Sciences*, 12 June 2007, pp. 10288–93; J. G. Canadell et al., "Contributions to Accelerating Atmospheric CO_2 Growth from Economic Activity, Carbon Intensity, and Efficiency of Natural Sinks," *Proceedings of the National Academy of Sciences*, 20 November 2007, pp. 18866–70.

2. M. Parry et al., "Squaring Up to Reality," *Nature Reports Climate Change*, 29 May 2008, pp. 68–71.

3. Box 2–1 from the following: United Nations, *United Nations Framework Convention on Climate Change* (UNFCCC), at unfccc.int/essential_background/convention/background/items/1349.php; R. Verheyen, *Climate Change Damage and International Law: Prevention, Duties and State Responsibility* (Martinus Nijhoff Publishers, 2005); A. D. Sagar and T. Banuri, "In Fairness to Current Generations: Lost Voices in the Climate Debate," *Energy Policy*, September 1999, pp. 509–14; P. Baer et al., "CLIMATE CHANGE: Equity and Greenhouse Gas Responsibility," *Science*, 29 September 2000, p. 2287; R. E. Zeebe et al., "OCEANS: Carbon Emissions and Acidification," *Science*, 4 July 2008, pp. 51–52; L. Harvey, "Dangerous Anthropogenic Interference, Dangerous Climatic Change, and Harmful Climatic Change: Non-trivial Distinctions with Significant Policy Implications," *Climatic Change*, May 2007, pp. 1–25; H.-H. Rogner et al., "Introduction," in IPCC, *Climate Change 2007: Mitigation* (Cambridge, U.K.: Cambridge University Press, 2007).

4. Raupach et al., op. cit. note 1; Canadell et al., op. cit. note 1; S. Rahmstorf et al., "Recent Climate Observations Compared to Projections," *Science*, 4 May 2007, p. 709.

5. P. Brohan et al., "Uncertainty Estimates in Regional and Global Observed Temperature Changes: A New Data Set from 1850," *Journal of Geophysical Research–Atmospheres*, 24 June 2006; M. E. Mann et al., "Proxy-based Reconstructions of Hemispheric and Global Surface Temperature Variations over the Past Two Millennia," *Proceedings of the National Academy of Sciences*, 9 September 2008, pp. 13252–57.

6. IPCC, op. cit. note 1; IPCC, *Climate Change 2007: The Physical Science Basis* (Cambridge, U.K.: Cambridge University Press, 2007).

7. S. Rahmstorf, "A Semi-Empirical Approach to Projecting Future Sea-Level Rise," *Science*, 19 January 2007, pp. 368–70; R. Horton et al., "Sea Level Rise Projections for Current Generation CGCMs Based on the Semi-empirical Method," *Geophysical Research Letters*, 26 January 2008.

8. IPCC, op. cit. note 6; IPCC, *Climate Change 2007: Impacts, Adaptation and Vulnerability* (Cambridge, U.K.: Cambridge University Press, 2007).

9. A. Sterl et al., "When Can We Expect Extremely High Surface Temperatures," *Geophysical Research Letters*, 19 July 2008.

10. IPCC, op. cit. note 8.

11. IPCC, op. cit. note 6; S. H. Schneider et al., "Assessing Key Vulnerabilities and the Risk from Climate Change," in IPCC, op. cit. note 8, pp. 779–810; A. E. Derocher, N. J. Lunn, and I. Stirling, "Polar Bears in a Warming Climate," in *Integrative and Comparative Biology*, April 2004, pp. 163–76; D. M. Lawrence et al., "Accelerated Arctic Land Warming and Permafrost Degradation during Rapid Sea Ice Loss," *Geophysical Research Letters*, 13 June 2008.

12. I. Eisenman, N. Untersteiner, and J. S. Wettlaufer, "On the Reliability of Simulated Arctic Sea Ice in Global Climate Models," *Geophysical Research Letters*, 18 May 2007, pp. 1–4; X. D. Zhang and J. E. Walsh, "Toward a Seasonally Ice-covered Arctic Ocean: Scenarios from the IPCC AR4 Model Simulations," *Journal of Climate*, May 2006, pp. 1730–47; A. Fischlin et al., "Ecosystems, Their Properties, Goods, and Services," in IPCC, op. cit. note 8, pp. 211–72; Stroeve et al., op. cit. note 1; J. Stroeve et al., "Arctic Sea Ice Decline: Faster than Forecast," *Geophysical Research Letters*, 1 May 2007, pp. 1–5.

13. T. P. Barnett, J. C. Adam, and D. P. Lettenmaier, "Potential Impacts of a Warming Climate on Water Availability in Snow-dominated Regions," *Nature*, 17 November 2005, pp. 303–09; W. J. Cai and T. Cowan, "Evidence of Impacts from Rising Temperature on Inflows to the Murray-Darling Basin," *Geophysical Research Letters*, 2 April 2008; B. Timbal and D. A. Jones, "Future Projections of Winter Rainfall in Southeast Australia Using a Statistical Downscaling Technique," *Climatic Change*, January 2008, pp. 165–87.

14. D. B. Lobell and C. B. Field, "Global Scale Climate–Crop Yield Relationships and the Impacts of Recent Warming," *Environmental Research Letters*, March 2007; M. Auffhammer, V. Ramanathan, and J. R. Vincent, "Integrated Model Shows that Atmospheric Brown Clouds and Greenhouse Gases Have Reduced Rice Harvests in India," *Proceedings of the National Academy of Sciences*, 26 December 2006, pp. 19668–72; W. Cramer, "Air Pollution and Climate Change Both Reduce Indian Rice Harvests," *Proceedings of the National Academy of Sciences*, 26 December 2006, pp. 19609–10.

15. Sterl et al., op. cit. note 9.

16. T. M. Lenton et al., "Inaugural Article: Tipping Elements in the Earth's Climate System," *Proceedings of the National Academy of Sciences*, 12 February 2008, pp. 1786–93.

17. Ibid.; Lawrence et al., op. cit. note 11.

18. M. Oppenheimer and A. Petsonk, "Article 2 of the UNFCCC: Historical Origins, Recent Interpretations," *Climatic Change*, December 2005, pp. 195–226; Rogner et al., op. cit. note 3; E. Kriegler, "On the Verge of Dangerous Anthropogenic Interference with the Climate System?" *Environmental Research Letters*, January-March 2007.

19. F. J. Rijsberman and R. J. Swart, eds., *Targets and Indicators of Climate Change* (Stockholm: Stockholm Environment Institute, 1990); A. K. Jain and W. Bach, "The Effectiveness of Measures to Reduce the Man-made Greenhouse Effect: The Application of a Climate-policy Model," *Theoretical and Applied Climatology*, June 1994, pp. 103–18; German Advisory Council on Global Change, *Scenario for the Derivation of Global CO_2 Reduction Targets and Implementation Strategies* (Berlin: 1995).

20. European Community, Climate Change—Council Conclusions 8518/96 (Presse 188-G) 25/26. VI.96. 1996; Council of the European Union, Presidency Conclusions—Brussels, 22 and 23 March 2005 IV. CLIMATE CHANGE (Brussels: European Commission, 2005), p. 39.

21. W. L. Hare, *Fossil Fuels and Climate Protection: The Carbon Logic* (Amsterdam: Greenpeace International, 1997), p. 90.

22. J. Hansen et al., "Dangerous Human-made Interference with Climate: A GISS ModelE Study," *Atmospheric Chemistry and Physics*, vol. 7, no. 9 (2007), pp. 2287–2312; J. Hansen et al., "Target Atmospheric CO_2: Where Should Humanity Aim?" *Open Atmospheric Science Journal*, 2008; J. Hansen et al., "Target Atmospheric CO_2: Supporting Material," *Open Atmospheric Science Journal*, 2008.

23. L. D. D. Harvey, "Allowable CO_2 Concentrations under the United Nations Framework Convention on Climate Change as a Function of the Climate Sensitivity Probability Distribution Function," *Environmental Research Letters*, January-March 2007; B. Hare and M. Meinshausen, "How Much Warming Are We Committed to and How Much Can be Avoided?" *Climatic Change*, March 2006, pp. 1–39.

24. Table 2–1 based on the following: IPCC, op. cit. note 1; Schneider et al., op. cit. note 11; Lobell and Field, op. cit. note 14; Hansen et al., "Where Should Humanity Aim?," op. cit. note 22; C. H. Sekercioglu et al., "Climate Change, Elevational Range Shifts, and Bird Extinctions," *Conservation Biology*, February 2008, pp. 140–50; J. A. Pounds et al., "Global Warming and Amphibian Losses; The Proximate Cause of Frog Declines? (Reply)," *Nature*, 30 May 2007, pp. E5–E6; C. Le Bohec et al., "King Penguin Population Threatened by Southern Ocean Warming," *Proceedings of the National Academy of Sciences*, 19 February 2008, pp. 2493–97; M. Boko et al., "Africa," in IPCC, op. cit. note 8, pp. 433–67; R. L. Naylor et al., "Assessing Risks of Climate Variability and Climate Change for Indonesian Rice Agriculture," *Proceedings of the National Academy of Sciences*, 8 May 2007, pp. 7752–57; R. Seager et al., "Model Projections of an Imminent Transition to a More Arid Climate in Southwestern North America," *Science*, 25 May 2007, pp. 1181–84; E. Rignot et al., "Recent Antarctic Ice Mass Loss from Radar Interferometry and Regional Climate Modelling," *Nature Geoscience*, 13 January 2008, pp. 106–10.

25. Quote from R. Tol et al., "Adaptation to Five Metres of Sea Level Rise," *Journal of Risk Research*, vol. 9, no. 5 (2006), pp. 467–82.

26. M. Meinshausen et al., "Multi-gas Emissions Pathways to Meet Climate Targets," *Climatic Change*, March 2006, pp. 1–44.

27. Box 2–2 from the following: J. M. Murphy et al., "Quantification of Modelling Uncertainties in a Large Ensemble of Climate Change Simulations," *Nature*, 12 August 2004, pp. 768–72; IPCC, op. cit. note 1; National Oceanic and Atmospheric Administration, "The NOAA Annual Greenhouse Gas Index," at www.esrl.noaa.gov/gmd/aggi, viewed 6 October 2008; Rogner et al., op. cit. note 3.

28. IPCC, op. cit. note 6.

29. Hare and Meinshausen, op. cit. note 23.

30. C. A. Senior and J. F. B. Mitchell, "The Time-Dependence of Climate Sensitivity," *Geophysical Research Letters*, vol. 27, no. 17 (2000), pp. 2685–88; G. Hooss et al., "A Nonlinear Impulse Response Model of the Coupled Carbon Cycle-Climate System (NICCS)," *Climate Dynamics*, December 2001, pp. 189–202; Hare and Meinshausen, op. cit. note 23; G. A. Meehl et al., "How Much More Global Warming and Sea Level Rise?" *Science*, 18 March 2005, pp. 1769–72.

31. D. Archer, H. Kheshgi, and E. MaierReimer, "Multiple Timescales for Neutralization of Fossil Fuel CO_2," *Geophysical Research Letters*, vol. 24, no. 4 (1997), pp. 405–08; D. Archer, "Fate of Fossil Fuel CO_2 in Geologic Time," *Geophysical Research Letters*, 10 May 2005, pp. 1–6.

32. M. Meinshausen, S. C. B. Raper, and T. M. L. Wigley, "Emulating IPCC AR4 Atmosphere-Ocean and Carbon Cycle Models for Projecting Global-mean, Hemispheric and Land/Ocean Temperatures: MAGICC 6.0," *Atmospheric Chemistry and Physics Discussion*, vol. 8, no. 2 (2008), pp. 6153–72.

33. G. J. M. Velders et al., "The Importance of the Montreal Protocol in Protecting Climate," *Proceedings of the National Academy of Sciences*, 20 March 2007, pp. 4814–19; Meinshausen, Raper, and Wigley, op. cit. note 32.

34. Hare and Meinshausen, op. cit. note 23.

35. D. P. V. Vuuren et al., "Stabilizing Greenhouse Gas Concentrations at Low Levels: An Assessment of Reduction Strategies and Costs," *Climatic Change*, March 2007, pp. 119–59; S. Rao et al., *IMAGE and MESSAGE Scenarios Limiting GHG Concentrations to Low Levels* (Laxenburg, Austria: International Institute for Applied Systems Analysis, 2008).

36. Vuuren et al., op. cit. note 35; Rao et al., op. cit. note 35; B. Knopf et al., "Deliverable M2.6: Report on First Assessment of Low Stabilisation Scenarios," Adaptation and Mitigation Strategies: Supporting European Climate Policy: Project co-funded by the European Commission within the Sixth Framework Programme (2002–2006) (Potsdam, Germany: Potsdam Institute for Climate Impact Research, 2008), p. 44.

37. Vuuren et al., op. cit. note 35; Rao et al., op. cit. note 35; Knopf et al., op. cit. note 36; C. Azar et al., "Carbon Capture and Storage From Fossil Fuels and Biomass—Costs and Potential Role in Stabilizing the Atmosphere," *Climatic Change*, January 2006, pp. 47–79.

38. Hare, op. cit. note 21; C. Azar and H. Rodhe, "Targets for Stabilization of Atmospheric CO_2," *Science*, 20 June 1997, pp. 1818–19; T. M. L. Wigley, R. Richels, and J. A. Edmonds, "Economic and Environmental Choices in the Stabilization of Atmospheric CO_2 Concentrations," *Nature*, 18 January 1996, pp. 240–43; P. Read and J. Lermit, "Bio-energy with Carbon Storage (BECS): A Sequential Decision Approach to the Threat of Abrupt Climate Change," *Energy*, November 2005, pp. 2654–71; D. W. Keith, M. Ha-Duong, and J. K. Stolaroff, "Climate Strategy with CO_2 Capture from the Air," *Climatic Change*, January 2006, pp. 17–45; J. Rhodes and D. Keith, "Biomass with Capture: Negative Emissions within Social and Environmental Constraints: An Editorial Comment," *Climatic Change*, April 2008, pp. 321–28; F. Kraxner, S. Nilsson, and M. Obersteiner, "Negative Emissions from BioEnergy Use, Carbon Capture and Sequestration (BECS)—The Case of Biomass Production by Sustainable Forest Management from Semi-natural Temperate Forests," *Biomass and Bioenergy*, April-May 2003, pp. 285–96; P. Read, "Biosphere Carbon Stock Management: Addressing the Threat of Abrupt Climate Change in the Next Few Decades: An Editorial Essay," *Climatic Change*, April 2008, pp. 305–20.

39. J. I. House, I. C. Prentice, and C. Le Quéré, "Maximum Impacts of Future Reforestation or Deforestation on Atmospheric CO_2," *Global Change Biology*, November 2002, pp. 1047–52; C. Müller et al., "Effects of Changes in CO_2, Climate, and Land Use on the Carbon Balance of the Land Biosphere during the 21st Century," *Journal of Geophysical Research*, 26 June 2007, pp. 1–14.

40. Vuuren et al., op. cit. note 35; Rao et al., op. cit. note 35; Knopf et al., op. cit. note 36; Azar et al., op. cit. note 37.

41. J. Koornneef et al., "Life Cycle Assessment of a Pulverized Coal Power Plant with Post-combustion Capture, Transport and Storage of CO_2," *International Journal of Greenhouse Gas Control*, October 2008, pp. 448–67.

42. Rao et al., op. cit. note 35; M. Meinshausen et al., "Multi-gas Emissions Pathways to Meet Climate Targets," *Climatic Change*, March 2006, pp. 151–94; P. L. Lucas et al., "Long-term Reduction Potential of Non-CO_2 Greenhouse Gases," *Environmental Science & Policy*, April 2007, pp. 85–103.

43. C. Jaeger, H. Schellnhuber, and V. Brovkin, "Stern's Review and Adam's Fallacy," *Climatic Change*, August 2008, pp. 207–18.

44. IPCC, op. cit. note 3.

45. J. E. Aldy, S. Barrett, and R. N. Stavins, *Thirteen Plus One: A Comparison of Global Climate Policy Architectures* (Milan: Fondazione Eni Enrico Mattei, 2003); N. Höhne et al., *Climate Change: Options for the Second Commitment Period of the Kyoto Protocol* (Berlin: Federal Environment Agency, 2005).

46. Velders et al., op. cit. note 33.

Chapter 3. Farming and Land Use to Cool the Planet

1. Delia C. Catacutan, *Scaling up Landcare in the Philippines: Issues, Methods and Strategies* (Bogor, Indonesia: Southeast Asia Regional Research Programme, World Agroforestry Centre, 2007); R.A. Cramb et al., "The 'Landcare' Approach to Soil Conservation in the Philippines: An Assessment of Farm-level Impacts," *Australian Journal of Experimental Agriculture*, vol. 47, no. 6 (2007), pp. 721–26.

2. "The New Face of Hunger," *The Economist*, 17 April 2008; climate footprint of food, fiber, and transportation from Intergovernmental Panel on Climate Change (IPCC), *Climate Change 2007: Synthesis Report* (Geneva: 2007).

3. H. Steinfeld et al., *Livestock's Long Shadow: Environmental Issues and Options* (Rome: Food and Agriculture Organization, 2006); N. Uphoff et al., "Understanding the Functioning and Management of Soil Systems," in N. Uphoff et al., eds., *Biological Approaches to Sustainable Soil Systems* (Boca Raton, FL: CRC Press, 2006), pp. 3–13.

4. Steinfeld et al., op. cit. note 3; Box 3–1 from ibid., from IPCC, op. cit. note 2, and from P. Smith et al., "Agriculture," in IPCC, *Climate Change 2007: Mitigation of Climate Change* (Cambridge, U.K.: Cambridge University Press, 2007); M. Santilli et al. "Tropical Deforestation and the Kyoto Protocol," *Climatic Change*, August, 2005, pp. 267–76.

5. Data on Mexico from Willem Janssen and Svetlana Edmeades, World Bank, unpublished, 2008; B. Jamieson, "Vermont Maple Syrup Hard Hit by Climate Change: Warmer Temperatures, Shorter Winters Could Move Industry North to Canada," *ABC News*, 24 March 2007.

6. Data on Gangotri glacier from L. R. Brown, "Melting Mountain Glaciers Will Shrink Grain Harvests in China and India," *Eco-Economy Update* (Washington, DC: Earth Policy Institute, 20 March 2008); population dependent on livestock from S. Anderson, "Animal Genetic Resources and Sustainable Livelihoods," *Ecological Economics*, July 2003, pp. 331–39.

7. IPCC, *Climate Change 2001: Impacts, Adaptation and Vulnerability. Contribution of Working Group II to the Third Assessment Report of the Intergovernmental Panel on Climate Change* (Cambridge, U.K.: Cambridge University Press, 2001), Box 5–10.

8. S. J. Scherr and J. A. McNeely, eds., *Farming with Nature: The Science and Practice of Ecoagriculture* (Washington, DC: Island Press, 2007).

9. Figure 3–1 based on sketch by Molly Phemister, Ecoagriculture Partners. See full scientific discussion of climate change mitigation opportunities in agriculture, by region, in IPCC, op. cit. note 4.

10. Uphoff et al., "Understanding the Functioning and Management of Soil Systems," op. cit. note 3.

11. U.N. Food and Agriculture Organization (FAO), "New Global Soil Database," press release (Rome: 21 July 2008).

12. Smith et al., op. cit. note 4; fertilizer use data for 2002 from FAO, *FAOSTAT Statistical Database*, at faostat.fao.org.

13. T. J. LaSalle and P. Hepperly, *Regenerative Organic Farming: A Solution to Global Warming* (Kutztown, PA: Rodale Institute, 2008).

14. N. Uphoff et al., "Issues for More Sustainable Soil System Management," in Uphoff et al., *Biological Approaches*, op. cit. note 3, pp. 715–27.

15. T. Goddard et al., eds., *No-Till Farming Systems*, Special Publication No. 3 (Bangkok: World Association of Soil and Water Conservation, 2008); P. Hobbs, R. Gupta, and C. Meisner, "Conservation Agriculture and Its Applications in South Asia," in Uphoff et al., *Biological Approaches*, op. cit. note 3, pp. 357–71.

16. A. Calegari, "No-tillage System in Parana State, South Brazil," in E. M. Bridges et al., eds., *Response to Land Degradation* (Enfield, NH: Science Publishers, 2001), pp. 344–45.

17. R. Derpsch, "No Tillage and Conservation Agriculture: A Progress Report," in Goddard et al., op. cit. note 15, pp. 7–42; D. C. Reicosky, "Carbon Sequestration and Environmental Benefits from No-Till Systems," in ibid., pp. 43–58.

18. J. Lehmann, J. Gaunt, and M. Rondon, "Bio-Char Sequestration in Terrestrial Ecosystems—A Review," *Mitigation and Adaptation Strategies for Global Change*, March 2006, pp. 395–419.

19. Ibid.

20. S. Wood, K. Sebastian, and S. J. Scherr, *Pilot Analysis and Global Ecosystems: Agrosystems* (Washington, DC: International Food Policy Research Institute and World Resources Institute (WRI), 2000); Smith et al., op. cit. note 4.

21. R. R. B. Leakey, "Domesticating and Marketing Novel Crops," in Scherr and McNeely, op. cit. note 8, pp. 83–102; J. D. Glover, C. M. Cox, and J. P. Reganold, "Future Farming: A Return to Roots?" *Scientific American*, August 2007.

22. L. R. DeHaan et al., "Perennial Grains," in Scherr and McNeely, op. cit. note 8, pp. 61–82.

23. T. S. Cox et al., "Prospects for Developing Perennial Grain Crops," *BioScience*, August 2006, pp. 649–59.

24. Ibid.

25. C. A. Palm et al., *Carbon Sequestration and Trace Gas Emissions in Slash-and-Burn and Alternative Land Uses in the Humid Tropics*, ASB Climate Change Working Group, Final Report, Phase II (Nairobi: Alternatives to Slash-and-Burn Programme Coordination Office, World Agroforestry Centre, 1999).

26. U.N. Environment Programme, "A Billion Tree Campaign," at www.unep.org/BILLION TREECAMPAIGN.

27. J. R. Smith, *Tree Crops: A Permanent Agriculture* (New York: Harcourt, 1929); Steinfeld et al., op. cit. note 3.

28. Leakey, op. cit. note 21.

29. M. A. Sanderson and P. R. Adler, "Perennial Forages as Second Generation Bioenergy Crops," *International Journal of Molecular Sciences*, May 2008, pp. 768–88; J. C. Milder et al., "Biofuels and Ecoagriculture: Can Bioenergy Production Enhance Landscape-scale Ecosystem Conservation and Rural Livelihoods?" *International Journal of Agricultural Sustainability*, vol. 6, no. 2 (2008), pp. 105–21.

30. John Holdren, Woods Hole Institute, presentation at Katoomba Group meeting, "Building an Infrastructure Fund for the Planet," Washington, DC, June 2008; meat consumption in China from FAO, op. cit. note 12; C. Delgado et al., *Livestock to 2020: The Next Food Revolution*, Food, Agriculture, and Environment Discussion Paper 28 (Washington, DC: International Food Policy Research Institute, 1999).

31. Steinfeld et al., op. cit. note 3.

32. C. L. Neely and R. Hatfield, "Livestock Systems," in Scherr and McNeely, op. cit. note 8, pp.121–42.

33. Al Rotz et al., *Grazing and the Environment* (University Park, PA: Pasture Systems and Watershed Management Research Unit, Agricultural Research Service, U.S. Department of Agriculture, 2008).

34. Steinfeld et al., op. cit. note 3.

35. Ibid.

36. Ibid.

37. Methane from P. Forster and V. Ramaswamy, "Changes in Atmospheric Constituents and in Radiative Forcing," in IPCC, *Climate Change 2007: The Physical Science Basis* (Cambridge, U.K.: Cambridge University Press, 2007), p. 212.

38. Department of Agricultural and Biological Engineering, "Penn England Farm Case Study," Penn State University, University Park, PA, January 2008.

39. WRI, Earth Trends Information Portal, at

earthtrends.wri.org, viewed September 2008; B.G. Mackey et al., *Green Carbon: The Role of Natural Forests in Carbon Storage* (Canberra, Australia: ANU E Press, 2008); 217 tons of carbon per hectare average for temperate forests is from IPCC, op. cit. note 2.

40. G. J. Nabuurs et al., "Forestry," in IPCC, op. cit. note 4.

41. Decision 2/CP.13, Conference of the Parties to the Framework Convention on Climate Change, Bali, 3–15 December 2007.

42. International Finance Corporation, "The Biodiversity and Agricultural Commodities Program," at www.ifc.org/ifcext/sustainability.nsf/Content/Biodiversity_BACP.

43. A. Pagdee, Y. Kim, and P. J. Daugherty, "What Makes Community Forest Management Successful: A Meta-Study from Community Forests Throughout the World," *Society and Natural Resources*, January 2006, pp. 33–52.

44. R. J. S. Beeton et al., *Australia State of the Environment 2006*, Independent Report to the Australian Government Minister for the Environment and Heritage (Canberra, Australia: 2006).

45. FAO, *Community-based Fire Management: Case Studies from China, The Gambia, Honduras, India, the Lao People's Democratic Republic and Turkey*, Forest Resources Development Service, Working Paper FFM/2 (Bangkok: Regional Office for Asia and the Pacific, 2003).

46. A. Molnar, S. J. Scherr, and A. Khare, *Who Conserves the World's Forests? Community-Driven Strategies that Protect Forests and Respect Rights* (Washington, DC: Forest Trends and Ecoagriculture Partners, 2004).

47. W. Cavendish, "Empirical Regularities in the Poverty-Environment Relationship in Rural Households: Evidence from Zimbabwe," *World Development*, November 2000, pp. 1979–2003.

48. M. Falkenmark, J. Rockström, and H. Savenije, *Balancing Water for Humans and Nature: The New Approach to Ecohydrology* (London: Earthscan, 2004).

49. A. E. Sidahmed, "Rangeland Development for the Rural Poor in Developing Countries: The Experience of IFAD," in Bridges et al., op. cit. note 16, pp. 455–65.

50. WRI et al., *World Resources 2008* (Washington, DC: WRI, 2008), pp. 142–57.

51. J. A. McNeely and S. J. Scherr, *Ecoagriculture: Strategies to Feed the World and Save Wild Biodiversity* (Washington, DC: Island Press, 2003), p. 145.

52. Conservation International, "Chocó-Manabí Corridor Project, Ecuador," at www.conservation.org/learn/forests/Pages/project_choco_manabi.aspx.

53. Sustainable Food Laboratory, Sustainability Institute, Hartland, VT, at www.sustainablefoodlab.org.

54. "Dole to Make Banana and Pineapple Supply Chain Carbon Neutral," *ClimateBiz News*, 10 August 2007; R. Bayon, A. Hawn, and K. Hamilton, *Voluntary Carbon Markets: An International Business Guide to What They Are and How They Work* (London: Earthscan, 2006); S. J. Scherr, J. C. Milder, and C. Bracer, *How Important Will Different Types of Compensation and Reward Mechanisms Be in Shaping Poverty & Ecosystem Services across Africa, Asia & Latin America over the Next Two Decades*, ICRAF Working Paper No. 40 (Nairobi: World Agroforestry Centre, 2007); BioCarbon Fund, Carbon Finance Unit, World Bank, at carbonfinance.org/Router.cfm?Page=BioCF; Michael Specter, "Big Foot: In Measuring Carbon Emissions, It's Easy to Confuse Morality and Science," *The New Yorker*, 25 February 2008.

55. Forest Stewardship Council, at www.fsc.org.

56. Box 3–2 from K. Hamilton et al., *Forging a Frontier: State of the Voluntary Carbon Markets 2008* (Washington, DC. Katoomba Group's Ecosystem Marketplace, 2008), and from L. Micol et al., *Redução das Emissões do Desmata-*

mento e da Degradação (REDD) Potencial de Aplicação em Mato Grosso (Cuiabá, Brazil: Instituto Centro da Vida, 2008).

57. Climate Neutral Network, "Costa Rica," at www.climateneutral.unep.org; Roberto Dobles Mora, "Costa Rica's Commitment on the Path to Becoming Carbon-Neutral," *UN Chronicle Online Edition*, issue 2 (2007).

58. *Environmental Performance Index 2008*, at epi.yale.edu; European Union's subsidy number from "Who Gets What from the Common Agricultural Policy," at www.farmsubsidy.org; Environmental Working Group, *Farm Subsidy Database*, at www.farm.ewg.org, viewed 14 October 2008.

59. Stephanie Hanson, "African Agriculture," *Council on Foreign Relations Backgrounder*, 28 May 2008.

60. G. Volpi, "Climate Change Mitigation, Deforestation and Human Development in Brazil," Occasional Paper No. 39, *Human Development Report 2007/2008* (New York: U.N. Development Programme, 2007); Achim Steiner, Executive Director, U.N. Environment Programme, speeches at international meetings, 2008, at www.unep.org.

61. Population Division, *World Population Prospects: The 2006 Revision* (New York: United Nations, 2008).

62. J. Skoet and K. Stamoulis, *The State of Food Insecurity in the World 2006* (Rome: FAO, 2006).

The Risks of Other Greenhouse Gases

1. Figure of 17 percent based on Intergovernmental Panel on Climate Change (IPCC), *Climate Change 2007: Synthesis Report* (Geneva: 2007), p. 4, and on IPCC, *Safeguarding the Ozone Layer and the Global Climate System* (Geneva: 2005), p. 135.

2. IPCC, *Safeguarding the Ozone Layer*, op. cit. note 1, Table 11.2.

3. The 100-year global warming potential (GWP) of various F-gases: CFC-11 at 4,750; CFC-12 at 10,900; HCFC-22 at 1,810; HCFC-141b at 725; HFC-23 at 14,800; HFC-125 at 3,500; HFC-134a at 1,430; HFC-152a at 124; PFCs at 6,500–12,200; and SF_6 at 22,800, according to IPCC, *Climate Change 2007: The Physical Science Basis* (Cambridge, U.K.: Cambridge University Press, 2007), pp. 212–13. Other GWPs are: HFC-404A at 3,922; HFC-410A at 2,088; and HFC-507 at 3,985, according to A. Cohr Pachai, "Phase Out of R22 and Then What?" in Prokima, *Natural Refrigerants: Sustainable Ozone- and Climate-Friendly Alternatives to HCFCs* (Eschborn, Germany: German Technical Cooperation, 2008), pp. 237–44.

4. IPCC, op. cit. note 3, pp. 212–13.

5. Alternative Fluorocarbons Environmental Acceptability Study, at www.afeas.org/about.html.

6. Figure of 0.6 percent was calculated based on IPCC, *Safeguarding the Ozone Layer*, op. cit. note 1, p. 135, and on "Summary for Policymakers," in IPCC, op. cit. note 3, p. 4.

7. More information on Greenfreeze can be found at www.greenpeace.org/usa/campaigns/global-warming-and-energy/green-solutions/greenfreeze.

8. More information on Refrigerants Naturally can be found at www.refrigerantsnaturally.com.

9. Unilever, "Europe: New Ice Cream Cabinets Cut Impact on Climate Change," at www.unilever.com; Ben and Jerry's, "Hydrocarbon Freezers: The New Cool! The Cleaner Greener Freezer," at benandjerrys.com; "Coca-cola's Olympic Coolers 100% HFC-free," *ACR News*, 19 September 2007.

10. "Retailers Opt for Natural Refrigerants," R744.com, 6 March 2007; Tesco, "Corporate Responsibility Review 2008: Climate Change," at www.tescoreports.com.

11. U.N. Environment Programme, *2006 Report of the Refrigeration, Air Conditioning and Heat*

Pumps Technical Options Committee: 2006 Assessment (Nairobi: 2007).

12. "Austria and Denmark Told to Justify F-Gas Bans," *EurActiv.com*, 5 April 2007; Blaise Horisberger, Federal Office of Environment, Switzerland, "The Swiss Approach on Climate Protection in Refrigeration," PowerPoint presentation, Co2ol Food, Berlin, 22–23 May 2007; "Regulation (EC) No 842/2006 of the European Parliament and of the Council of 17 May 2006 on Certain Fluorinated Greenhouse Gases," and "Directive 2006/40/EC of the European Parliament and of the Council of 17 May 2006 Relating to Emissions from Air-conditioning Systems in Motor Vehicles," *Official Journal of the European Union*; California Air Resources Board, Climate Change Program, at www.arb.ca.gov/cc/ccea/ccea.htm.

13. David Doniger, NRDC Climate Center, "A Sectoral Agreement for HFCs?" PowerPoint presentation in Accra, Ghana, 26 August 2008.

Reducing Black Carbon

1. V. Ramanathan and G. Carmichael, "Global and Regional Climate Changes due to Black Carbon," *Nature Geoscience*, 23 March 2008; Tami C. Bond and Haolin Sun, "Can Reducing Black Carbon Emissions Counteract Global Warming?" *Environmental Science & Technology*, 15 August 2005, pp. 5921–26; Mark Z. Jacobson, "Control of Fossil-Fuel Particulate Black Carbon and Organic Matter, Possibly the Most Effective Method of Slowing Global Warming," *Journal of Geophysical Research*, 15 October 2002.

2. Figure from Tami Bond, Testimony before Hearing on Black Carbon and Climate Change, U.S. House Committee on Oversight and Government Reform, Washington, DC, 18 October 2007.

3. Charles S. Zender, Testimony before Hearing on Black Carbon and Climate Change, U.S. House Committee on Oversight and Government Reform, Washington, DC, 18 October 2007; V. Ramanathan and Y. Feng, "On Avoiding Dangerous Anthropogenic Interference with the Climate System: Formidable Challenges Ahead," *Proceedings of the National Academy of Sciences*, 23 September 2008, pp. 14245–50.

4. Intergovernmental Panel on *Climate Change, Climate Change 2007: The Physical Science Basis* (Cambridge, U.K.: Cambridge University Press, 2007), Chapter 2; Ramanathan and Carmichael, op. cit. note 1.

5. Ramanathan and Feng, op. cit. note 3.

6. Clean Air Task Force, *Diesel Engines: Emissions Controls and Retrofits* (Boston: April 2005).

7. Felicity Barringer, "Maritime Organization Seeks to Cut Air Pollution from Oceangoing Ships," *New York Times*, 9 October 2008.

8. International Network for Environmental Compliance and Enforcement, "Jump-Starting Climate Protection: INECE Targets Compliance with Laws Controlling Black Carbon," Washington, DC, 12 June 2008.

Women and Climate Change: Vulnerabilities and Adaptive Capacities

1. Intergovernmental Panel on Climate Change, *Summary for Policymakers. Climate Change 2001: Impacts, Adaptation, and Vulnerability* (Geneva: 2001).

2. Table based on the following: Morocco drought from V. Nelson et al., "Uncertain Predictions, Invisible Impacts, and the Need to Mainstream Gender in Climate Change Adaptations," *Gender and Development*, July 2002, pp. 51–59; E. Neumayer and T. Plümper, "The Gendered Nature of Natural Disasters: The Impact of Catastrophic Events on the Gender Gap in Life Expectancy, 1981–2002," *Annals of the Association of American Geographers*, vol. 97, no. 3 (2007), pp. 551–66; X. Rodo et al., "ENSO and Cholera: A Nonstationary Link Related to Climate Change?" *Proceedings of the National Academy of Sciences*, 1 October 2002, pp. 12901–06; G. Zhou et al., "Association Between Climate Variability and Malaria Epidemics in the East African Highlands," *Proceedings of the National Academy of Sciences*, 24 February 2004, pp.

2375–80; medical services and workload from Nelson et al., op. cit. this note; C. Thomas et al., "Extinction Risk from Climate Change," *Nature*, 8 January 2004, pp. 145–48; L. C. Stige et al., "The Effect of Climate Variation on Agro-pastoral Production in Africa," *Proceedings of the National Academy of Sciences*, 28 February 2006, pp. 3049–53; rural women's crop production from U.N. Food and Agriculture Organization (FAO), "Gender and Food Security: Agriculture," at www.fao.org/Gender/en/agri-e.htm; climate-related crop changes from FAO, *Gender and Food Security: Synthesis Report of Regional Documents* (Rome: undated). Malaria deaths of women and infants from "Pregnant Women and Infants," Global Malaria Programme, World Health Organization, at www.who.int/malaria/pregnant womenandinfants.html.

3. Oxfam, *Millennium Development Goals Gender Quiz*, at www.oxfam.org.uk:80/gener ationwhy/do_something/campaigns/healthand education/quiz/index.htm; C. Moser and D. Satterthwaite, "Pro-poor Climate Change Adaptation in the Urban Centers of Low and Middle-Income Countries," presented at Workshop on Social Dimension of Climate Change, World Bank, Washington, DC, 5 March 2008.

4. Women's more limited access to assets is documented by different institutions throughout the U.N. system.

5. Inter-American Development Bank, *Hurricane Mitch: Women's Needs and Contributions* (Washington, DC: 1999).

6. The Equilibrium Fund, "Reforestation," at www.theequilibriumfund.org/page.cfm?pageid=5494.

7. The Green Belt Movement, "Green Belt Movement and the World Bank Sign Emission Reductions Purchase Agreement (ERPA)," press release (Nairobi: 15 November 2006).

8. FAO, "Women—Users, Preservers and Managers of Agrobiodiversity," fact sheet (Rome: Sustainable Development Department, December 2001).

9. "New from Bali: Launch of the Global Gender and Climate Alliance," *WEDO News & Views*, January 2008.

The Security Dimensions of Climate Change

1. Robert T. Watson, Marufu Zinyowera, and Richard H. Moss, eds., *IPCC Special Report: The Regional Impacts of Climate Change: An Assessment of Vulnerability, Summary for Policymakers* (Geneva: Intergovernmental Panel on Climate Change (IPCC), 1997), p. 3; IPCC, *Climate Change 2007: Synthesis Report* (Geneva: 2007), pp. 46, 50.

2. See, for example, Ken Conca, *Governing Water: Contentious Transnational Politics and Global Institution Building* (Cambridge, MA: The MIT Press, 2006); U.N. Food and Agriculture Organization, *Summary of World Food and Agricultural Statistics 2005* (Rome: 2005).

3. Richard G. Lugar, "Opening Statement for Hearing on National Security Implications of Climate Change," Senate Committee on Foreign Relations, Washington, DC, 9 May 2007.

4. CNA Corporation, Military Advisory Board, *National Security and the Threat of Climate Change* (Alexandria, VA: 2007), p. 7; George W. Bush, "National Security Strategy of the United States of America," Washington, DC, 16 March 2006; Council of the European Union, "A Secure Europe in a Better World: European Security Strategy," Brussels, 12 December 2003.

5. Ragnhild Nordås and Nils Petter Gleditsch, "Climate Change and Conflict," *Political Geography*, August 2007, pp. 627–38; Thomas Homer-Dixon, "Environmental Scarcities and Violent Conflict: Evidence from Cases," *International Security*, summer 1994, pp. 5–40.

6. Clionadh Raleigh and Henrik Urdal, "Climate Change, Environmental Degradation and Armed Conflict," *Political Geography*, August 2007, pp. 674–94.

7. Nicholas Stern, *The Economics of Climate*

Change: The Stern Review (Cambridge, U.K.: Cambridge University Press, 2007), p. 56; Rafael Reuveny, "Environmental Change, Migration and Conflict: Theoretical Analysis and Empirical Explorations," presented at Workshop on Human Security and Climate Change, Asker, Norway, 21–23 June 2005.

8. Adrian Martin, "Environmental Conflict Between Refugee and Host Communities," *Journal of Peace Research*, May 2005, p. 333.

9. U.N. Environment Programme (UNEP), *Sudan: Post-Conflict Environmental Assessment* (Geneva: 2007), pp. 58, 84.

10. Ibid., p. 87.

11. U.N. Development Programme, *Human Development Report 2007/2008* (New York: 2007), p. 88.

12. Kurt M. Campbell et al., *The Age of Consequences: The Foreign Policy and National Security Implications of Global Climate Change* (Washington, DC: Center for Strategic and International Studies and the Center for A New American Security, 2007), p. 57; Sudha Ramachandran, "The Threat of Islamic Extremism to Bangladesh," *The Power and Interest News Report*, 27 July 2005.

Climate Change's Pressures on Biodiversity

1. T. Lovejoy and L. Hannah, eds., *Climate Change and Biodiversity* (New Haven, CT: Yale University Press, 2005).

2. T. D. Prowse, "River-ice Hydrology," in M. G. Anderson, ed., *Encyclopedia of Hydrological Sciences* (West Sussex: John Wiley and Sons Ltd., 2005); L. G. Thompson et al., "Abrupt Tropical Climate Change: Past and Present," *Proceedings of the National Academy of Sciences*, 11 July 2006, pp. 10536–43.

3. Kenneth Weiss, "Polar Bear Is Listed as Threatened Species," *Los Angeles Times*, 15 May 2008.

4. Table from the following: flowers from Sandra Bell, "Climate Change: Signs of the Times," *Kew Magazine*, summer 2007, pp. 22–27; tree swallows, Edith's Checkerspot, Golden Toads, and coral reefs from Lovejoy and Hannah, op. cit. note 1; eel grass from K. Blankenship, "Underwater Grasses at the Tipping Point?" *Bay Journal* (Alliance for the Chesapeake Bay), September 2006; American ash tree from M. Davy, "A Beetle and Balmy Weather May Bench a Baseball Staple," *New York Times*, 11 July 2007.

5. Intergovernmental Panel on Climate Change (IPCC), *Climate Change 2007: Synthesis Report. Summary for Policymakers* (Geneva: 2007).

6. G. M. Hewitt, "Post-glacial Re-colonization of European Biota," *Biological Journal of the Linnean Society*, September 1999, pp. 87–112.

7. E. E. Berg et al., "Spruce Beetle Outbreaks on the Kenai Peninsula, Alaska, and Kluane National Park and Reserve, Yukon Territory: Relationship to Summer Temperatures and Regional Differences in Disturbance Regimes," *Forest Ecology and Management*, 1 June 2006, pp. 219–32.

8. E. Salati and P. B. Vose, "Amazon Basin: A State of Equilibrium," *Science*, 13 July 1984, pp. 129–38.

9. P. M. Cox et al., *Amazon Dieback under Climate-Carbon Cycle Projections for the 21st Century*, Hadley Centre Technical Note 42 (Bracknell, Berks, U.K.: Hadley Centre, Met Office, 2003).

10. J. A. Kleypas et al., *Impacts of Ocean Acidification on Coral Reefs and Other Marine Calcifiers: A Guide for Future Research*, Report of a workshop in St. Petersburg, FL, 18–20 April 2005, sponsored by the National Science Foundation, the National Oceanic and Atmospheric Association, and the U.S. Geological Survey.

11. R. A. Feely et al., "Evidence for Upwelling of Corrosive 'Acidified' Water onto the Continental Shelf," *Science*, 13 June 2008, pp. 1490–92; V. J. Fabry et al., "Ocean Acidification's Effects on Marine Ecosystems and Biogeochemistry," Ocean Carbon and Biogeochemistry Scoping Workshop on Ocean Acidification Research, La Jolla, CA, 9–11

October 2007, *Eos, Transactions* (American Geophysical Union), vol. 89, no. 15 (2008), p. 143.

12. IPCC, op. cit. note 5.

13. G. F. Midgely et al., "Assessing the Vulnerability of Species Richness to Anthropogenic Climate Change in a Biodiversity Hotspot," *Global Ecology and Biogeography*, vol. 11, no. 6 (2002), pp. 445–51.

14. S. H. Pearsall, III, "Managing for Future Change on the Albemarle Sound," in Lovejoy and Hannah, op. cit. note 1, pp. 359–61.

15. IPCC, op. cit. note 5.

Small Island Developing States at the Forefront of Global Climate Change

1. New Economics Foundation, *Up in Smoke? Latin America and the Caribbean: The Threat from Climate Change to the Environment and Human Development* (London: 2006); Oxfam, *Climate Alarm: Disasters Increase as Climate Change Bites* (Oxford, U.K.: November 2007).

2. Pew Center on Global Climate Change, *Coral Reefs: Potential Contributions of Climate Change to Stresses on Coral Reef Ecosystems and Global Climate Change* (Arlington, VA: February 2004); Kent Carpenter et al., "One-Third of Reef-Building Corals Face Elevated Extinction Risk from Climate Change and Local Impacts," *Science*, 25 July 2008, pp. 560–63.

3. "Summary for Policymakers," in Intergovernmental Panel on Climate Change, *Climate Change 2007: The Physical Science Basis* (Cambridge, U.K.: Cambridge University Press, 2007), p. 21.

4. World Health Organization, *Climate Change and Human Health: Risks and Response* (Geneva: 2003).

5. Ministry of Planning and National Development of the Maldives, *Maldives Key Indicators* (Malé: date unknown); Maumoon Abdul Gayoom, speech before the Forty-second Session of the U.N. General Assembly, New York, 19 October 1987.

6. Dunya Maumoon, Deputy Minister for Foreign Affairs, Republic of Maldives, presentation to World Bank Workshop on the Social Dimensions of Climate Change, March 2008; Government of the Republic of Maldives and the Asian Development Bank, *Poverty Reduction Partnership Agreement*, 2002; World Bank, *World Development Indicators 2008* (Washington, DC: 2008).

7. Maldives Ministry of Environment, Energy and Water, *National Adaptation Programme of Action (NAPA): Republic of Maldives* (Malé: 2006).

8. Ibid.

9. Alliance of Small Island States, *No Island Left Behind*, Negotiating Brief for the UNFCCC Conference in Bali, November 2007.

10. United Nations Human Rights Council, Resolution 7/23: Human Rights and Climate Change, 28 March 2008.

The Role of Cities in Climate Change

1. Sources using 75–80 percent include the Clinton Climate Initiative, the United Nations, and the official web sites for the Mayors' offices of London and New York. For these sources and a more detailed explanation of their inherent errors, see David Satterthwaite, "Cities' Contribution to Global Warming: Notes on the Allocation of Greenhouse Gas Emissions," *Environment and Urbanization*, October 2008, pp. 539–49.

2. Shobhakar Dhakal, *Urban Energy Use and Greenhouse Gas Emissions in Asian Mega-Cities: Policies for a Sustainable Future* (Kitakyushu, Japan: Institute for Global Environmental Strategies, 2004); Mayor of London, *Action Today to Protect Tomorrow: The Mayor's Climate Change Action Plan* (London: Greater London Authority, 2007); Office of Long-term Planning and Sustainability, *Inventory of New York City Greenhouse Gas Emissions* (New York: Mayor's Office of Operations, 2007); C. Dubeux and E. La Rovere, "Local Perspectives in the Control of Greenhouse Gas Emissions–The Case of Rio de Janeiro," *Cities*, October 2007, pp. 353–64; J. VandeWeghe and C. Kennedy, "A Spatial Analysis of Residential Greenhouse Gas Emissions in the Toronto Census Metropolitan Area," *Journal of Industrial Ecology*, spring 2007, pp. 133–44.

3. David Dodman, "Blaming Cities for Climate Change? An Analysis of Urban Greenhouse Gas Emissions Inventories," *Environment and Urbanization*, forthcoming; Secretaria Municipal do Verde e do Meio Ambiente de São Paulo, *Inventário de Emissões de Efeito Estufa do Município de São Paulo* (São Paulo: Centro de Estudos Integrados sobre Meio Ambiente e Mudanças Climáticas da Coordenação dos Programas de Pós-graduação de Engenharia da Universidade Federal do Rio de Janeiro, 2005); Department of Health, Air Quality Division, *District of Columbia Greenhouse Gas Emissions Inventories and Preliminary Projections* (Washington DC: 2005).

4. Dodman, op. cit. note 3.

5. Peter Newman, "The Environmental Impact of Cities," *Environment and Urbanization*, October 2006, pp. 275–96; Bloomberg, op. cit. note 2.

6. Satterthwaite, op. cit. note 1; Office of Long-term Planning and Sustainability, op. cit. note 2; Mayor of London, op. cit. note 2.

7. David Satterthwaite et al., *Adapting to Climate Change in Urban Areas: The Possibilities and Constraints in Low- and Middle-income Nations* (London: International Institute for Environment and Development, 2007).

8. Luz Stella Velásquez, "The Bioplan: Decreasing Poverty in Manizales, Colombia, through Shared Environmental Management," in Steve Bass et al., eds., *Reducing Poverty and Sustaining the Environment* (London: Earthscan Publications, 2005), pp. 44–72.

Climate Change and Health Vulnerabilities

1. Intergovernmental Panel on Climate Change (IPCC), *Climate Change 2007: Impacts, Adaptation and Vulnerability* (Cambridge, U.K.: Cambridge University Press, 2007).

2. P. R. Epstein and E. Mills, eds., *Climate Change Futures: Health, Ecological and Economic Dimensions* (Boston: Center for Health and the Global Environment, Harvard Medical School, 2005); IPCC, *Climate Change 2007: The Physical Science Basis* (Cambridge, U.K.: Cambridge University Press, 2007).

3. P. R. Epstein, "Climate and Health," *Science*, 16 July 1999, pp. 347–48.

4. P. R. Epstein, "Climate Change and Public Health: Emerging Infectious Diseases," *Encyclopedia of Energy* (Amsterdam: Elsevier Science, 2004), pp. 381–92.

5. P. S. Moore and C. V. Broome, "Cerebrospinal Meningitis Epidemics," *Scientific American*, November 1994, pp. 38–45.

6. J. Almendares et al., "Critical Regions, A Profile of Honduras," *Lancet*, 4 December 1993, pp. 1400–02; Epstein and Mills, op. cit. note 2.

7. Dr Martin Ivan Sinclair, "Community and Local Government Participation in Malaria Control in Honduras," V Reunion del Comite Directivo Proyecto DDT/GEF/OPS, Mexico City, July 2008.

8. ACTIVIDADES Escuela de la Sustentabilidad Amigos de la Tierra América Latina, at www.censat.org/ambientalistas/60/Actividades.html; F. Magkos, F. Arvaniti, and A. Zampelas, "Organic Food: Nutritious Food or Food for Thought? A Review of the Evidence," *International Journal of Food Sciences and Nutrition*, September 2003, pp. 357–71.

9. P. R. Epstein, W. Moomaw, and C. Walker, eds., *Healthy Solutions for the Low Carbon Economy: Guidlines for Investors, Insurers and Policy Makers* (Boston: Harvard Medical School, 2008).

India Starts to Take on Climate Change

1. "Special Report, The World's Billionaires," *Forbes Magazine*, 8 March 2007; "India World's

Third Largest Economy: OECD," *Financial Express*, 6 December 2007; Netherlands Environmental Assessment Agency, "China Contributing Two Thirds to Increase in CO_2 Emissions," press release (Amsterdam: 13 June 2008).

2. Intergovernmental Panel on Climate Change (IPCC), *Climate Change 2007: Synthesis Report. Summary for Policymakers* (Geneva: 2007), p. 13.

3. Ministry of Environment and Forests, *India: Addressing Energy Security and Climate Change* (Delhi: October 2007); "India More Vulnerable to Climate Change: UNIDO," *Times of India*, 3 September 2008.

4. European Union, *India: Country Strategy Paper 2007–2013 & Multi-annual Indicative Programme 2007–2010* (Brussels: 2007); "India Becomes a Billionaire," U.N. Population Division, at www.un.org/esa/population/pubs archive/india/ind1bil.htm.

5. International Energy Agency, *World Energy Outlook 2007: China and India Insights* (Paris: 2007).

6. Amit Garg, P.R. Shukla, and Manmohan Kapshe, "The Sectoral Trends of Multigas Emissions Inventory of India," *Atmospheric Environment*, 40 August 2006, pp. 4608–20.

7. Quotes from Government of India, *National Five Year Plan* (Delhi: June 2008), p. 205; Prime Minister's Council on Climate Change, *National Action Plan on Climate Change* (Delhi: Government of India, June 2008), p. 2.

8. IPCC quote from "India More Vulnerable to Climate Change: UNIDO," op. cit. note 3.

9. Government of India, op. cit. note 7, p. 207.

10. Prime Minister's Council on Climate Change, op. cit. note 7.

11. Climate Challenge India, "Initial Assessment of India's National Action Plan on Climate Change," Centre for Social Markets, London, 3 July 2008.

12. Prime Minister's Council on Climate Change, op. cit. note 7, p. 45.

13. Ministry of Environment and Forests, op. cit. note 3, pp. 1, 6.

14. KPMG India, *Climate Change: Is India Inc. Prepared?* (Delhi: August 2008), pp. 7–11.

15. Anand Mahindra, discussion with author, July 2008.

16. "The Green Trade," *Businessworld*, 22 September 2008, p. 36.

17. Chatham House, *Changing Climates: Interdependencies on Energy and Climate Security for China and Europe* (London: November 2007), p. vi.

18. "The Green Trade," op. cit. note 16.

19. Environmental Planning & Coordination Organisation, Madhya Pradesh, discussion with author.

A Chinese Perspective on Climate and Energy

1. Netherlands Environmental Assessment Agency, "China Contributing Two Thirds to Increase in CO_2 Emissions," press release (Bilthoven: 13 June 2008).

2. "China's Energy Consumption Structure Changes, Energy Saving and Emission Reduction Face Challenges," *China Security News*, 1 March 2007; State Council, *China Energy White Paper* (Beijing: December 2007).

3. China Council for International Cooperation on Environment and Development (CCICED), "2006 Policy Recommendations by the China Council for International Cooperation on Environment and Development," at www.cciced.org/2008-02/19/content_10192515_3.htm.

4. City dwellers from China Population Info

Web site, at www.cpirc.org.cn/tjsj/tjsj_cy_1.asp, viewed August 2008; urban energy use from Qi Zhongxu, "Expert Opinion: Three Challenges China Facing in Energy Strategies," *Xinhua News Agency*, 1 June 2006.

5. CCICED, op. cit. note 3; Christopher L. Weber et al., "The Contribution of Chinese Exports to Climate Change," *Energy Policy*, September 2008, pp. 3572–77.

6. National People's Congress, "The 11th Five-year Plan for National Economic and Social Development," 16 March 2006.

7. "Central Finance Added 10 Billion Yuan for Energy Saving and Emission Reduction Special Funds," Chinese Embassy in the US, at www.china-embassy.org/chn/gyzg/t345508.htm, 27 July 2007; "Central Policies Meeting Local Challenges, Controls on Steel Export Increase Calls for Caution," *China Business Times,* 14 January 2008; Jia Hepeng, "Challenges to Energy Conservation and Emission Reduction" (draft), *China Insights*, September 2008.

8. "Central Finance Added 10 Billion Yuan," op. cit. note 7; Jia, op. cit. note 7.

9. Liu Zheng, "National Development and Reform Commission: Aborting Local Preferential Policies for Energy-intensive Industries," *Xinhua News Agency*, 26 May 2007; Zhu Jianhong, "Policy Analysis: One-vote Veto on Energy Saving and Emission Reduction," *People's Daily*, 30 November 2007.

10. Wang Youling and Zhou Wei, "To Break the Bottleneck on Energy and Environment, China Making Overhaul of Energy Conservation Law," *Xinhua News Agency*, 28 October 2007.

11. Liu Zheng and An Pei, "China Unit GDP Energy Intensity Down 1.23% in 2006, Failing Annual Targets," *Xinhua News Agency*, 28 February 2007; Zhou Yingfeng and Liu Zheng, "National Unit GDP Energy Intensity Down 3.66% in 2007," *Xinhua News Agency*, 14 July 2008.

12. Urbanization data from "Urbanization," in World Bank, *2007 World Development Indicators* (Washington, DC: 2007); Gerard Wynn and Emma Graham-Harrison, "China's First Climate Change Steps Too Small," *Reuters*, 23 April 2007.

13. Eric Martinot and Li Junfeng, *Special Report: Powering China's Development, the Role of Renewable Energy* (Washington, DC: Worldwatch Institute, November 2007).

14. Ibid.; Li Junfeng, "China's Wind Power Development Exceeds Expectations," *e2–Eye on Earth*, Worldwatch Institute, 2 June 2008.

15. Data provided by Li Junfeng of China Renewable Energy Industries Association.

16. Ibid.

17. Martinot and Li, op. cit. note 13.

Trade, Climate Change, and Sustainability

1. B. R. Copeland and M. S. Taylor, *Trade and the Environment: Theory and Evidence* (Princeton, N.J.: Princeton University Press, 2003); A. W. Wyckoff and J. M. Roop, "The Embodiment of Carbon in Imports of Manufactured Products: Implications for International Agreements on Greenhouse Gas Emissions," *Energy Policy*, March 1994, pp. 187–94.

2. Exports from multinationals from Bo Xilai, Minister of Commerce of China, speech at ASEAN Economic Ministers meeting, 26 August 2007.

3. New Economics Foundation (NEF), *Chinadependence: The Second UK Interdependence Report* (London: NEF and the Open University, 2007); Y. Li and C. N. Hewitt, "The Effect of Trade between China and the UK on National and Global Carbon Dioxide Emissions," *Energy Policy*, forthcoming; R. Reinvang and G. Peters, *Norwegian Consumption, Chinese Pollution: An Example of How OECD Imports Generate CO2 Emissions in Developing Countries* (Oslo and Trondheim: WWF Norway, WWF China Programme Office, and Norwegian University of Science and

Technology, 2008); initial assessment and Table from T. Wang and J. Watson, *Who Owns China's Carbon Emissions?* Tyndall Centre Briefing Note No. 23 (Norwich, U.K.: Tyndall Centre for Climate Change Research, 2007); N. Ahmad and A. Wyckoff, *Carbon Dioxide Emissions Embodied in International Trade of Goods* (Paris: Organisation for Economic Co-operation and Development, 2003); Chinese Academy of Social Sciences, in "WWF: Don't Ignore the Net Export of Embodied Energy from China," 11 December 2007, at www.wwfchina.org/english/loca.php?loca=496; International Energy Agency, *World Energy Outlook 2007* (Paris: 2007); G. P. Peters and E. G. Hertwich, "CO_2 Embodied in International Trade with Implications for Global Climate Policy," *Environmental Science & Technology*, 1 March 2008, pp. 1401–07; G. P. Peters et al., "China's Growing CO_2 Emissions—A Race between Increasing Consumption and Efficiency Gains," *Environmental Science & Technology*, 1 September 2007, pp. 5939–44; C. L. Weber et al., "The Contribution of Chinese Exports to Climate Change," *Energy Policy*, September 2008, pp. 3572–77.

4. G. P. Peters, "From Production-based to Consumption-based National Emission Inventories," *Ecological Economics*, 15 March 2008, pp. 13–23.

5. T. Houser et al., "Trade Measures," in T. Houser et al., eds., *Leveling the Carbon Playing Field: International Competition and US Climate Policy Design* (Washington, DC: Peterson Institute for International Economics, 2008).

6. For arguments of industrial lobbyists, see, for example, Carbon Trust, *EU ETS Impacts on Profitability and Trade: A Sector by Sector Analysis* (London: 2008).

Adaptation in Locally Managed Marine Areas in Fiji

1. This article is based on the authors' work and fieldwork in Fiji.

2. Quote from "What Works: Adaptive Management," at www.lmmanetwork.org.

Building Resilience to Drought and Climate Change in Sudan

1. See ReliefWeb, *Sudan: Drought Appeal No. 11/2001*, at www.reliefweb.int/w/rwb.nsf/vID/7FC68B9789E06194C1256A0900590AFC?OpenDocument.

2. B. Osman-Elasha, "Climate Variability and Change: Impacts on Peace and Stability in Sudan and the Region," Nile Development Forum, Khartoum, Sudan, January 2008.

3. B. Osman-Elasha and A. Sanjak, "Livelihoods and Drought in Sudan," in N. Leary et al., eds., *Climate Change and Vulnerability* (London: Earthscan, 2007), pp. 239–56; Government of Sudan, *First National Communications under the UNFCCC* (Khartoum, Sudan: 2002).

4. B. Osman-Elasha et al., "Community Development and Coping with Drought in Rural Sudan," in N. Leary et al., eds., *Climate Change and Adaptation* (London: Earthscan, 2007), pp. 90–108; B. Osman-Elasha, *Environmental Strategies to Increase Human Resilience to Climate Change: Lessons for Eastern and Northern Africa*, Final report, Assessments of Impacts and Adaptations to Climate Change (AIACC) (Washington, DC: AIACC Project Office, International START Secretariat, 2006).

5. Sustainable Livelihoods Unit, *Operationalizing the Sustainable Livelihoods Approach: The Civil Society Alternative* (New York: U.N. Development Programme, 1999); K. Pasteur, *Tools for Sustainable Livelihoods: Policy Analysis* (Brighton, U.K.: Institute of Development Studies, 2001).

6. B. Dougherty, A. Abusuwar, and K. Abdel Razik, *Community Based Rangeland Rehabilitation for Carbon Sequestration and Biodiversity*, Report of the Terminal Evaluation, April/May 2001 (New York: UNDP, unpublished); E. Spanger-Siegfried et al., *The Role of Community-Based Rangeland Rehabilitation in Reducing Vulnerability to Climate Impacts: Summary of a Case Study from Drought-Prone Bara Province, Sudan*, prepared for the International Institute for Sustainable Development, International Union for Conservation of Nature, and Stockholm Environment Institute project on Climate Change, Vulnerable Communities and Adaptation, 2005.

7. S. Zaki-Eldeen and A. Hanafi, *Environmental Strategies for Increasing Human Resilience in Sudan: Lessons for Climate Change Adaptation in North and East Africa*, Case Study (Washington, DC: AIACC Project Office, International START Secretariat, 2004).

8. Ibid.

9. H. Abdellatti et al., *Khor Arba'at Livelihoods and Climate Variability*, AIACC Case Study (Khartoum, Sudan: Higher Council for Environment and Natural Resources, 2003); S. Bashier, *Surface Runoff in the Red Sea Province*, RESAP Technical Paper No. 5 (Khartoum, Sudan: Red Sea Area Programme, 1991); H. Abdel Ati, "The Red Sea State: Drought Cycles and Human Adaptation," presented to the IIRR Drought Cycle Management Writeshop, Nairobi, 28 October –7 November 2003.

10. Abdellatti et al., op. cit. note 9.

11. E Sanjak, *Water Harvesting Technique as a Coping Mechanism to Climate Variability and Change (Drought)*, North Darfur State Case Study (Washington, DC: AIACC Project Office, International START Secretariat, 2003).

12. Ibid.

13. B. Osman et al., *Sustainable Livelihood Approach for Assessing Community Resilience to Climate Change: Case Studies from Sudan*, AIACC Working Paper No. 17 (Washington, DC: AIACC Project Office, International START Secretariat, 2005).

Geoengineering to Shade Earth

1. P. J. Rasch et al., "An Overview of Geoengineering of Climate Using Stratospheric Sulfate Aerosols," *Philosophical Transactions of the Royal Society Part A*, 29 August 2008.

2. Ibid.

3. Intergovernmental Panel on Climate Change, *Climate Change 2001: The Scientific Basis* (Cambridge, U.K.: Cambridge University Press, 2001), p. 944.

4. "Geoengineering," in National Academy of Sciences (NAS), *Policy Implications of Greenhouse Warming: Mitigation, Adaptation and the Science Base* (Washington, DC: National Academy Press, 1992), pp. 433–64.

5. K. Caldeira and L. Wood, "Global and Arctic Climate Engineering: Numerical Model Studies," *Philosophical Transactions of the Royal Society Part A*, 29 August 2008.

6. J. T. Early, "The Space-based Solar Shield to Offset Greenhouse Effect," *Journal of the British Interplanetary Society*, December 1989, pp. 567–69; B. Govindasamy, K. Caldeira, and P. B. Duffy, "Geoengineering Earth's Radiation Balance to Mitigate Climate Change from a Quadrupling of CO_2," *Global and Planetary Change*, 10 June 2003, pp. 157–68.

7. J. Latham et al., "Global Temperature Stabilization via Controlled Albedo Enhancement of Low-level Maritime Clouds," *Philosophical Transactions of the Royal Society Part A*, 29 August 2008; P. J. Crutzen, "Albedo Enhancement by Stratospheric Sulfur Injections: A Contribution to Resolve a Policy Dilemma?" (editorial essay), *Climatic Change*, August 2006, pp. 211–19.

8. NAS, op. cit. note 4.

9. L. Lane et al., eds., *Workshop Report on Managing Solar Radiation* (Hanover, MD: National Aeronautics and Space Administration, 2007); T. M. L. Wigley, "A Combined Mitigation/Geoengineering Approach to Climate Stabilization," *Science*, 20 October 2006, pp. 452–54.

10. D. M. Matthews and K. Caldeira, "Transient Climate-Carbon Simulations of Planetary Geoengineering," *Proceedings of the National Academy of Sciences*, 12 June 2007; Crutzen, op. cit. note 7; Caldeira and Wood, op. cit. note 5.

11. Lane et al., op. cit. note 9.

Carbon Capture and Storage

1. G. Göttlicher, *The Energetics of Carbon Dioxide Capture in Power Plants* (Washington, DC: trans. by National Energy Technology Laboratory, U.S. Department of Energy, 1999).

2. Mineralization process from Intergovernmental Panel on Climate Change, *IPCC Special Report on Carbon Dioxide Capture and Storage*, prepared by Working Group III (Cambridge, U.K.: Cambridge University Press, 2005), Chapter 7; for algae, see, for example, algae@work, at www.algaeatwork.com.

3. National Energy Technology Laboratory, *Carbon Sequestration Technology Roadmap and Program Plan* (Washington, DC: U.S. Department of Energy, 2007).

4. For predictions on large-scale applications, see European Technology Platform for Zero Emission Fossil Fuel Power Plants, "CO_2 Capture and Storage," at www.zero-emissionplatform.eu, and Greenpeace International, *False Hope: Why Carbon Capture and Storage Won't Save the Climate* (Amsterdam: 2008).

5. Quote from J. Goodell, "Coal's New Technology: Panacea or Risky Gamble?" *Yale Environment 360*, posted 14 July 2008; Ecofys in cooperation with TNO, *Global Carbon Dioxide Storage Potential and Costs* (Utrecht, Netherlands: 2004); uncertainty about suitability from Massachusetts Institute of Technology, *The Future of Coal* (Cambridge, MA: 2007); liability issue from D. Spreng, G. Marland, and A. M. Weinberg, "CO_2 Capture and Storage: Another Faustian Bargain?" *Energy Policy*, February 2007, pp. 850–54.

6. P. Viebahn et al., "Comparison of Carbon Capture and Storage with Renewable Energy Technologies regarding Structural, Economic, and Ecological Aspects," *International Journal of Greenhouse Gas Control*, April 2007, p. 126; Wuppertal Institute for Climate, Environment and Energy et al., *RECCS – Ecological, Economic and Structural Comparison of Renewable Energy Technologies (RE) with Carbon Capture and Storage (CCS)*, Final Report for German Ministry for the Environment (Wuppertal, Germany: 2008), p. 111. Results confirmed by J. Koornneef et al., "Life Cycle Assessment of a Pulverized Coal Power Plant with Post-combustion Capture, Transport and Storage of CO_2," *International Journal of Greenhouse Gas Control*, October 2008, pp. 448–67.

7. Erik Shuster, *Estimating Freshwater Needs to Meet Future Thermoelectric Generation Requirements* (Pittsburgh, PA: National Energy Technology Laboratories, 2007), p. 60; Greenpeace International, op. cit. note 4, p. 19.

8. Wuppertal Institute et al., op. cit. note 6, pp. 109, 110.

9. Capture costs from Wuppertal Institute et al., op. cit. note 6, p. 135ff; renewables' cost advantages from Viebahn et al., op. cit. note 6, p. 128.

10. Quote from H. Fernando et al., *Capturing King Coal: Deploying Carbon Capture and Storage Systems in the U.S. at Scale* (Washington, DC: World Resources Institute, 2008).

11. A. Stangeland et al., *How to Combat Global Warming* (Oslo: Bellona, 2008); J. Hansen et al., "Target Atmospheric CO_2: Where Should Humanity Aim?" *Open Atmospheric Science Journal*, 2008.

12. Two examples of scenarios are Greenpeace International and European Renewable Energy Council (EREC), *Energy (R)evolution: A Sustainable World Energy Outlook* (Amsterdam and Brussels: 2007) for the global level, and German Ministry for the Environment, Nature Conservation and Nuclear Safety, *Lead Study 2007—Update and Reassessment of the Strategy to Increase the Use of Renewable Energies up until the Years 2020 and 2030, plus an Outlook to 2050* (Berlin: 2007) for the national level.

13. European Commission, "Proposal for a Directive of the European Parliament and of the Council on the Geological Storage of Carbon Dioxide and Amending Council Directives 85/337/EEC, 96/61/EC, Directives

2000/60/EC, 2001/80/EC, 2004/35/EC, 2006/12/EC and Regulation (EC) No 1013/2006," Brussels, 2008; fears about financing from Greenpeace and EREC, op. cit. note 12, from Wuppertal Institute et al., op. cit. note 6, and from European Renewable Energy Research Centres Agency, "Input to the Public Consultation on the Strategic Energy Technology Plan," Brussels, 2007, p. 15.

Using the Market to Address Climate Change

1. Intergovernmental Panel on Climate Change, *Climate Change 2007: The Physical Science Basis* (Cambridge, U.K.: Cambridge University Press, 2007).

2. Energy Information Agency, *Monthly Energy Review* (Washington, DC: U.S. Department of Energy, 2008).

3. Emission rates for 2006, the most recent year data are available, are from www.eia.doe.gov/environment.html.

4. Paul L. Joskow, Richard Schmalensee, and Elizabeth M. Bailey, "The Market for Sulfur Dioxide Emissions," *The American Economic Review*, September 1998, pp. 669–85.

Technology Transfer for Climate Change

1. Romina Picolotti, "A Tale of Two Cities: Lessons for Climate Negotiators," *MEA Bulletin* (IISD Linkages), 29 November 2007.

2. U.N. Framework Convention on Climate Change (UNFCCC), *Investment and Financial Flows to Address Climate Change* (Bonn: October 2007), pp. 86–91.

3. Stephen O. Andersen, K. Madhava Sarma, and Kristen N. Taddonio, *Technology Transfer for the Ozone Layer: Lessons for Climate Change* (London: Earthscan 2007); D. Hunter, J. Salzman, and D. Zaelke, *International Environmental Law & Policy*, 3rd ed. (West Publishing Co., 2007).

4. Romina Picolotti, "Fast and Furious: Early Agreement on Fair and Equitable Financing is Key to Post-2012 Treaty," *MEA Bulletin* (IISD Linkages), 12 June 2008; Romina Picolotti, "Rethinking Climate Strategies," in Donald Kaniaru, ed., *The Montreal Protocol: Celebrating 20 Years of Environmental Progress—Ozone Layer and Climate Protection* (London: Cameron May, 2007), p. 155; K. Madhava Sarma, "Technology Transfer Mechanism for Climate Change," (draft), 24 March 2008, at www.igsd.org/docs/SarmaTT%2024Mar08.pdf; G. J. M. Velders et al., "The Importance of the Montreal Protocol in Protecting Climate," *Proceedings of the National Academy of Sciences*, 20 March 2007, pp. 4814–19; K. Madhava Sarma and Durwood Zaelke, "Start and Strengthen: The Importance of Immediate Action for Climate Mitigation," *MEA Bulletin* (IISD Linkages), 27 June 2008.

5. U.N. Environment Programme (UNEP), "Bali Strategic Plan for Technology Support and Capacity Building," 25 February 2005.

6. Andersen, Sarma, and Taddonio, op. cit. note 3.

7. UNEP, "Report of the Fourth Meeting of the Parties to the Montreal Protocol on Substances that Deplete the Ozone Layer," 25 November 1992, Annex V, p. 1.

8. UNFCCC, op. cit. note 2.

9. Montreal Protocol on Substances that Deplete the Ozone Layer, Articles 5(1) and 10(1), 16 September 1987.

Electric Vehicles and Renewable Energy Potential

1. M. Duvall, "Plug-In Hybrids on the Horizon: Building a Business Case," *EPRI Journal*, spring 2008, pp. 8–15.

2. J. G. Dorn, "Solar Thermal Power Coming to a Boil," *Earth Policy Update* (Washington, DC: Earth Policy Institute, July 2008).

3. Share of world vehicle fleet based on U.S. vehicle fleet figure from U.S. Department of Trans-

portation, Bureau of Transportation Statistics, at www.bts.gov/publications/transportation_statistics
_annual_report/2007/html/chapter_02/table_a_06.html, viewed 9 September 2008, and on world vehicle fleet figure from Michael Renner, "Vehicle Production Rises, But Few Cars Are 'Green'," *Vital Signs Online* (Washington, DC: Worldwatch Institute, May 2008); 63,000 wind turbine figure based on average capacity factor of 33 percent, from American Wind Energy Association, "Wind Web Tutorial: Wind Energy Basics," at www.awea.org/faq; U.S. Department of Energy, *20% Wind Energy by 2030: Increasing Wind Energy's Contribution to U.S. Electricity Supply* (Washington, DC: July 2008), p. 40; Mark Duvall, Electric Power Research Institute, testimony before Subcommittee on Energy, Science Committee, U.S. House of Representatives, Washington, DC, 17 May 2006, p. 4; vast renewable energy resources from U.N. Development Programme, *World Energy Assessment: Energy and the Challenge of Sustainability* (New York: 2000), and from T. B. Johansson et al., "The Potentials of Renewable Energy: Thematic Background Paper," International Conference for Renewable Energies, Bonn, Germany, 2004.

4. Michael Renner, "Jobs in Renewable Energy Expanding," *Vital Signs Online* (Washington, DC: Worldwatch Institute, July 2008); Jianxiang Yang, "China Speeds Up Renewable Energy Development," *China Watch* (Washington, DC: Worldwatch Institute, October 2006); Jonathan Watts, "Energy in China: 'We Call It the Three Gorges of the Sky. The Dam There Taps Water, We Tap Wind'," (London) *Guardian*, 25 July 2008.

5. Wind power in Denmark from European Wind Energy Association, "Strategic Overview of the Wind Energy Sector," at www.ewea.org/index.php?id=195; T. V. Holm, DONG Energy A/S, Hearing on $4 Gasoline and Fuel Economy: Auto Industry at a Crossroads, Select Committee on Energy Independence and Global Warming, U.S. House of Representatives, Washington, DC, 26 June 2008.

6. Project Better Place, at www.projectbetterplace.com, viewed 23 July 2008; T. Hamilton, "Toronto Could Become Electric Transportation Hub," *Toronto Star*, 10 September 2008.

7. W. Kempton and J. Tomic, "Vehicle-to-Grid Power Fundamentals: Calculating Capacity and Net Revenue," *Journal of Power Sources*, 1 June 2005, pp. 268–79; Holm, op. cit. note 5.

8. "RechargeIT.org: A Google.org Project," at www.google.org/recharge/index.html.

9. Alan L. Madian et al., *US Plug-In Hybrid and US Light Vehicle Data Book*, prepared for Plug-In Electric Vehicles 2008: What Role for Washington? conference (Washington, DC: LECG, June 2008).

10. Ibid.

11. Energy Information Administration, *Electric Power Annual* (Washington, DC: 2007); Holm, op. cit. note 5.

12. Madian et al., op. cit. note 9.

13. W. Kempton and J. Tomic, "Vehicle-to-Grid Power Implementation: From Stabilizing the Grid to Supporting Large-scale Renewable Energy," *Journal of Power Sources*, 11 April 2005, pp. 280–94.

Employment in a Low-Carbon World

1. This text is based on Worldwatch Institute, with technical assistance by the Global Labor Institute, Cornell University, *Green Jobs: Towards Decent Work in a Sustainable, Low-Carbon World* (Nairobi: 2008), commissioned by the U.N. Environment Programme (UNEP) as part of its joint Green Jobs Initiative with the International Labour Organization, International Trade Union Confederation, and International Organisation of Employers.

2. Job estimates are the result of a wide-ranging literature review conducted for the report commissioned by UNEP; see Worldwatch Institute with Global Labor Institute, op. cit. note 1.

3. Government Accountability Office, *Wind*

Power's Contribution to Electric Power Generation and Impact on Farms and Rural Communities (Washington, DC: 2004); Ben Block, "The Afterlife of German Coal Mining," *Eye on Earth* (Worldwatch Institute), 30 July 2008; Dipal Chandra Barua, *Grameen Shakti: Pioneering and Expanding Green Energy Revolution to Rural Bangladesh* (Dhaka, Bangladesh: Grameen Bank Bhaban, 2008).

4. Projections from Greenpeace International and Global Wind Energy Council, *Global Wind Energy Outlook 2006* (Amsterdam and Brussels: 2006), and from European Photovoltaic Industry Association and Greenpeace International, *Solar Generation IV–2007* (Brussels and Amsterdam: 2007); labor intensity from Daniel M. Kammen, Kamal Kapadia, and Matthias Fripp, *Putting Renewables to Work: How Many Jobs Can the Clean Energy Industry Generate?* (Berkeley, CA: Renewable and Appropriate Energy Laboratory, University of California, Berkeley, 2004).

5. Worldwatch Institute with Global Labor Institute, op. cit. note 1; German experience from Werner Schneider, German Trade Union Confederation, presentation at Trade Union Assembly on Labour and Environment, Nairobi, Kenya, 15–17 January 2006.

6. Fastest-rising carbon emissions from "Transport Sector Must Lead in the Climate Change Fight, UN Official Says," *UN News Service*, 30 May 2008; estimate of green jobs in auto manufacturing from Worldwatch Institute with Global Labor Institute, op. cit. note 1; retrofitting from Alana Herro, "Retrofitting Engines Reduces Pollution, Increases Incomes," *Eye on Earth* (Worldwatch Institute), 1 August 2007.

7. Europe from European Commission, *Panorama of Transport, 2007 Edition* (Brussels: Eurostat Statistical Books, 2007), p. 64, and from Eurostat, *European Business Facts and Figures*, 2007 edition (Luxembourg: Office for Official Publications of the European Communities, 2007), p. 195; China and India from Clel Harral, Jit Sondhi, and Guang Zhe Chen, "Highway and Railway Development in India and China, 1992–2002," Transport Note No. TRN-32 (Washington, DC: World Bank, 2006); bus rapid transit from "The BRT Model," *Sustainable Mobility*, October 2006, p. 11; cleaner buses from European Commission, op. cit. this note, p. 154, and from International Association of Public Transit, *Better Urban Mobility in Developing Countries* (Brussels: 2003), p. 19; New Delhi from "Capital Gets a 'Green' Diwali Gift," *The Hindu*, 7 November 2007.

8. Aimee McKane, Lynn Price, and Stephane de la Rue du Can, "Policies for Promoting Industrial Energy Efficiency in Developing Countries and Transition Economies," prepared for the U.N. Industrial Development Organization (Berkeley, CA: Lawrence Berkeley National Laboratory, 2007); scrap-based production and associated energy savings from International Iron and Steel Institute, *Steel and You: The Life of Steel* (Brussels: 10 January 2008), p. 3, from Subodh Das and Weimin Yin, "Trends in the Global Aluminium Fabrication Industry," *JOM*, February 2007, p. 84 (steel), and from Gary Gardner, "Aluminum Production Continues Upward," in Worldwatch Institute, *Vital Signs 2007–2008* (New York: W. W. Norton & Company, 2007), p. 59; jobs in secondary steel production is an estimate from Worldwatch Institute with Global Labor Institute, op. cit. note 1; jobs in secondary aluminum production from Japan Aluminum Association, "Outline of the Japanese Aluminum Industry," undated, at www.aluminum.or.jp/english/common/pdf/e_industry.pdf, from U.S. Census Bureau, *Secondary Smelting and Alloying of Aluminum: 2002* (Washington, DC: 2004), Table 1, and from European Aluminium Association and Organisation of European Aluminium Refiners and Remelters, *Aluminium Recycling: The Road to High Quality Products* (Brussels: 2004), p. 16.

9. U.S. recycling jobs from "Special Report: Investing in Recycling!" *Progressive Investor*, February/March 2008; Wael Salah Fahmi, "The Impact of Privatization of Solid Waste Management on the Zabaleen Garbage Collectors of Cairo," *Environment and Urbanization*, October 2005, pp. 155–70; "Brazil's Recycling Map Shows Close to 2,500 Firms Working in the Sector," *Brazil Magazine*, 4 October 2005; China from Christina

Reiss, "WRF in Shanghai: Bridging the Gap," *Recycling Magazine*, No. 16 (2007), p. 10.

10. Worldwatch Institute with Global Labor Institute, op. cit. note 1; U.K. and Ireland study from James Morison, Rachel Hine, and Jules Pretty, "Survey and Analysis of Labour on Organic Farms in the U.K. and the Republic of Ireland," *International Journal of Agricultural Sustainability*, vol. 3, no. 1 (2005).

11. Worldwatch Institute with Global Labor Institute, op. cit. note 1; people involved in agroforestry from World Bank, *Sustaining Forests: A Development Strategy* (Washington, DC: 2004), p. 16.

12. Recycling jobs from Andreas Manhart, *Key Social Impacts of Electronics Production and WEEE-Recycling in China* (Freiburg, Germany: Öko-Institut, June 2007), p. 15; plantation working conditions from Oxfam International, "Bio-fuelling Poverty," Oxfam Briefing Note (Oxford: 1 November 2007), from Rachel Smolker et al., *The Real Cost of Agrofuels: Food, Forest and the Climate* (Amsterdam: Global Forest Coalition, 2007), and from International Labour Organization, "Indonesian Plantation Workers Still Face Lack of Labour Rights," press release (Jakarta: 26 August 2005); impact on rural livelihoods from Friends of the Earth, LifeMosaic, and Sawit Watch, *Losing Ground: The Human Rights Impacts of Oil Palm Plantation Expansion in Indonesia* (London, Edinburgh, and Bogor: 2008).

Climate Justice Movements Gather Strength

1. Nigeria from "Women Taking the Lead in Reversing Climate Change," *WRM Bulletin* (World Rainforest Movement), October 2006; Friends of the Earth Australia, "Worldwide Protests Wednesday 24/10 Against Oil Pipeline in Amazon Headwaters," press release (Fitzroy, Australia: 24 October 2001); La Via Campesina (International Peasant Movement), "Peasants Denounce Impacts of Climate Change in Rural Areas and its Most Severe Consequences," press release (Jakarta: 12 December 2007).

2. U.N. Development Programme Indonesia, *The Other Half of Climate Change: Why Indonesia Must Adapt to Protect its Poorest People* (Jakarta: 2007).

3. Joshua Karliner, "Climate Justice Summit Provides Alternative Vision," 21 November 2000, at www.corpwatch.org.

4. Ibid.

5. International Indian Treaty Council, "Declaration of the First International Forum of Indigenous People's on Climate Change," Lyon, France, 4–6 September 2000; World Rainforest Movement, "Mount Tamalpais Declaration" (San Francisco, May 2000); CorpWatch, "Mount Tamalpais Declaration on Climate and Forests" (Oakland, CA: May 2000).

6. British Council Indonesia, "Asian Young Leaders Climate Forum," at www.britishcouncil.org/indonesia-society-aylcf-2007.htm.

7. Friends of the Earth International, "What's Missing from the Climate Talks?" press release (Bali, Indonesia: 14 December 2007).

8. Jorge Daniel Taillant, *Human Rights, Development and Climate Change Negotiations*, prepared for Climate Negotiations in Bali, December 2007 (Córdoba, Argentina: Center for Human Rights and Environment, 2007).

9. U.N. Development Group, *Guidelines on Indigenous Peoples' Issues* (New York: February 2008); Edward Cameron, director, Human Rights and Climate Change Program, World Bank, e-mail to author, 22 August 2008.

10. Tearfund, "Tearfund Submission on the Bali Action Plan (Para 1)" (Teddington, U.K.: 21 July 2008); Global Forest Coalition, "Effective Strategies to Reduce Emissions from Deforestation in Developing Countries (REDD) Must Address Leakage and Incorporate Social Impact Criteria," submitted to the Secretariat of the U.N. Framework Convention on Climate Change (Amsterdam, Netherlands, 20 March 2008); Climate Action Network International, "Views Regarding the Work Program for Issues under the Bali Action Plan," submitted to the Secretariat of the U.N. Framework Convention on Climate Change (Washington, DC: 22 February 2008).

Shifting Values in Response to Climate Change

1. C. Holst, ed., *Get Satisfied: How Twenty People Like You Found the Satisfaction of Enough* (Westport, CT: Easton Studio Press, 2007), pp. 31, 32.

2. Ibid., p. 8.

3. Associations with biophilia (the love of all life) from S. Saunders and D. Munro, "The Construction and Validation of a Consumer Orientation Questionnaire (SCOI) Designed to Measure Fromm's (1955) 'Marketing Character' in Australia," *Social Behavior and Personality*, vol. 28, no. 3 (2000), pp. 219–40; environmental attitudes and behaviors from M. L. Richins and S. Dawson, "A Consumer Values Orientation for Materialism and Its Measurement: Scale Development and Validation," *Journal of Consumer Research*, December 1992, pp. 303–16, and from P. W. Schultz et al., "Values and Their Relationship to Environmental Concern and Conservation Behavior," *Journal of Cross-cultural Psychology*, July 2005, pp. 457–75; ecological footprints from K. W. Brown and T. Kasser, "Are Psychological and Ecological Well-being Compatible? The Role of Values, Mindfulness, and Lifestyle," *Social Indicators Research*, November 2005, pp. 349–68; timber company example from K. M. Sheldon and H. McGregor, "Extrinsic Value Orientation and the Tragedy of the Commons," *Journal of Personality*, April 2000, pp. 383–411.

4. "Intrinsic" and "extrinsic" goals first described in T. Kasser and R. M. Ryan, "Further Examining the American Dream: Differential Correlates of Intrinsic and Extrinsic Goals," *Personality and Social Psychology Bulletin*, March 1996, pp. 280–87.

5. R. G. Tedeschi and L. G. Calhoun, "Posttraumatic Growth: Conceptual Foundations and Empirical Evidence," *Psychological Inquiry*, January 2004, pp. 1–18; K. Ring, *Heading Toward Omega: In Search of the Meaning of the Near-death Experience* (New York: Morrow, 1984).

6. P. J. Cozzolino et al., "Greed, Death, and Values: From Terror Management to Transcendence Management Theory," *Personality and Social Psychology Bulletin*, March 2004, pp. 278–92; E. L. B. Lykins et al., "Goal Shifts Following Reminders of Mortality: Reconciling Posttraumatic Growth and Terror Management Theory," *Personality and Social Psychology Bulletin*, August 2007, pp. 1088–99.

7. T. Kasser, *The High Price of Materialism* (Cambridge, MA: The MIT Press, 2002); T. Kasser, "Materialism and Its Alternatives," in M. Csikszentmihalyi and I. S. Csikszentmihalyi, eds., *A Life Worth Living: Contributions to Positive Psychology* (Oxford: Oxford University Press, 2006), pp. 200–14.

8. Effects of death thoughts on consumption from T. Kasser and K. M. Sheldon, "Of Wealth and Death: Materialism, Mortality Salience, and Consumption Behavior," *Psychological Science*, July 2000, pp. 348–51; effects of death thoughts on attitudes toward wilderness from S. L. Koole and A. E. Van den Berg, "Lost in the Wilderness: Terror Management, Action Orientation, and Nature Evaluation," *Journal of Personality and Social Psychology*, June 2005, pp. 1014–28; effects of economically hard times from P. R. Abramson and R. Inglehart, *Value Change in Global Perspective* (Ann Arbor: University of Michigan Press, 1995), from T. Kasser et al., "The Relations of Maternal and Social Environments to Late Adolescents' Materialistic and Prosocial Aspirations," *Developmental Psychology*, November 1995, pp. 907–14, and from K. M. Sheldon and T. Kasser, "Psychological Threat and Extrinsic Goal Striving," *Motivation and Emotion*, March 2008, pp. 37–45.

9. J. Schor, *Born to Buy* (New York: Scribner, 2004); S. Linn, *Consuming Kids* (New York: The New Press, 2004).

10. Quote from D. Rosnick and M. Weisbrot, *Are Shorter Work Hours Good for the Environment? A Comparison of U.S. and European Energy*

Consumption (Washington, DC: Center for Economic Policy and Research, 2006), p. 1; overview of time poverty in J. deGraaf, ed., *Take Back Your Time: Fighting Overwork and Time Poverty in America* (San Francisco: Berrett-Koehler, 2003).

11. N. Marks et al., *The Happy Planet Index: An Index of Human Well-being and Environmental Impact* (London: New Economics Foundation, 2006); problems with the business case for environmental change from T. Crompton, *Weathercocks and Signposts: The Environment Movement at a Crossroads* (Godalming, U.K.: WWF-UK, 2008).

12. Facilitating post-traumatic growth from R. A. Niemeyer, "Fostering Posttraumatic Growth: A Narrative Elaboration," *Psychological Inquiry*, January 2004, pp. 53–59, and from J. L. Pals and D. P. McAdams, "The Transformed Self: A Narrative Understanding of Posttraumatic Growth," *Psychological Inquiry*, January 2004, pp. 65–69.

Not Too Late to Act

1. Michael Pollan, "Farmer in Chief," *New York Times Magazine*, 12 October 2008; Stephen Heintz, "Time for an End to Business as Usual," *Alliance Magazine*, September 2007; European Climate Foundation at www.europeanclimate.org; interview with Uday Khemka, Alliance for Philanthropy and Social Investment Worldwide, September 2007; Debra Kahn, "Champion of Climate Change Fights Girds for a New Battleground," ClimateWire, 14 October 2008; Andy Darrell, "New York City's Sustainability Plan: A Bold Greenprint…," *Climate 411* (blog), Environmental Defense, 23 April 2007.

2. Robert Pollin et al., *Green Recovery: A Program to Create Good Jobs and Start Building a Low-Carbon Economy* (Washington, DC, and Amherst, MA: Center for American Progress and Political Economy Research Institute, September 2008).

3. Lester Brown, *Time for Plan B: Cutting Carbon Emissions 80 Percent by 2020* (Washington, DC: Earth Policy Institute, 2008); Clayton B. Cornell, "The World's Most Fuel Efficient Car: 285 MPG, Not a Hybrid," *Gas 2.0*, 12 March 2008.

4. McKinsey Global Institute, *Productivity of Growing Global Energy Demand: A Microeconomic Perspective* (November 2006); Joshua Zumbrun, "The Most Energy-Efficient Countries," *Forbes.com*, 7 July 2008.

5. Nick Hodge, "Renewable Energy in Africa: Some Like It Hot," *Energy & Capital*, 13 August 2007; "Solar Thermal Electric Power Plants Throughout MENA" (blog), *After Gutenberg*, 28 September 2007.

6. Hank Green, "China's Wind Power Set to Hit 100 Gigawatts," *EcoGeek*, 18 July 2008; Global Wind Energy Council, "U.S, China and Spain Lead Global Wind Markets in 2007," press release (Brussels: 6 February 2008); E. B. Boyd, "Tomorrowland: An Eco-smart Urban Design Competition Turns 'What Ifs' into 'What Is'," *Utne Reader*, September-October 2008; Jeffery Greenblatt, "Clean Energy 2030," *Google.com*, 1 October 2008.

7. U.N. Environment Programme (UNEP), "The Billion Tree Campaign Enters a Second Wave," at www.unep.org/billiontreecampaign; UNEP, "Communities, Corporations and Countries Deliver on Planting of One Billion Trees," press release (Nairobi: 28 November 2007); "Planting a Billion Trees Campaign Is Launched by United Nations Environment Programme," Horizon Solutions Web site, 9 November 2006.

8. Paul Hawken, *Blessed Unrest* (New York: Viking Press, 2008).

Chapter 4. An Enduring Energy Future

1. Austrian Federal Ministry for Transport, Innovation and Technology, "Model Region Güssing," *Forshungsforum 1/2007* (Vienna: 2007); Jonathan Tirone, "'Dead-end' Austrian Town Blossoms with Green Energy," *Bloomberg News*, 28 August 2007; Elizabeth Kolbert, "The Island in the Wind: A Danish Community's Victory Over Carbon Emissions," *The New Yorker*, 7 July 2008.

2. Larger cities from, for example, Janet L.

Sawin and Kristen Hughes, "Energizing Cities," in Worldwatch Institute, *State of the World 2007* (New York: W.W. Norton & Company, 2007), pp. 90–107; tenfold rise in price of oil from U.S. Energy Information Administration (EIA), "Petroleum Navigator," at tonto.eia.doe.gov/dnav/pet/hist/rwtcM.htm, viewed 27 September 2008.

3. International Energy Agency (IEA), *World Energy Outlook 2007* (Paris: 2007); "Summary for Policy Makers," in Intergovernmental Panel on Climate Change (IPCC), *Climate Change 2007: Mitigation of Climate Change* (Cambridge, U.K.: Cambridge University Press, 2007), p. 13; Jan Hamrin, Holmes Hummel, and Rachel Canapa, *Review of Renewable Energy in Global Scenarios*, prepared for the IEA (San Francisco: Center for Resource Solutions, 2007), p. i.

4. IEA, *Energy Technology Perspectives 2008: Scenarios and Strategies to 2050* (Paris: 2008), p. 39.

5. REN21, *Renewables 2007 Global Status Report* (Paris and Washington, DC: REN21 Secretariat and Worldwatch Institute, 2008), p. 6.

6. Growth rates from ibid., p. 7. Figure 4–1 from the following: renewables based on U.N. Development Programme (UNDP), *World Energy Assessment: Energy and the Challenge of Sustainability* (New York: 2000), and on T. B. Johansson et al., "The Potentials of Renewable Energy: Thematic Background Paper," International Conference for Renewable Energies, Bonn, Germany, 2004; global energy use is 2005 data from IEA, op. cit. note 3, p. 4.

7. U.N. Environment Programme (UNEP), International Environmental Technology Centre, *Energy and Cities: Sustainable Building and Construction* (Osaka, Japan: 2003), p. 1.

8. Half or less from U.S. Department of Energy (DOE), Energy Efficiency and Renewable Energy, "Technology Fact Sheet: Resources for Whole Building Design," GHG Management Workshop, 25–26 February 2003, p. 11; 80 percent from James Read, associate director, Arup Communications, on "Deeper Shades of Green," *Design E2*, U.S. Public Broadcasting System, summer 2006; information technology from Kurt Yeager, executive director, Galvin Electricity Initiative, e-mail to Janet Sawin, 2 September 2008.

9. Delhi green development from William Moomaw and Charles Bralber, *Scaling Alternative Energy: The Role of Emerging Markets, Dialogue Synthesis Report* (Medford, MA: The Fletcher School, Tufts University, 2008).

10. Developing countries from Chris Goodall, "The Rebound Effect," *Carbon Commentary*, 11 November 2007, at www.carboncommentary.com/2007/11/11/51, and from Steve Sorrell, "The Rebound Effect: Overview of Existing Research," UK Energy Research Centre, University of Strathclyde, Glasgow, powerpoint presentation, 28 February 2008; wealthier markets from Frank Gottron, *Energy Efficiency and the Rebound Effect: Does Increasing Efficiency Decrease Demand?* (Washington, DC: Congressional Research Service, 30 July 2001); U.S. case studies from Jonah Bea-Taylor, "Beating the Energy Efficiency Paradox" (blog), Rocky Mountain Institute, May 2008, at www.rmi.org.

11. Department for Environment, Food, and Rural Affairs, *UK Climate Change Programme: Annual Report to Parliament* (London: July 2008), p. 11.

12. Cheaper for building facades from Steven Strong, "Solar Electric Buildings: PV as a Distributed Resource," *Renewable Energy World*, July-August 2002, p. 171; peak demand times from IEA, *Renewables for Power Generation: Status and Prospects* (Paris: 2003), p. 19; cheaper than diesel from Satish Kumar, U.S. Agency for International Development, ECO-III Project in India, presentation at Energy Efficiency Global Forum, Washington, DC, 14 November 2007.

13. "Information on Passive Houses," at www.passivhaustagung.de/Passive_House_E/passivehouse.html, viewed 30 July 2008.

14. Richard Perez et al., "Solution to the Summer Blackouts? How Dispersed Solar Power-Generating Systems Can Help Prevent the Next Major

Outage," *Solar Today*, July/August 2005; Richard Perez, Atmospheric Sciences Research Center, State University of New York at Albany, e-mail to Janet Sawin, 3 October 2006; Denmark from Janet Sawin's personal experience, December 2000.

15. U.K. government from Fiona MacDonald, "Turning Our Homes into Power Stations," *Sunday Metro*, 4 August 2008.

16. Box 4–1 from the following: Kurt Yeager, executive director, Galvin Electricity Initiative, e-mails to Janet Sawin, 3 August 2006, 2 September 2008, and 12 September 2008, and discussion with Janet Sawin, 25 May 2007; John Carey, "A Smarter Electrical Grid," *Business Week*, 11 January 2008; Mark MacCracken, president, CALMAC Manufacturing Corporation, presentation at Energy Efficiency Global Forum & Exposition 2007, 11–14 November 2007; Energy Future Coalition, "What Is a Smart Grid, and Why Is It Important?" at www.energyfuturecoalition.org/preview.cfm?catID=57, viewed 29 July 2008; "A Wise Grid," *EnergyBiz Insider*, 11 May 2007; Pacific Gas and Electric from Marc Gunther, "Making the Dumb Grid Smarter," *RenewableEnergyWorld.com*, 29 October 2007; Netherlands from James Griffin, "Smart Metering: Slowly, but Surely," *UtiliPoint International*, 4 April 2007; Michael Setters, "Focus on European Smart Grids," *RenewableEnergyWorld.com*, 9 April 2008.

17. Technical issues addressed in Europe from Xavier Lemaire, "Regulation and Distributed Generation," Sustainable Energy Regulation Network/Renewable Energy and Energy Efficiency Partnership, presentation for ERRA Integration Workshop, Budapest, 6 July 2007, at www.rec.org/REEEP/workshops/distributed_generation/lemaire_regulation_and_distributed_generation.pdf; no payment for power from "Combined Heat and Power Petition," The Construction Centre, August 2008, at www.theconstructioncentre.co.uk/news/latest-news/combined-heat-and-power-petition.html.

18. B. Metz et al., "Energy Supply," in IPCC, op. cit. note 3.

19. Shares of energy use and emissions from ibid.; one-third conversion from EIA, DOE, "How Electricity is Generated," November 2007, at www.eia.doe.gov/kids/energyfacts/sources/electricity.html#Generation; losses from World Bank, *World Development Report 1997* (New York: Oxford University Press, 1997), and from Seth Dunn, *Micropower The Next Electrical Era*, Worldwatch Paper 151 (Washington, DC: Worldwatch Institute, 2000), p. 46; incandescent bulb from D. Cleland, *Sustainable Energy Use and Management, Proceedings of the Conference, People and Energy, How Do We Use It?* (Christchurch: Royal Society of New Zealand, 2005), pp. 91–102.

20. Vijaya Ramachandran, "Power and Roads for Africa," essay (Washington, DC: Center for Global Development, March 2008), p. 9; Sahara from Schott Solarthermie GmbH, cited in Ryan O'Keefe, vice president, Solar Development, FPL Energy, LLC, Texas Solar Forum, Austin, 24 April 2008; Middle East and elsewhere from Fred Morse, Senior Advisor, U.S. Operations, Abengoa Solar, presentation for "Concentrating Solar Power: What Can Solar Thermal Electricity Deliver, and at What Price?" Webcast, 26 June 2008; China's wind from Worldwatch calculation based on 3,200 gigawatts of potential from China Meteorological Administration, cited in Zijun Li, "China's Wind Energy Potential Appears Vast," *Eye on Earth* (Worldwatch Institute), 2 November 2005, and on "Installed Electric Capacity Reaches 713m Kilowatts," *ChinaDaily.com*, 14 January 2008; U.S. wind from Janet L. Sawin et al., *American Energy: The Renewable Path to Energy Security* (Washington, DC: Worldwatch Institute and Center for American Progress, 2006), p. 26; vast resources from UNDP, op. cit. note 6, and Johansson et al., op. cit. note 6.

21. REN21, op. cit. note 5, p. 6; European Wind Energy Association (EWEA), "Wind Energy Leads EU Power Installations in 2007, But National Growth Is Inconsistent," press release (Brussels: 4 February 2008); Ryan Wiser and Mark Bolinger, "Annual Report on U.S. Wind Power Installation, Cost, and Performance Trends: 2007," prepared for DOE, May 2008, p. 4; photovoltaics (PV) is Worldwatch calculation based on data from Paul Maycock and Prometheus Institute,

PV News, various issues, and from Travis Bradford, Prometheus Institute, discussion with Janet Sawin, 29 April 2008; Michael Kanellos, "Can Wind Energy be Cheaper than Regular Power?" *CNET.com*, 14 August 2007; Stephanie Busari, "U.S., China Lead Way in Tapping Wind Power," *CNN.com*, 29 July 2008; solar thermal in California from Rainer Aringhoff, president, Solar Millennium LLC, presentation for "Concentrating Solar Power: What Can Solar Thermal Electricity Deliver, and at What Price?" Webcast, 26 June 2008; China and India from David R. Mills and Robert G. Morgan, "A Solar-Powered Economy: How Solar Thermal Can Replace Coal, Gas and Oil," *Renewable Energy World*, 3 July 2008; DOE expects solar PV to be cost-competitive with baseload power in the United States by 2015, according to David Rodgers, Deputy Assistant Secretary for Energy Efficiency, DOE, presentation on panel "New Approaches to Environmentally Conscious Building Envelope Design and Technologies," at Energy Efficiency Global Forum & Exposition 2007, 11–14 November 2007; PV cost in Europe and elsewhere from Ashley Seager, "Solar Future Brightens as Oil Soars," (London) *The Guardian*, 16 June 2008.

22. DOE, Energy Efficiency and Renewable Energy, *20% Wind Energy by 2030: Increasing Wind Energy's Contribution to U.S. Electricity Supply* (Washington, DC: 2008), p. 93; EWEA, *Large-Scale Integration of Wind Energy into the European Power Supply: Analysis, Issues and Recommendations* (Brussels: December 2005), p. 6.

23. Innovative technologies from IEA, *Variability of Wind Power and Other Renewables: Management Options and Strategies* (Paris: June 2005), p. 41; for direct current see, for example, New Energy, "Grid Connection of Offshore Windparks," at www.newenergy.org.cn/english/guide/offshore-1.htm; Robin McKie, "How Africa's Desert Sun Could Bring Europe Power," (London) *Observer*, 2 December 2007.

24. Great Plains from study by Midwest Independent Service Operator, cited in Ken Silverstein, "Cleaning the Transmission Process," *RenewableEnergyWorld.com*, 14 July 2008, and from Electric Reliability Council of Texas, cited in Michael Goggin, "Texas Study: Benefits of Wind Transmission Outweigh Costs," *RenewableEnergyWorld.com*, 11 April 2008; solar in U.S. Southwest from Aringhoff, op. cit. note 21; offshore resources from OCS Alternative Energy and Alternate Use Programmatic EIS Information Center, U.S. Department of the Interior, "Offshore Wind Energy," at ocsenergy.anl.gov/guide/wind/index.cfm.

25. Ken Zweibel, James Mason, and Vasilis Fthenakis, "A Solar Grand Plan: By 2050 Solar Power Could End U.S. Dependence on Foreign Oil and Slash Greenhouse Gas Emissions," *Scientific American*, January 2008, p. 69; ocean energy from Sawin et al., op. cit. note 20, p. 33.

26. Richard Baxter, "A Call for Back-Up: How Energy Storage Could Make a Valuable Contribution to Renewables," *Renewable Energy World*, September 2007; "Cities to Store Wind Power for Later Use: Iowa Project Would Compress Air in Underground Caverns," *Associated Press*, 4 January 2006; Electricity Storage Association, "Technologies & Applications: CAES," at www.electricitystorage.org/tech/technologies_technologies_caes.htm; David Marcus, "Moving Wind to the Mainstream: Leveraging Compressed Air Energy Storage," *RenewableEnergyWorld.com*, 1 October 2007; compressed air and wind from Roger Peters with Linda O'Malley, *Storing Renewable Power* (Calgary, Canada: Pembina Institute, 2008), pp. 11–12; José Etcheverry, *Developing Renewable Energy to their Full Potential Through Smart Grids and Storage Options* (Vancouver, Canada: David Suzuki Foundation, undated).

27. DOE, op. cit. note 22, p. 77; Edgar A. DeMeo et al., "Wind Plant Integration: Advances in Insights and Methods" (draft 5), 7 July 2007; IEA, op. cit. note 23, p. 42; Edgar A. DeMeo et al., "Accommodating Wind's Natural Behavior: Advances in Insights and Methods for Wind Plant Integration," *IEEE Power & Energy Magazine*, November/December 2007, p. 60.

28. EWEA, op. cit. note 22, p. 18; DeMeo et al., "Accommodating Wind's Natural Behavior," op. cit. note 27; Hannele Holttinen et al., "State-of-the-Art Design and Operation of Power Systems with Large Amounts of Wind Power, Summary of IEA Wind Collaboration," presented at European Wind Energy Conference 2007, Milan, Italy, 7–10 May 2007, p. 10.

29. EWEA, "Strategic Overview of the Wind Energy Sector," 2008, at www.ewea.org/index.php?id=195; James Kanter, "Denmark Leads the Way in Green Energy—To a Point," *International Herald Tribune*, 21 March 2007; German Wind Energy Institute, "Windenergie in Deutschland—Aufstellungszahlen fur das Jahr 2007," at www.dewi.de/dewi/fileadmin/pdf/publications/Statistics%20Pressemitteilungen/31.12.07/press eanhang_2007%20.pdf; Pacific Gas and Electric, "2008 Renewables," at www.pge.com/renewableRFO; DeMeo et al., "Accommodating Wind's Natural Behavior," op. cit. note 27, p. 67; J. Charles Smith and Brian Parsons, "What Does 20% Look Like? Developments in Wind Technology and Systems," *IEEE Power & Energy Magazine*, November/December 2007, p. 24.

30. Sheryl Carter, Devra Wang, and Audrey Chang, "The Rosenfeld Effect in California: The *Art* of Energy Efficiency," Natural Resources Defense Council, 2006, at www.energy.ca.gov/commissioners/rosenfeld_docs/rosenfeld_effect/presentations/NRDC.pdf.

31. "BPL Global Enters Joint Venture with BPL Africa to Expand Smart Grid Footprint," press release (Pittsburgh, PA: 29 May 2008); India from Phillip Bane, "Major Smart Grid Projects Announced," *SmartGridNews.com*, 17 June 2008; Europe from James Griffin, "Smart Metering: Slowly, but Surely," UtiliPoint International, 4 April 2007; Arc Innovations, "Arc Innovations Begins Christchurch Network Rollout," press release (Christchurch, New Zealand: 28 June 2007); "Xcel Energy Announces Smart Grid Plan for City of Boulder," *Clean Edge News*, 13 March 2008; smooth integration from Kurt Yeager, executive director, Galvin Electricity Initiative, "Facilitating the Transition to a Smart Electric Grid," testimony to House Subcommittee on Energy and Air Quality, 3 May 2007; controlling the flow from Yeager, op. cit. note 8.

32. Leila Abboud, "Thar She Blows: DONG's Wind Woes" (blog), *Wall Street Journal*, 11 March 2008.

33. Figure 4–2 from Wolfram Krewitt et al., *Renewable Energy Deployment Potentials in Large Economies*, prepared for REN21 (Stuttgart, Germany: Institute of Technical Thermodynamics, German Aerospace Center (DLR), 2008), pp. 8, 18–37; REN21, *Renewable Energy Potentials, Summary Report* (Paris: 2008), pp. 14–15; Zweibel, Mason, and Fthenakis, op. cit. note 25, pp. 64–73.

34. Enercon GmbH, SolarWorld AG and Schmack Biogas AG, "The Combined Power Plant–The First Stage in Providing 100% Power from Renewable Energy," press release (Berlin: 9 October 2007); "Background Paper: The Combined Power Plant" (undated), at www.kombikraftwerk.de/fileadmin/downloads/Background_Information_Combined_power_plant.pdf; Dave Gilson, "Power Q&A: S. David Freeman," *Mother Jones*, 21 April 2008, at www.motherjones.com/interview/2008/05/interview-s-david-freeman.html.

35. O. Langniss et al., *Renewables for Heating and Cooling: Untapped Potential* (Paris: Renewable Energy Technology Deployment, IEA, 2007), p. 15.

36. John Perlin, *A Forest Journey: The Role of Wood in the Development of Civilization* (Cambridge, MA: Harvard University Press, 1991); Pompeii from Geothermal Education Office, "Geothermal Energy," slideshow, funded by U.S. Department of Energy, undated; John Perlin, "Solar Evolution: The History of Solar Energy," California Solar Center, 2005, at www.californiasolarcenter.org/history_passive.html.

37. Solar heating rank from REN21, op. cit. note 5, p. 6; China and top per capita countries from Werner Weiss, Irene Bergmann, and Gerhard Faninger, *Solar Heat Worldwide: Markets and Contribution to the Energy Supply 2005*, prepared for the Solar Heating and Cooling Programme (Paris: IEA, 2007), p. 17; Israel from European Solar Thermal Industry Federation (ESTIF), *Solar*

Thermal Action Plan for Europe: Heating & Cooling from the Sun (Brussels: 2007), p. 20; "First of Its Kind: Hybrid SolarWall PV/T System in Olympic Village," *SolarWall News*, 24 July 2008.

38. Industrial heat from Werner Weiss, "Untapped Potential: Solar Heat for Industrial Applications," *Renewable Energy World*, January-February 2006, pp. 68–74, from Langniss et al., op. cit. note 35, p. 29, and from ESTIF, op. cit. note 37, pp. 6–7; Claudia Vannoni, Riccardo Battisti, and Serena Drigo, "Potential for Solar Heat in Industrial Processes," prepared within Task 33 "Solar Heat for Industrial Processes" of the IEA Solar Heating and Cooling Programme and Task IV of the IEA SolarPACES Programme (Madrid: CIEMAT, 2008), pp. 1–3, 12.

39. Pellet stoves from Christian Rakos, "Time for Stability: An Update on International Wood Pellet Markets," *Renewable Energy World*, January/February 2008; biofuels from "Germany Starts Testing Bio Oil for Heat," *RenewableEnergyWorld.com*, 26 November 2007; Lena Sommestad, Director, Swedish District Heating Association, "Swedish District Energy—Innovation for Sustainable Development," May 2008, at cdea.ca/events/de-presentations/1_Sommestad_L.pdf, viewed 23 July 2008; Austria from Jane Burgermeister, "Biomass Heat and Electricity Plants on the Rise in Europe," *RenewableEnergyWorld.com*, 6 May 2008; Svend Brandstrup Hansen, "Bioenergy in Denmark," in *Nordic Bioenergy 2007: Invitation and Programme*, Stockholm, 11–13 June 2007, p. 3; "Poland Country Profile," European Bank for Reconstruction and Development, at www.ebrdrenewables.com/sites/renew/countries/Poland/profile.aspx; efficiency from Heinz Kopetz, "Biomass—A Burning Issue," *Renewable Energy Focus*, March/April 2007, pp. 52–58.

40. Geothermal uses and countries from Langniss et al., op. cit. note 35, p. 51; Wilson Rickerson et al., "An Overview of Renewable Heating in the United States: Policy and Market Trends," in Heinrich Böll Foundation, *The Missing Piece in Climate Policy: Renewable Heating and Cooling in Germany and the U.S.* (Washington, DC: 2008), p. 19.

41. Efficiencies from IEA, *Combined Heat and Power: Evaluating the Benefits of Greater Global Investment* (Paris: 2008), p. 10; traditional fossil fuel units from "Cogeneration: More Energy, Less Pollution From Fossil Fuels," *eJournalUSA*, 9 May 2008; residential-scale units from Mark Clayton, "It Heats. It Powers. Is it the Future of Home Energy?" *Christian Science Monitor*, 14 November 2006.

42. Christopher M. Looney and Stephen K. Oney, "Seawater District Cooling and Lake Source District Cooling," *Energy Engineering*, vol. 104, no. 5 (2007), pp. 34–45; "Deep Lake Water Cooling: Chilled Water for Cooling Toronto's Buildings," at www.enwave.com/dlwc.php, viewed 16 August 2008; A. Pongtornkulpanich et al., "Experience with a Fully Operational Solar-Driven 10-ton LiBr/H2O Single-Effect Absorption Cooling System in Thailand," *Renewable Energy*, May 2008, pp. 943–49.

43. Wolfram Krewitt, "Integration of Renewable Energy into Future Energy Systems," paper presented at IPCC Scoping Meeting Special Report on Renewable Energy and Climate Change, Lübeck, Germany, 21–25 January 2008, p. 132; European Geothermal Energy Council, *Geothermal Heating & Cooling Action Plan for Europe* (Brussels: 2007), p. 5.

44. IEA from Langniss et al., op. cit. note 35, p. 16; Kristin Seyboth et al., "Recognising the Potential for Renewable Energy Heating and Cooling," *Energy Policy*, July 2008, pp. 2460–63.

45. Barriers from Li Hua, "China's Solar Thermal Industry: Threat or Opportunity for European Companies?" *Renewable Energy World*, July-August 2002, p. 107, and from ESTIF, op. cit. note 37, pp. 6–7.

46. German market from Werner Weiss, Irene Bergmann, and Gerhard Faninger, "Solar Heat Worldwide: Markets and Contribution to the Energy Supply 2006," IEA Solar Heating and Cooling Programme, Gleisdorf, Austria, May 2008; subsidies from Friedrich Ebert Stiftung, *Solar Energy in Germany* (London: March 2006); Baden Württemberg from "First Heating Law for

Renewable Energy in Germany," *Energy Server* (newsletter for Renewable Energy and Energy Efficiency), 2 August 2007; federal regulations from Arne Jungjohann, Heinrich Böll Foundation, e-mail to Janet Sawin, 25 July 2008; Spain from REN21, *Renewables Global Status Report 2006 Update* (Paris and Washington, DC: REN21 Secretariat and Worldwatch Institute, 2006), p. 2; "Hawaii Solar Mandate First in the Nation: Bill's Introducer Senator Gary L. Hooser Considers Groundbreaking Action 'Vital,'" by Hawaii State Senate Majority, 30 June 2008, at www.hawaiireporter.com/story.aspx?a6d06dbe-0f6b-461f-89b0-2d26d6be30b9.

47. Figure 4–3 from Krewitt et al., op. cit. note 33, pp. 8, 10, 18–37; projection for 2050 from ibid. and from REN21, op. cit. note 33, pp. 15–16.

48. Two thirds released as heat from IEA, op. cit. note 41, p. 5; European losses from Sommestad, op. cit. note 39.

49. Owen Bailey and Ernst Worrell, *Clean Energy Technologies: A Preliminary Inventory of the Potential for Electricity Generation* (Berkeley, CA: Environmental Energy Technologies Division, Ernest Orlando Lawrence Berkeley National Laboratory, 2005), p. 2.

50. Northern Europe from IEA, op. cit. note 41, p. 8; 75 percent of Finland's district heat came from CHP in 2001, from ibid., p. 16; District Energy St. Paul, *Annual Report: Mapping a Course Toward Saint Paul's Energy Independence* (Saint Paul, MN: 2006).

51. Thomas R. Casten and Richard Munson, "Recycled Energy Can Power Industry," *Energy Management*, September 2007, p. 20.

52. Industry share from China Council for International Cooperation on Environment and Development, "2006 Policy Recommendations by the China Council for International Cooperation on Environment and Development," at www.cciced.org/2008-02/19/content_10192515_3.htm; steel and cement from "Current Situation and Challenges for Waste Heat Utilization," Shanghai Energy Saving Information Portal, 4 January 2007, at www.365jn.cn/html/2007/0104/4181.htm; Saudi Arabia quote from Roger Ballentine, in Michael Kanellos, "Waste Heat: The Next Frontier for Clean-Tech Companies," *CNetNews.com*, 21 April 2008; Baosteel generation from "Shaangu TRT: Assisting Energy Saving and Emission Reduction of China's Steel Industry," *China Industry News*, 12 March 2008; average Chinese is Worldwatch calculation based on 1,600 kilowatt-hours a year per person consumption, from Quirin Schiermeier et al., "Electricity without Carbon," *Nature*, 13 August 2008, p. 817; Eastern China from Zhang Yange, "Combined Heat and Power: Survival of An Energy Conservation Industry," *Economic Observer*, 2 July 2006.

53. Appropriate Infrastructure Development Group, "Biodigesters," at www.aidg.org/biodigesters.htm; James Kanter, "Sweden Turning Sewage into a Gasoline Substitute," *International Herald Tribune*, 27 May 2008.

54. Marty Weil, "Animal Planet: Another Avenue to Renewable Fuels," *RenewableEnergyWorld.com*, 17 June 2008; Matthew L. Wald, "Gassing Up with Garbage," *New York Times*, 24 July 2008; Stephen Lacey, "Part 1: What to Do with All This Waste?" *RenewableEnergyWorld.com*, 26 November 2007.

55. Martin Bensmann, "Slime from Neptune's Garden," *New Energy*, April 2008, p. 73; Graham Jesmer, "WMU Researchers Create Biofuels from Waste Oil & Algae," *RenewableEnergyWorld.com*, 15 April 2008.

56. Martin Goetzeler, Osram, "Towards a New Culture of Lighting," presentation for Worldwatch Institute Efficient Lighting Symposium, Washington, DC, 28 May 2008.

57. Capacity in North America from EIA, DOE, *Electric Power Annual 2007: Overview*, at www.eia.doe.gov/cneaf/electricity/epa/epa_sum.html; Jun Ishii, *Technology Adoption and Regulatory Regimes: Gas Turbine Electricity Generators from 1980 to 2001* (Berkeley, CA: Center for the Study of Energy Markets, University of California

Energy Institute, 2004).

58. EWEA, "Wind Energy Leads EU Power Installations in 2007, But National Growth is Inconsistent," press release (Brussels: 4 February 2008); Prometheus Institute and Greentech Media, *PV News*, April 2008, p. 6; International Atomic Energy Agency (IAEA), *Power Reactor Information System*, at www.iaea.org/programmes/a2; Jessica Hunt, "Nanosolar's Breakthrough—Solar Now Cheaper than Coal," undated, at www.celsias.com/article/nanosolars-breakthrough-technology-solar-now-cheap.

59. Indian manufacturers' purchases from Gokul Subramaniam, "Suzlon Energy Posts 97 Percent Jump in Net Profit in Q1 FY09 Despite Forex Loss," *International Business Times*, 31 July 2008; Global Wind Energy Council (GWEC), "Top 10 Total Installed Capacity" and "Top 10 New Capacity,"at www.gwec.net, viewed 4 April 2008; GWEC, "US, China & Spain Lead World Wind Power Market in 2007," press release (Brussels: 15 February 2008); Travis Bradford, Prometheus Institute, "World PV Market Update and Photovoltaic Markets, Technology, Performance, and Cost to 2015," presentation, March 2008; projections for China from Junfeng Li, China Renewable Energy Industries Association, discussion with Yingling Liu, Worldwatch Institute.

60. Photovoltaic status based on annual (and cumulative) sales from Paul Maycock and Prometheus Institute, *PV News*, various issues, and on Travis Bradford, Prometheus Institute, discussion with Janet Sawin, 29 April 2008; wind status and projections based on historical annual and cumulative installation data from BTM Consult, EWEA, and American Wind Energy Association (AWEA), on GWEC, "Global Wind Energy Markets Continue to Boom—2006 Another Record Year," press release (Brussels: 2 February 2007), and on GWEC, "Global Installed Wind Power Capacity (MW)—Regional Distribution," at www.gwec.net, viewed 4 April 2008; nuclear power projections based on data from Worldwatch Institute database and from IAEA, op. cit. note 58. Figure 4–4 from the following: Worldwatch calculations based on fossil fuel data from BP, *BP Statistical Review of World Energy* (London: 2008); on nuclear power from Worldwatch Institute database compiled from statistics from IAEA, op. cit. note 58; on Janet L. Sawin, "Another Sunny Year for Solar Power," *Vital Signs Online* (Washington, DC: Worldwatch Institute, 2008); on Janet L. Sawin, "Wind Power Continues Rapid Rise," *Vital Signs Online* (Washington, DC: Worldwatch Institute, 2008); and on Joe Monfort, "Despite Obstacles, Biofuels Continue Surge," *Vital Signs Online* (Washington, DC: Worldwatch Institute, 2008).

61. U.S. trucks from Paul Gipe, "Can the U.S. Reach 100 Percent Renewable Electricity in 10 Years?" *RenewableEnergyWorld.com*, 17 July 2008.

62. Box 4–2 from the following: "IT Services Recommendations on When to Replace Aging Computers," Information Technology Services, Stanford University, at www.stanford.edu/services/ess/adminapps/recommended.html, modified 15 July 2008; D. V. Spitzley et al., *Automotive Life Cycle Economics and Replacement Intervals* (Ann Arbor, MI: Center for Sustainable Systems, University of Michigan, 2004); California Air Resources Board, "Estimation of Average Vehicle Lifetime Miles Traveled," Sacramento, CA, 2004; U.S. power plants efficiency and age from EIA, *Annual Energy Review 2007* (Washington, DC: DOE, 2007), and from "Existing U.S. Coal Plants," *Sourcewatch* (Center for Media and Democracy), at www.sourcewatch.org/index.php?title=Existing_U.S._Coal_Plants; similar situation elsewhere from IEA, *Key World Energy Statistics 2006* (Paris: 2006).

63. Wind payback from AWEA, "The Most Frequently Asked Questions About Wind Energy," 2002, p. 11, at www.awea.org/pubs/documents/FAQ2002%20-%20web.PDF; solar payback from Energy Efficiency and Renewable Energy, DOE, "Solar FAQs—Photovoltaics—Financial Considerations," updated 8 February 2007, and from National Renewable Energy Laboratory, "PV FAQs—What is the Energy Payback for PV?" December 2004.

64. Manfred Fischedick, Wuppertal Institute, "The German Renewable Energy Act: Success and Ongoing Challenges," ICORE 2004 Conference for Renewable Energies, Bangalore, January 2004; Bundesministerium für Umwelt, Naturschutz und Reaktorsicherheit (BMU, German Federal Ministry for the Environment, Nature Conservation and Nuclear Safety), "Big Boost for Renewable Energies: Share in Electricity Supply Has Gone up to 14 Per Cent," Berlin, 22 January 2008; revised German targets from "German Parliament Adopts Climate Package to Reduce CO_2 Emissions by 2020," *Thomson Financial News Limited*, 6 June 2008.

65. Denmark's economy quoted in EWEA, "With Increased Research, Renewable Energy Can Supply More than 20% of Europe's Energy Demand," press release (Brussels: 3 April 2008); renewable share in 1980 from Ministry of Foreign Affairs of Denmark, "The Danish Example—Towards an Energy Efficient and Climate Friendly Economy; COP 15 United Nations Climate Change Conference, Copenhagen, 2009," at www.cop15.dk/en/menu/About-Denmark/The-Danish-Example; 2008 share and 2011 and 2020 goals from Karl Larsen, "Denmark Continues its Renewable Tradition," *Renewable Energy Focus*, July/August 2008, p. 66; Geoffrey Lean and Bryan Kay, "Four Nations in Race to be First to Go Carbon Neutral," (London) *The Independent*, 30 March 2008; John Vidal, "Sweden Plans to be World's First Oil-free Economy," *The Guardian*, 8 February 2006.

66. Worldwatch calculation based on 2005 data from EIA, *International Energy Outlook 2008* (Washington, DC: DOE, June 2008), p. 1, and from EIA, "Table 1.3 Primary Energy Consumption by Source, 1949–2007," at www.eia.doe.gov/emeu/aer/txt/ptb0103.html, viewed 12 September 2008.

67. American Institute of Architects and economic benefits from 2030, Inc./Architecture 2030, *The 2030 BluePrint: Solving Climate Change Saves Billions* (Santa Fe, NM: 2008), pp. 2, 4.

68. DOE, op. cit. note 22, pp. 1–19.

69. Charles F. Kutscher, ed., *Tackling Climate Change in the U.S.: Potential Carbon Emissions Reductions from Energy Efficiency and Renewable Energy by 2030* (Washington, DC: American Solar Energy Society, 2007), pp. 3, 10, 34–35.

70. Chinese Renewable Energy Industries Association, "Sector Review of Renewable Energy in China and Its Potential for CDM Projects," undated, at cdm.ccchina.gov.cn/english/UpFile/File161.DOC; Peter Meisen and Eléonore Quéneudec, *Overview of Renewable Energy Potential in India* (San Diego, CA: Global Energy Network Institute, 2006); Brazil from Fernando R. Martins et al., "Solar and Wind Resources Database to Support Energy Policy and Investments in South America," in E. Ortega and S. Ulgiati, eds., *Proceedings of IV Biennial International Workshop, Advances in Energy Studies*, Unicamp, Campinas, Brazil, 16–19 June 2004, pp. 419–27.

71. Denmark's experience from Monica Prasad, "On Carbon, Tax and Don't Spend" (op ed), *New York Times*, 25 March 2008.

72. J. L. Míguiz et al., "Review of the Energy Rating of Dwellings in the European Union as a Mechanism for Sustainable Energy," *Renewable & Sustainable Energy Reviews*, February 2006, pp. 24–45.

73. End of technical lifetimes from Alexander Ochs, *AICGS Policy Report: Overcoming the Lethargy: Climate Change, Energy Security, and the Case for a Third Industrial Revolution* (Washington, DC: American Institute for Contemporary German Studies, Johns Hopkins University, 2008); Ontario Ministry of Energy and Infrastructure, "Moving Forward on Coal Replacement," press release (Toronto, Canada: 16 May 2008).

74. Jobs from Mariah Blake, "Germany's Key to Green Energy," *Christian Science Monitor*, 20 August 2008; new industries and 79 million tons from BMU, *Electricity from Renewable Energy Sources: What Does it Cost Us?* (Berlin: 2008); cost from ibid. and from Daniel Argyropoulos, BMU, "Renewable Energy Source Act in Germany: Current Status and Perspectives," presentation at Strategy Workshop on Feed-in Tariffs and Their Application in the United States, Washington, DC, 2 March 2008; 9-million-ton reduction from Alex Dunnin, "Emissions Trading Won't Fix Climate Change: Expert," *Financial Standard*, 23 May 2008; most effective and efficient from Nicholas Stern, *Stern Review: The Economics of Climate Change* (Cambridge, U.K.: Cambridge University Press, 2006), p. 366; laws in other countries from REN21, op. cit. note 5, p. 7.

75. IEA from UNEP, "Breaking Down the Barriers to a Green Economy: UNEP Launches Year Book 2008," press release (Monaco: 20 February 2008); Carl Levesque, "In Energy Sector, Renewables Get Less Federal Support," *RenewableEnergyWorld.com*, 19 November 2007.

76. "Revealing the High Cost of Energy Subsidies: An Interview with Trevor Morgan," *Subsidy Watch* (Global Subsidies Initiative), September 2008; UNEP, Division of Technology, Industry and Economics, *Reforming Energy Subsidies: Opportunities to Contribute to the Climate Change Agenda* (Geneva: 2008); 96 percent from Keith Bradsher, "Fuel Subsidies Overseas Take a Toll on U.S.," *New York Times*, 28 July 2008.

77. Worldwatch Institute, with technical assistance by the Global Labor Institute, Cornell University, *Green Jobs: Towards Decent Work in a Sustainable, Low-Carbon World* (Nairobi: 2008), commissioned by UNEP as part of its joint Green Jobs Initiative with the International Labour Organization, International Trade Union Confederation, and International Organisation of Employers; Bonn from Janet Sawin's attendance at International Conference for Renewable Energies, Bonn, Germany, 1–4 June 2004.

78. Skeptical comments from *Famous Authoritative Pronouncements*, at www.av8n.com/physics/ex-cathedra.htm; 8 percent of U.S. homes from "Remarkable Progress in Electrical Development: Notable Features in the Increase of the Use of Electricity in Small Plants and Households," *New York Times*, 8 January 1905, and from Edison Electric Institute, "Historical Statistics of the Electric Utility Industry Through 1970," at www.eia.doe.gov/cneaf/electricity/page/electric_kid/append_a.html; 3,000 vehicles from Ritz Site, "Early Ford Models," at www.ritzsite.net/FORD_1/02_eford.htm.

79. Non-hydro renewables from REN21, op. cit. note 5.

Chapter 5. Building Resilience

1. Quote from S. Vermeulen et al., *Springing Back: Climate Resilience at Africa's Grassroots*, Sustainable Development Opinion (London: International Institute for Environment and Development (IIED), 2008).

2. J. M. Scheuren et al., *Annual Disaster Statistical Review: The Numbers and Trends 2007.* (Belgium: Center for Research on the Epidemiology of Disasters, Université Catholique de Louvain, 2008); heaviest impact from S. Huq and J. Ayers, *Critical List: The 100 Nations Most Vulnerable to Climate Change*, Sustainable Development Opinion (London: IIED, 2007).

3. Intergovernmental Panel on Climate Change (IPCC), *Climate Change 2007: Impacts, Adaptation and Vulnerability. Contribution of Working Group II to the Fourth Assessment Report of the Intergovernmental Panel on Climate Change* (Cambridge, U.K.: Cambridge University Press, 2007), p. 883.

4. S. H. Schneider et al., "Assessing Key Vulnerabilities and the Risk from Climate Change," in IPCC, op. cit. note 3, p. 781.

5. Table 5–1 adapted from IPCC, *Climate Change 2007: Synthesis Report. Contribution of Working Groups I, II and III to the Fourth Assessment Report of the Intergovernmental Panel on Climate Change* (Geneva: 2007); Cavite City from D. Satterthwaite et al., *Adapting to Climate Change in Urban Areas: The Possibilities and Constraints in Low- and Middle-Income Nations*, Human Settlements Discussion Paper Series, Climate Change and Cities 1 (London: IIED, 2007).

6. IPCC, op. cit. note 3, p. 880.

7. Poverty levels and malaria from U.N. Development Programme, *Human Development Report 2007/2008* (New York: Palgrave Macmillan, 2007); water and sanitation from D. Satterthwaite and G. McGranahan, "Providing Clean Water and Sanitation," in Worldwatch Institute, *State of the World 2007* (New York: W. W. Norton & Company, 2007), p. 27.

8. P. Blaikie et al., *At Risk: Natural Hazards, People's Vulnerability, and Disasters* (London: Routledge, 1994).

9. B. H. Pandit and S. Barsila, *Impact Evaluation of Practical Action Implemented Program on "Increasing the Resilience of Poor Communities to Cope with the Impacts of Climate Change"* (Kathmandu, Nepal: Practical Action, 2007).

10. Ibid.

11. Ecological footprint concept from M. Wackernagel and W. Rees, *Our Ecological Footprint: Reducing Human Impact on the Earth* (Gabriola Island, BC: New Society Publishers, 1996); N. W. Adger et al., "Social-Ecological Resilience to Coastal Disasters," *Science*, 12 August 2005, pp. 1036–39.

12. Department for Environment, Food and Rural Affairs, *Conserving Biodiversity in a Changing Climate: Guidance on Building Capacity to Adapt* (London: 2007); A. Fischlin et al., "Ecosystems, Their Properties, Goods, and Services," in IPCC, op. cit. note 3, pp. 211–72.

13. L. J. Hansen, J. L. Biringer, and J. R. Hoffman, eds., *Buying Time: A User's Manual for Building Resistance and Resilience to Climate Change in Natural Systems* (Berlin: WWF Climate Change Program, 2003); Adger et al., op. cit. note 11; L. Burke, E. Selig, and M. Spalding, *Reefs at Risk in Southeast Asia* (Washington, DC: World Resources Institute, 2002).

14. K. Dow, "The Extraordinary and the Everyday in Explanations of Vulnerability to an Oil Spill," *Geographical Review*, vol. 89, no. 1 (1999), pp. 74–93.

15. Number surviving on less than $1 per day from International Fund for Agricultural Development (IFAD), *Rural Poverty Report 2001: The Challenge of Ending Rural Poverty* (Rome: 2001); W. E. Easterling and P. K. Aggarwal, "Food, Fibre and Forest Products." in IPCC, op. cit. note 3, pp. 273–313; R. A. Chambers, A. Pacey, and L. A. Thrupp, *Farmer First: Farmer Innovation and Agricultural Research* (London: Intermediate Technology Publications, 1989).

16. Easterling and Aggarwal, op. cit. note 15; IPCC, op. cit. note 3; IFAD, *Climate Change and the Future of Smallholder Agriculture*, Discussion Paper for Round Table on Climate Change, Thirty-first session of IFAD Governing Council, 14 February 2008 (Rome: 2008); M. Boko et al., "Africa," in IPCC, op. cit. note 3, pp. 433–67.

17. Bangladesh from S. Huq and J. Ayers, *Climate Change Impacts and Responses in Bangladesh*, briefing note prepared for the European Parliament (London and Brussels: IIED and Policy Department, Economic and Scientific Policy, DG Internal Policies of the Union, 2008); study cited in Working Group on Climate Change and Development, *Up in Smoke? Asia and the Pacific* (London: New Economics Foundation and IIED, 2007).

18. Working Group on Climate Change and Development, op. cit. note 17; Huq and Ayers, op. cit. note 17.

19. Easterling and Aggarwal, op. cit. note 15.

20. IFAD, op. cit. note 16.

21. U.N. Food and Agriculture Organization, *Community Forestry: Herders' Decision-Making in Natural Resources Management in Arid and Semi-arid Africa*, Community Forestry Notes 4 (Rome: 1990).

22. "Rainwater Harvesting in Zimbabwe," *Practical Action*, at practicalaction.org/?id=climatechange_rainwater.

23. Ibid.

24. D. Brown, T. Slaymaker, and N. K. Mann, *Access to Assets: Implications of Climate Change for Land and Water Policies and Management* (London: Overseas Development Institute, 2007).

25. World Bank, "Rural Institutions and Climate Change: Discussion Forum," at www-esd.worldbank.org/ricc.

26. M. E. Hellmuth et al., eds., *Climate Risk Management in Africa: Learning from Practice* (Palisades, NY: International Research Institute for Climate and Society, Earth Institute at Columbia University, 2007).

27. Ibid.

28. Urban dwellers living in poverty and their needs from UN-HABITAT, *The Challenge of Slums: Global Report on Human Settlements 2003* (London: Earthscan, 2003), and from D. Satterthwaite, *The Under-Estimation of Urban Poverty in Low and Middle Income Nations*, Poverty Reduction in Urban Areas Working Paper 14 (London: IIED, 2004).

29. M. Alam and M. Rabbani, "Vulnerabilities and Responses to Climate Change for Dhaka," *Environment and Urbanization*, April 2007, pp. 81–97.

30. Khulna from unpublished data in Capacity Strengthening in Least Developed Countries for Adaptation to Climate Change project.

31. Box 5–1 from the following: Albert F. Appleton, "How New York City Used an Ecosystem Services Strategy Carried Out Through an Urban-Rural Partnership to Preserve the Pristine Quality of Its Drinking Water and Save Billions of Dollars and What Lessons It Teaches about Using Ecosystem Services," presented at The Katoomba Conference, Tokyo, November 2002; Pablo Lloret, Informe, Ecuador, PowerPoint presentation, 29 April 2008; author's observations.

32. D. Dodman and D. Satterthwaite, "Institutional Capacity, Climate Change Adaptation, and the Urban Poor," *Institute for Development Studies Bulletin*, October 2008.

33. D. Roberts, "Thinking Globally, Acting Locally: Institutionalizing Climate Change at the Local Government Level in Durban, South Africa," *Environment and Urbanization*, October 2008, pp. 521–38.

34. L. S. Velásquez B., "Agenda 21; A Form of Joint Environmental Management in Manizales, Colombia," *Environment and Urbanization*, October 1998, pp. 9–36; L. S. Velásquez B., "The Bioplan: Decreasing Poverty in Manizales, Colombia, through Shared Environmental Management," in S. Bass et al., eds., *Reducing Poverty and Sustaining the Environment* (London: Earthscan, 2005), pp. 44–72.

35. R. Sanchez-Rodriguez, M. Fragkias, and W. Solecki, *Urban Responses to Climate Change: A Focus on the Americas*, Workshop Report, Urbanization and Global Environmental Change, June 2008; Satterthwaite et al., op. cit. note 5.

36. World Bank, *Clean Energy and Development: Towards an Investment Framework* (Washington, DC: 2006); ActionAid USA, *Compensating for Climate Change: Principles and Lessons for Equitable Adaptation Funding* (Washington DC: 2007); Oxfam International, *Adapting to Climate Change: What's Needed in Poor Countries, and Who Should Pay*, Oxfam Briefing Paper 104 (Washington, DC: 2007).

37. Details from Global Environment Facility Web site, at www.gefweb.org, and from Climate Change Convention Web site, at unfccc.int.

38. ActionAid USA, op. cit. note 36; Clean Development Mechanism description from N. Taiyab, *Exploring the Market for Voluntary Carbon Offsets* (London: IIED, 2006); B. Müller, "The Nairobi Climate Change Conference: A Breakthrough for Adaptation Funding," *Oxford Energy & Environment Comment* (Oxford: Oxford Institute for Energy Studies, 2007).

39. ActionAid USA, op. cit. note 36; Oxfam International, op. cit. note 36; I. Burton, E. Diringer, and J. Smith, *Adaptation to Climate Change: International Policy Options* (Arlington, VA: Pew Center on Global Climate Change, 2006).

40. S. Huq and J. Ayers, "Streamlining Adaptation to Climate Change into Development Projects

at the National and Local Level," in European Parliament, *Financing Climate Change Policies in Developing Countries* (Brussels: 2008); E. Levina, *Adaptation to Climate Change: International Agreements for Local Needs*, prepared for the Annex I Expert Group on the UNFCCC (Paris: International Energy Agency, Organisation for Economic Co-operation and Development, 2007); Burton, Diringer, and Smith, op. cit. note 39.

41. Huq and Ayers, op. cit. note 40; M. Thompson and S. Rayner, "Cultural Discourses," in S. Rayner and E. L. Malone, eds., *Human Choice and Climate Change, Volume 1: The Societal Framework* (Columbus, OH: Battelle Press, 1998); Oxfam, op. cit. note 36; ActionAid USA, op. cit. note 36.

42. Huq and Ayers, op. cit. note 40.

43. R. J. T. Klein et al., "Inter-Relationships Between Adaptation and Mitigation," in IPCC, op. cit. note 3, pp. 745–77.

44. John Vidal, "UK Gives £50m to Bangladesh Climate Change Fund," (London) *Guardian*, 8 September 2008.

45. IIED, *Adaptation by the Poorest* (London: forthcoming).

46. Sattherthwaite et al., op. cit. note 5.

47. T. J. Wilbanks et al., "Toward an Integrated Analysis of Mitigation and Adaptation: Some Preliminary Findings," *Mitigation and Adaptation Strategies for Global Change*, June 2007, p. 714; R. Swart and F. Raes, "Making Integration of Adaptation and Mitigation Work: Mainstreaming into Sustainable Development Policies?" *Climate Policy*, vol. 7, no. 4 (2007), pp. 288–303; N.H. Ravindranath, "Mitigation and Adaptation Synergy in Forest Sector," *Mitigation and Adaptation Strategies for Global Change*, June 2007, pp. 843–53; J. Ayers and S. Huq, "The Value of Linking Mitigation and Adaptation: A Case Study of Bangladesh," accepted for publication in *Environmental Management.*

48. T. J. Wilbanks et al., "Possible Responses to Global Climate Change: Integrating Mitigation and Adaptation," *Environment*, vol. 45, no. 5 (2003), pp. 28–38; T. J. Wilbanks, J. Sathaye, and R. J. T. Klein, "Introduction," *Mitigation and Adaptation Strategies for Global Change*, June 2007, pp. 639–41; Wilbanks et al., op. cit. note 47; H. H. Dang, A. Michaelowa, and D. D. Tuan, "Synergy of Adaptation and Mitigation Strategies in the Context of Sustainable Development: The Case of Vietnam," *Climate Policy*, vol. 3, supp. 1 (2003), pp. S81–S96; Klein et al., op. cit. note 43.

49. S. Huq and M. Grubb, "Preface," *Mitigation and Adaptation Strategies for Global Change*, June 2007, pp. 645–49; Dang, Michaelowa, and Tuan, op. cit. note 48; C. Rosenzweig and F. N. Tubiello, "Adaptation and Mitigation Strategies in Agriculture: An Analysis of Potential Synergies," *Mitigation and Adaptation Strategies for Global Change*, June 2007, pp. 855–73.

50. H. D. Venema and M. Cisse, *Seeing the Light: Adapting to Climate Change with Decentralized Renewable Energy in Developing Countries* (Winnipeg, MN: International Institute for Sustainable Development, 2004); H. D. Venema and I. H. Rehman, "Decentralized Renewable Energy and the Climate Change Mitigation-Adaptation Nexus," *Mitigation and Adaptation Strategies for Global Change*, June 2007, pp. 875–900; Ayers and Huq, op. cit. note 47.

51. Ravindranath, op. cit. note 47; Venema and Rehman, op. cit. note 50; Ayers and Huq, op. cit. note 47.

52. Venema and Rehman, op. cit. note 50; Huq and Grubb, op. cit. note 49.

Chapter 6. Sealing the Deal to Save the Climate

1. Population data from Population Division, *World Population Prospects: The 2006 Revision* (New York: United Nations, 2008).

2. Box 6–1 based on the following: Lasse Ringius, Asbjørn Torvanger, and Arild Underdal, "Burden Sharing and Fairness Principles in International Climate Policy," *International Environ-*

mental Agreements: Politics, Law, and Economics, March 2002, pp. 1–22; "Policies, Instruments, and Cooperative Arrangements," in Intergovernmental Panel on Climate Change (IPCC), *Climate Change 2007: Mitigation of Climate Change* (Cambridge, U.K.: Cambridge University Press, 2007), p. 770; "Decision-making Frameworks," in IPCC, *Climate Change 2001: Mitigation* (Cambridge, U.K.: Cambridge University Press, 2001), p. 669; Egalitarian Principle from Carlo Carrera, *Efficiency and Equity in Climate Change Policy* (Berlin: Springer, 2000), pp. 333–35; population sizes from U.S. Bureau of the Census, *International Data Base*, electronic database, Suitland, MD, viewed 28 September 2008; carbon dioxide emissions from Carbon Dioxide Information Analysis Center (CDIAC), "Ranking of the World's Countries by 2005 Total CO_2 Emissions from Fossil-fuel Burning, Cement Production, and Gas Flaring," Oak Ridge National Laboratory (ORNL), at cdiac.ornl.gov/trends/emis/top2005.tot, viewed 29 September 2008; emissions since 1950 are a Worldwatch calculation based on data in CDIAC, ORNL, at cdiac.ornl.gov/ftp/ndp030/global.1751_2005.ems, viewed 25 September 2008; gross domestic products from International Monetary Fund, *World Economic Outlook Database* (April 2008 update), at www.imf.org/external/pubs/ft/weo/2008/01/weodata/index.aspx, viewed 29 September 2008; survey from Andreas Lange, Carsten Vogt, and Andreas Ziegler, *On the Importance of Equity in International Climate Policy: An Empirical Analysis*, Discussion Paper No. 06-042 (Mannheim, Germany: Centre for European Economic Research, 2006).

3. United Nations, *Kyoto Protocol to the United Nations Framework Convention on Climate Change*, at unfccc.int/resource/docs/convkp/kpeng.pdf.

4. Emissions prevented from Karan Capoor and Philippe Ambrosi, *State and Trends of the Carbon Market 2008* (Washington: World Bank, 2008), p. 1. Data could not be found for carbon dioxide (CO_2)-equivalent emissions in 2006 and 2007, but the IPCC estimates that global greenhouse gas emissions totaled 49 billion tons CO_2 equivalent in 2004 (see IPCC, *Climate Change 2007: Synthesis Report, Summary for Policymakers* (Geneva: 2007), p. 5). And the Global Carbon Project estimates that global CO_2 emissions rose by 7.8 percent from 2004 to 2007, suggesting a likely comparable increase in CO_2 equivalent emissions (see Global Carbon Project, at lgmacweb.env.uea.ac.uk/lequere/co2/carbon_budget.htm, viewed 2 October 2008).

5. Carbon-related financial flows to developing countries from Capoor and Ambrosi, op. cit. note 4; development assistance data from Organisation for Economic Co-operation and Development, "Aid Statistics, Donor Aid Charts," at www.oecd.org; International Fund for Agricultural Development, *Sending Money Home: Worldwide Remittance Flows to Developing and Transition Countries* (Rome: 2007).

6. Data on 2007 carbon market from Capoor and Ambrosi, op. cit. note 4.

7. Estimate on China's electric power sector from Center for Global Development, "China Passes U.S., Leads World in Power Sector Carbon Emissions–CGD" (Washington, DC: 27 August 2008); Kevin A. Baumert, Timothy Herzog, and Jonathan Pershing, *Navigating the Numbers: Greenhouse Gas Data and International Climate Policy* (Washington, DC: World Resources Institute, 2005), figure 6–1 and p. 32.

8. Total of $19.5 billion from Capoor and Ambrosi, op. cit. note 4.

9. Juliette Jowitt and Patrick Wintour, "Cost of Tackling Climate Change Has Doubled, Warns Stern," (London) *Guardian*, 26 June 2008; global spending on oil from Steven Mufson, "This Time, It's Different," *Washington Post*, 27 July 2008; World Health Organization, "Spending on Health: A Global Overview," fact sheet No. 319 (Geneva: February 2007).

10. Stephen Castle and Mark Landler, "After 7 Years, Talks Collapse on World Trade," *New York Times*, 30 July 2008.

11. Michel den Elzen and Malte Meinshausen, *Meeting the EU 2°C Target: Global and Regional Emissions Implications* (Biltoven, Netherlands:

Netherlands Environmental Assessment Agency, 2005); U.N. Development Programme, *Human Development Report 2007/2008* (New York: 2007).

12. *Vienna Convention for the Protection of the Ozone Layer*, at www.unep.org/ozone/vc-text.shtml; *Montreal Protocol on Substances that Deplete the Ozone Layer*, at www.unep.org/OZONE/pdfs/Montreal-Protocol2000.pdf.

13. Funding to date from Multilateral Fund for the Implementation of the Montreal Protocol Web site, at www.multilateralfund.org, viewed 2 October 2008.

14. United Nations, *United Nations Framework Convention on Climate Change* (UNFCCC), at unfccc.int/essential_background/convention/background/items/1349.php.

15. United Nations, op. cit. note 3.

16. Reductions of 5 percent from UNFCCC Web site, at unfccc.int/kyoto_protocol/items/2830.php, viewed 2 October 2008.

17. Criticism from Charles Forelle, "French Firm Cashes In Under U.N. Warming Program," *Wall Street Journal*, 23 July 2008.

18. Environmentalist suit from court notice of application, at www.ecojustice.ca/media-centre/media-release-files/notice_of_application07_05_29.pdf.

19. "Byrd-Hagel Resolution," Senate Resolution 98, 105th Congress, 1st Session, 25 July 1997.

20. Population Division, op. cit. note 1; Jay S. Gregg, Robert J. Andress, and Gregg Marland, "China: Emissions Pattern of the World Leader in CO_2 Emissions from Fossil Fuel Consumption and Cement Production," *Geophysical Research Letters*, 24 April 2008.

21. Figure 6–1 from Global Carbon Project, op. cit. note 4.

22. For information on carbon markets, see Zöe Chafe and Hilary French, "Improving Carbon Markets," in Worldwatch Institute, *State of the World 2007* (New York: W. W. Norton & Company, 2007), pp. 91–106.

23. Industrial emissions estimate from Ottmar Edenhofer, deputy director, Potsdam Institute for Climate Impact Research, discussion with author, 26 June 2008.

24. Jonathan Fowlie and Fiona Anderson, "B.C. Introduces Carbon Tax," *Vancouver Sun*, 19 February 2008; City of Boulder, "Boulder Voters Pass First Energy Tax in the Nation," press release (Boulder, CO: 8 November 2006); Chicago carbon exchange from Chafe and French, op. cit. note 22; U.S. regional initiative from Robin Shulman, "Carbon Sale Raises $40 Million," *Washington Post*, 30 September 2008; New South Wales, "Greenhouse Gas Reduction Scheme," at www.greenhousegas.nsw.gov.au.

25. UNFCCC, "Report of the Conference of the Parties on Its Thirteenth Session, Held in Bali from 3 to 15 December 2007—Part Two: Action Taken by the Conference of the Parties at Its Thirteenth Session," at unfccc.int/resource/docs/2007/cop13/eng/06a01.pdf#page=3.

26. UNFCCC, "Summary Report of the Co-Chairs of the In-Session Thematic Workshop," at unfccc.int/files/meetings/intersessional/awg-lca_1_and_awg-kp_5/agendas/application/pdf/bkk_workshop_sumreport.pdf; Yvo de Boer, executive secretary, UNFCCC, address to the Sustainability Luncheon, World Petroleum Congress, Madrid, 3 July 2008.

27. Industrial sector approach from Rob Bradley et al., *Slicing the Pie: Sector-Based Approaches to International Climate Agreements* (Washington, DC: World Resources Institute, 2007); "China to Back Japan's Plan on Post-Kyoto Emissions," *Asahi Shimbun*, 3 May 2008.

28. Calculation of industrial process contribution to global emissions based on Baumert, Herzog, and Pershing, op. cit. note 7, pp. 4–5.

29. Proportion of CO_2 from land use from IPCC,

op. cit. note 4, p. 5; U.S. and Chinese proportion of global emissions from CDIAC, op. cit. note 2.

30. IPCC, op. cit. note 4, p. 20.

31. Ian Traynor and David Gow, "EU Promises 20% Carbon Reduction by 2020," (London) *Guardian*, 21 February 2007.

32. "Lieberman-Warner Climate Security Act of 2008," S. 3036, U.S. Senate, 20 May 2008; David M. Herszenhorn, "After Verbal Fire, Senate Effectively Kills Climate Change Bill," *New York Times*, 7 June 2008; activist criticism from Vivian Buckingham, "The Lieberman-Warner Bill: A Post Mortem," *The Skywriter* (blog for 1Sky), 9 June 2008; position of Senator John McCain from campaign Web site, at www.johnmccain.com/Informing/Issues/da151a1c-733a-4dc1-9cd3-f9ca5caba1de.htm, viewed 3 October 2008; position of Senator Barack Obama from campaign Website at my.barackobama.com/page/content/newenergy, viewed 3 October 2008.

33. Peter Christoff, "Post-Kyoto? Post-Bush? Towards an Effective Climate Coalition of the Willing," *International Affairs*, September 2006, pp. 831–60.

34. Box 6–2 based on the following: G-77 and China, "Proposal: Financial Mechanism for Meeting Financial Commitments under the Convention," Submission to the UNFCCC, undated; "Proposal by the G-77 and China for a Technology Transfer Mechanism under the UNFCCC," Submission to the UNFCCC, undated; "World Climate Change Fund: A Proposal by Mexico" presentation by Ambassador Juan Manuel Gómez Robledo, Workshop on Investment and Financial Flows, Bonn, 5 June 2008; "Submission by Mexico," Submission to the UNFCCC, 13 August 2008; Switzerland and India proposals from "Developing Countries Ask for New UNFCCC Financial Architecture" (7 June 2008), in Third World Network, *Bonn News Updates and Climate Briefings* (Penang, Malaysia: 2008), pp. 19–20; South Africa proposal from "G77/China Reaffirms Climate Funds Should be Within UNFCCC" (11 June 2008), in ibid., p. 25; Matthew Stilwell, "G77-China Propose 'Enhanced Financial Mechanism' for the UNFCCC," Info Service on Finance and Development, Third World Network, 29 August 2008; Norway proposal from Sven Harmeling and Christoph Bals, "The Need for Adaptation Financing and the International Debate" (draft), Germanwatch, 2008, and from "Norway's Submission on Auctioning Allowances" Submission to the UNFCCC, date unknown; "Developing Countries Submit Proposals for Comprehensive Adaptations Framework," *Accra News Update* (Third World Network), 27 August 2008. For other proposals, see Harvard Project on International Climate Agreements, "Climate Proposals: Overview of the Six Frameworks," at belfercenter.ksg.harvard.edu/project/56/harvard_project_on_international_climate_agreements.html?page_id=82; Daniel Bodansky, *International Climate Efforts Beyond 2012: A Survey of Approaches* (Arlington, VA: Pew Center on Global Climate Change, 2004); Paul Baer and Tom Athanasiou, *Frameworks & Proposals: A Brief, Adequacy and Equity-Based Evaluation of Some Prominent Climate Policy Frameworks and Proposals*, Global Issue Paper No. 30 (Berlin: Heinrich Böll Foundation, 2007).

35. Stern quote from Jowitt and Wintour, op. cit. note 9.

36. India and Germany from Claudia Kade, "Merkel Backs Climate Deal Based on Population," *Reuters*, 31 August 2007.

37. Luiz Pinguelli Rosa, Maria Silvia Muylaert, and Christiano Pires de Campos, *The Brazilian Proposal and its Scientific and Methodological Aspects: Working Draft* (Winnipeg, Canada: International Institute for Sustainable Development, 2003); Baumert, Herzog, and Pershing, op. cit. note 7, pp. 31–33.

38. EcoEquity and Stockholm Environment Institute, "Executive Summary," *The Right to Development in a Climate Constrained World: The Greenhouse Development Rights Framework*, June 2008, at www.ecoequity.org/GDRs/GDRs_ExecSummary.html.

39. Ibid.

40. Feed-in laws from REN21, *Renewables 2007 Global Status Report* (Paris and Washington, DC: REN21 Secretariat and Worldwatch Institute, 2008), p. 7; Björn Larsen, *World Fossil Fuel Subsidies and Global Carbon Emissions in a Model with Interfuel Substitition*, Policy Research Working Paper No. 1256 (Washington, DC: World Bank, 1994).

Climate Change Reference Guide

1. "Global Anthropogenic GHG Emissions," in Intergovernmental Panel on Climate Change (IPCC), *Climate Change 2007: Synthesis Report* (Geneva: 2007), p. 36.

2. Table based on "World GHG Emissions Flow Chart," World Resources Institute, Washington, DC; Figure from IPCC, op. cit. note 1, p. 36.

3. "Couplings Between Changes in the Climate System and Biogeochemistry," in IPCC, *Climate Change 2007: The Physical Science Basis* (Cambridge, U.K.: Cambridge University Press, 2007), p. 515; graphic based on Oak Ridge National Laboratory, "The Global Carbon Cycle," *Integrated Assessment Briefs* (Oakridge, TN: 1995).

4. "Changes in Atmospheric Constituents and in Radiative Forcing," in IPCC, op. cit. note 3, pp. 212–13. The scientific understanding about global warming potential has changed over time; these numbers, from the latest IPCC assessment, are based on the current understanding.

5. Carbon Dioxide Information Analysis Center (CDIAC), "Ranking of the World's Countries by 2005 Total CO_2 Emissions from Fossil-fuel Burning, Cement Production, and Gas Flaring" and "Ranking of the World's Countries by 2005 Per Capita Fossil-fuel CO_2 Emission Rates," at cdiac.ornl.gov/trends, viewed 3 October 2008.

6. Worldwatch calculations based on data from CDIAC, "National CO_2 Emissions from Fossil-Fuel Burning, Cement Manufacture, and Gas Flaring: 1751–2005," at cdiac.ornl.gov/ftp/ndp030/nation1751_2005.ems, viewed 1 October 2008; Russian figure based on 1951–91 data for USSR and 1992–2005 data for Russian Federation, with 1951–91 emissions adjusted to 55 percent of USSR totals (the ratio of 1992 Russian emissions to 1991 USSR emissions).

7. A. Neftal et al., "Historical CO_2 Record from the Siple Station Ice Core," CDIAC, at cdiac.ornl.gov/ftp/trends/co2/siple2.013; D. M. Etheridge et al., "Historical CO_2 Record Derived from a Spline Fit (20 Year Cutoff) of the Law Dome DE08 and DE08-2 Ice Cores," CDIAC, at cdiac.ornl.gov/ftp/trends/co2/lawdome.smoothed.yr20; National Oceanic and Atmospheric Administration, "Use of NOAA ESRL Data," NOAA dataset, at ftp.cmdl.noaa.gov/ccg/co2/in-situ/mlo/mlo_01C0_day.co2, viewed 13 October 2008.

8. J. Hansen et al., "Global Land-Ocean Temperature Index in .01 C, base period 1951–1980 (January-December)," Goddard Institute for Space Studies, at data.giss.nasa.gov/gistemp/tabledata/GLB.Ts+dSST.txt, viewed 1 October 2008; "Technical Summary," op. cit. note 5, pp. 36, 70.

9. National Climatic Data Center, "Global Temperatures," at www.ncdc.noaa.gov/oa/climate/research/2007/ann/global.html, viewed 3 October 2008.

10. Timothy M. Lenton et al., "Tipping Elements in the Earth's Climate System," *Proceedings of the National Academy of Sciences*, 12 February 2008, pp. 1786–93.

11. IPCC, *Climate Change 2007: Impacts, Adaptation and Vulnerability* (Cambridge, U.K.: Cambridge University Press, 2007).

12. "Risks of Climate Change Damages Would Be Reduced by Stabilizing CO_2 Concentration," in IPCC, *Climate Change 2001. Synthesis Report* (Cambridge, U.K.: Cambridge University Press, 2001); IPCC, op. cit. note 1, p. 67; James Hansen et al., "Target Atmospheric CO_2: Where Should Humanity Aim?" *Open Atmospheric Science Journal*, 2008; "Towards a Goal for Climate Change Policy," in Nicholas Stern, *The Economics of Climate Change: The Stern Review* (Cambridge, U.K.:

Index